Space and Society

Editor-in-Chief

Douglas A. Vakoch, METI International, San Francisco, CA, USA

Series Editors

Anthony Milligan, King's College London, London, UK

Beth O'Leary, Department of Anthropology, New Mexico State University, Las Cruces, NM, USA

The Space and Society series explores a broad range of topics in astronomy and the space sciences from the perspectives of the social sciences, humanities, and the arts. As humankind gains an increasingly sophisticated understanding of the structure and evolution of the universe, critical issues arise about the societal implications of this new knowledge. Similarly, as we conduct ever more ambitious missions into space, questions arise about the meaning and significance of our exploration of the solar system and beyond. These and related issues are addressed in books published in this series. Our authors and contributors include scholars from disciplines including but not limited to anthropology, architecture, art, environmental studies, ethics, history, law, literature, philosophy, psychology, religious studies, and sociology. To foster a constructive dialogue between these researchers and the scientists and engineers who seek to understand and explore humankind's cosmic context, the Space and Society series publishes work that is relevant to those engaged in astronomy and the space sciences, while also being of interest to scholars from the author's primary discipline. For example, a book on the anthropology of space exploration in this series benefits individuals and organizations responsible for space missions, while also providing insights of interest to anthropologists. The monographs and edited volumes in the series are academic works that target interdisciplinary professional or scholarly audiences. Space enthusiasts with basic background knowledge will also find works accessible to them.

More information about this series at http://www.springer.com/series/11929

Margaret Boone Rappaport · Konrad Szocik
Editors

The Human Factor in the Settlement of the Moon

An Interdisciplinary Approach

 Springer

Editors
Margaret Boone Rappaport
The Human Sentience Project, LLC
Tucson, AZ, USA

Konrad Szocik
Yale University
New Haven, CT, USA

Department of Social Sciences
University of Information Technology
and Management in Rzeszow
Rzeszow, Poland

ISSN 2199-3882   ISSN 2199-3890   (electronic)
Space and Society
ISBN 978-3-030-81387-1   ISBN 978-3-030-81388-8   (eBook)
https://doi.org/10.1007/978-3-030-81388-8

Cover design: Paul Duffield

This Springer imprint is published by the registered company Springer Nature Switzerland AG
The registered company address is: Gewerbestrasse 11, 6330 Cham, Switzerland

Preface

The Lunar Tour, 2096 A.D.

Come, take a tour with us around a Moon settlement that's seventy-five years old. For just a moment, we step into the future and look around, using a tour book that will become familiar...

Welcome. Welcome to the Moon! We're going to explore a whole new world. We'll begin your tour here at the Admin Dome, where you've just seen our Historical Exhibit, complete with the map of the Apollo landing sites. Be careful getting on the rail car! I'm guessing most of you don't yet have your Moon Legs. It takes a different step, doesn't it?

Now, look at Chap. 2 in your tour book. You will read that people have had all kinds of ideas about Earth's Moon for a very long time. In Chap. 2, Dr. Murray gives you a sampling—in culture, society, health, and ritual! Earth's Moon figures prominently in many religions and virtually all types of literature—oral and written. The Moon has shone brightly on each and every human culture on Earth, and sometimes as an animate figure!

Now, let's stop the rail car, and we'll take a look a Dr. Haviland's exhibit on the surface of the Moon. Look at your book: There is a good image of the Moon's surface a little earlier in Fig. 1.1. It captures an important visit by Apollo 14 leaving to return to Earth in 1971, with the Earthrise in the background. Beautiful! Haviland has done us all a favor by introducing the idea that the Moon's surface is not always stationary. It changes. It's always changing! Interesting!

Okay, as we get back on our rail car, glance over to the right and you'll see two important installations: Our battery bank for the Admin Dome and a little further to your right, our field of solar panels begins, and then it stretches about halfway to the rim of that crater. In your tour book, Drs. Lumbreras and Pérez Grande give you a good summary of power options for the Moon. We're using them all, and we're trying a few new ones, too!

Let's get back on the rail car and stop over there at a special set of domes that house our first greenhouses. Of course, now, just over the edge of that crater rim to your left is a mammoth, new set of greenhouses that stretch for about three miles.

It's the biggest greenhouse in the Solar System! ...except, maybe... I heard about a new one on Mars. [shrugs] Those Martians—they're always trying to top us...

In your tour book, Dr. Cazalis describes different types of agricultural installations—it's a very tricky business! It's taken new construction materials and new piping, and a couple years to build that big one. Our ability to feed ourselves has grown exponentially! And, of course, our replicators can now deliver banana flambé... The last chapter in your tour book by Drs. Corbally and Rappaport gives you a nice summary of some of their cultural prognostications 75 years ago. It's a light review of art, architecture, cuisine, and outdoor recreation. ...Yes, we do sponsor hiking trips twice a week for our visitors.

Oh, before we leave this view of the older greenhouses, remember that almost all of the hardscape is 3D-printed right on site. It's amazing to watch that operation... Dr. Braddock, in Chaps. 6 and 9 of your tour book, describes some of the dangers of the regolith that were just beginning to be understood, back, 75 years ago. It's harmless, of course, if you're inside a space suit, but getting anywhere near those 3D printers without one turned out to be dangerous! The stuff is bad for the respiratory system. ...Almost as bad as smoking, they say. [shrugs] We didn't know it at first.

Now watch your step, and once you're on the train, we'll head for that group of domes over there. That's our Med Center. Your tour book has Dr. Shelhamer's early survey of human body systems, and what the concerns were back 75 years ago. They were right about most of it. We still have to exercise more than we'd like to [swiveling and pointing] ...Oh, that's the gym over there. Everybody uses it. They send you back to Earth if you don't use it. Exercise on the Moon is like taxes back on Earth... Can't avoid it!

As we come up to the Med Center... We won't go in. It's off limits because of all the research going on there... Anyway, let me mention Dr. Newberg's review of the effects of lunar living on the human nervous system in your tour book. We've learned ways of countering some of that, too. But... they still keep an eye on us!

Now, as we get going again, we're heading for the Judicial Center, where we have our court and the Moon's only jail, which you can see off there to the left. We never thought we'd ever have to use it! ...Of course, sadly... Well, that's another story... The Judicial Center also has a reception room for dignitaries and heads of state. They've signed two treaties in that very room. It's a beautiful room. Fanciest place on the Moon! In your tour book, there's a list of Moon treaties with descriptions, by the early lunar diplomat Jurado Ripoll. He's been up here, you know. So has Newberg. A bunch of the authors in your tour book have paid the Moon a visit!

We won't go there today, but there is a military contingent in that separate dome over there. People fought it hard, but they always knew there would be a military presence, or two... or three or four... here on the Moon, just as Stewart and Rappaport describe in your tour book. We never go over there. We don't know what they do over there, but—they all eat with us in the Mess Hall! So, we know the guys and gals in uniform, but... shhhh... We've no idea what they do! Lots of antennae, more than usual. [nodding]

Of course, some people don't like their new life here on the Moon. They find it too confining, too uncomfortable, too inconvenient, and some people can't live with the low gravity and they go back to Earth. Dr. Schwartz describes the so-called Right to Return to Earth. It's written into every contract signed for a work tour. It's simply not for some people. There are a few every year.

The philosophers in your tour book like Dr. Szocik speculate on what they call a Space Refuge, where a better quality of life is more possible than on Earth. [shrugs] It's true, Earth is just beginning to rid itself of centuries of pollution, and not many who come out here want to return. A few. We started out right, with a good environmental plan—an "ethos" Norman and Reiss call it in your tour book. They were right! And so was Dr. Kendal. We took her advice and now there are specific workplace protections for all our major occupations. Of course, we can't protect our Governor from worrying! I think this moon base *runs* on his worry. We English-speaking Moon residents all clustered around this particular moon base, so he's got his hands full.

Careful! Watch your step. I want you to visit a special place. This is our one and only church on the Moon. It's staffed by a multi-denominational group of pastors, priests, and rabbis. There are others here, like the Bahá'í group, who meet on their own. If anybody thought the settlers here on the Moon would leave religion behind on Earth—they were mistaken!

And that is the point of three chapters in your tour book, by Drs. Impey, Oviedo, and Murray. They are different sorts of experts, but each one speculates on faith and religion away from the Earth. They were mostly right! If anything, they played down religion, at least, compared with what I've seen. Of course, I don't know all the moon bases well, like the Chinese. It's gigantic! It's over on the farside. I'm not sure what they do there, in the area of religion, but here, it's solid.

Now, look to the left, and you'll see our new Luna University. It's brand new and not quite open yet. They'll offer degree programs, and the military folk are already signed up. You know... Sociology and Mathematics and English classes. Speaking of Sociology, your tour book gives a good look at how the social scientists 75 years ago saw living on the Moon. Dr. Campa was right about a lot of things. We have to watch out for depression, that kind of thing, but we have a program now, and if someone seems a little blue, we all step in and lend a hand. It's *pro*-active! We don't just ignore it.

Now, we're coming around again to the Admin Dome. That's where you can sign up for the beginners mountain hiking group that'll leave in about 30 minutes—just like Corbally and Rappaport describe in your tour book. It's fun. It will also help you get your Moon Legs! We're at only 1/6th of Earth's gravity here, and it takes different muscles! Have a good time!

And watch your step as you get off the train.

Tucson, AZ, USA   Margaret Boone Rappaport
Rzeszów, Poland/New Haven,   Konrad Szocik
USA
March 2021

Contents

List of Figures

List of Tables

Chapter 1
Practical Planning Commences: Next Steps in the Settlement of Earth's Moon

Margaret Boone Rappaport⍟ and Konrad Szocik⍟

Abstract The Editors welcome readers with a perspective on human off-world existence that will be useful in the following decades of early settlement of Earth's Moon, Mars, the asteroids, and the moons of the gas giants Jupiter and Saturn. They emphasize the unity of the Earth and its Moon—as an astronomical reality, a research dimension, and a collective focus for environmental stewardship. The overarching goal of this book is to inject a sense of realism into the identification of human problems and the crafting of solutions over the coming decades of adaptation to an initially inhospitable, airless, and dusty location at one-sixth of Earth's gravity. The Editors identify five themes that emerged spontaneously, from the twenty chapters: (1) expansion of research models; (2) the multi-disciplinary and pragmatic nature of space research; (3) siting moon bases according to human requirements; (4) the unexpected emergence of the importance of religious expression in off-world settlement; and (5) the critical role of environmental protection in cis-lunar space, and on the Moon, as on the Earth.

1.1 The Human Factor in Space

Humans are in space and they have been for some time, since the early 1960s. As of this writing, almost 600 people have experienced conscious human life outside the protective envelope of Earth's atmosphere. More humans will join these pioneers at an increasing pace, and therefore, it is important to understand how our species, with its own unique traits and preferences, will survive and ultimately thrive in this initially

M. B. Rappaport (✉)
The Human Sentience Project, LLC, Tucson, AZ 85704, USA
e-mail: msbrappaport@aol.com

K. Szocik
Yale University, New Haven, CT, USA

University of Information Technology and Management in Rzeszow, Rzeszow, Poland

K. Szocik
e-mail: konrad.szocik@yale.edu

dangerous new place. Humans are what the philosophers and anthropologists call "social problem solvers". We solve our problems in groups of people, and we shall continue to solve an untold array of problems in space, in groups where our ideas blossom and ricochet in the process of sharing ideas with other human problem solvers—be they nearby, or far away in Mission Control. Joint problem solving occurred on the early Apollo missions, and it will continue in spite of the expected time lapses between Earth, Earth orbit, the ISS, the Moon, Mars, the asteroids, and the moons of the gas giants, Jupiter and Saturn.

Our first step toward a semi-permanent presence in space has been accomplished by the International Space Station, host to a variety of people from many of Earth's nations. Their variety reminds us that by settling the Moon, other humans will arise and be identified by new names. There will be humans who are not born on their home planet, and the Moon may be the first place that this happens, if not during a long space journey.

Eventually, the ISS will be followed by orbital hotels, research stations, convalescent centers, and a new type of structure, the space elevator, which will make ascent to orbit less costly and more convenient for non-specialists. Those accomplishments lie in our future, but before that, our task is to conceive of the human animal as the inhabitant of a nearby planetary body, the Moon. Mars will come soon thereafter, but the Moon is relatively close. It gives our species a footing, at 1/6th Earth's gravity, for stepping into outer space and venturing very far away from Earth.

We will likely launch our species from a relatively stable, lunar orbital platform, out into the rest of the Solar System, and beyond. Voyages to distant points in our Solar System will often originate on the Moon, partly because it is easier and cheaper to build spaceships that are already beyond much of Earth's gravitational pull. It is enormously costly to lift large weights from the surface. In addition, the lift-off of many ships will eventually be polluting.

The social scientists among us will watch closely to see how humans fare on the Moon (Chaps. 17–20). We will conduct research—much of it anticipated in the chapters of this volume. The Moon is in many ways a testing ground for human life on Mars, and the commercial mining of the asteroids.

This book is full of anticipation of great discoveries and riches, and not a little dread. We shall succeed, but it will be hard going, and in these pages, we try to conceive of human problems and their solutions. If we do not address the natural requirements of our species, we shall fail. Space exploration is not a simple technical exercise, but a task of making ourselves comfortable, productive, alert, and ready for action in a completely new and dangerous environment. Humans must be ready and able to exercise their extraordinary decision-making skills, drawing as they do on the skills of our artificial companions. We should accustom ourselves to their ability to make decisions faster, while humans are left with the ultimate decision, "Does it feel right?" The artificial intelligence of today cannot address this last question, although perhaps it will in the future. Space exploration may be that venture where human and artificial human learn to work together. We shall have to teach them how to do that.

In this volume, we try to imbed in our vision a renewed concern for environmental protection. It would be a sad and pitiful course if we turn our Moon into a trash heap or a toxic waste dump. We must protect it because it has, for millennia, given each human culture a theme for imaginative stories, songs, gods, and rituals. Apparently, our Moon is large as moons go, and that size equals its importance to humans, all of us. There is only one Moon. There is only one Earth. Their relationship is pictured anew in Fig. 1.1 (NASA 1971a), which was taken by Apollo 14 while it headed back home to Earth, appearing in the background as an Earthrise. The beauty in these vistas will not escape our descendants.

This is the next, and the third volume in our series of collections on the "Human Factor in Space" for Springer's larger series on "Space and Society." However, this is the first group of contributions to fully and consistently accept the notion that humans are leaving Earth. That is an enormous step for many humans. Previous

Fig. 1.1 Apollo 14 heads for home in 1971. Image Credit Apollo 14, NASA, JSC, ASU (Image Reprocessing: Andy Saunders)

volumes considered a variety of scientists' and philosophers' questions on whether humans should inhabit planetary bodies other than Earth. At this point in time, it seems a foregone conclusion: Humans are off the planet, and in a big way, as soon as they can afford it and figure out how to survive. It is a mental and physical chore, an individual and a collective mission plan for humanity, a technical challenge, and the repeated testing of realistic scenarios. Humans must use all their many cognitive attributes to survive where we are going. Living on the Moon will include an evaluation of our species in a uniquely new environment. Watch, as that species—our species—ventures off world. Watch us leave and settle new places. In these pages, read something about the challenges that engineers, scientists, and eventually celebrities and politicians will face. We have not thought of everything, but we have thought of a lot.

This book is dedicated to all the human missions to the Moon, in the past and in the future, but we focus here on the processes of establishing the first lunar settlements. The initial function of these nascent communities is probably as a place and a lifeway to support work, much of it, research. Other chapters extend into problems and solutions for later stages of settlement, to functions of fully supportive human communities involving work, recreation, government, self-sufficiency with industrial agriculture, and trade between Earth and other settlements in our Solar System. Previous volumes focused on challenges of a mission to Mars (Szocik 2019) and on the broad notion of human enhancements for all space missions, to the Moon, to Mars, and farther (Szocik 2020).

Our purpose is to create a sense of realism about this enormous human adventure. Our goal is to outline challenges that sound more concrete and practical than the questions and problems of the previous two volumes in our series. It is not unimportant that humans have already landed on the Moon. That challenge lies in front of us again, while we plan for Mars. Unlike a mission to Mars, we know we can fly to the Moon even today, and we know some of the precautions. In this volume, we discuss the rationale for a mission to *settle* the Moon, and the environmental issues of protecting both Earth and Moon, as inextricably linked, which they are. As a group of contributors, we champion the goal of sustainable development of Earth's Moon, and the social and ethical issues related to human health and well-being. There is a precaution in our realism: We need to remember that while we shall be the first to make the Moon home, we shall be far from the last. The whole of humanity needs an approach to conservation, waste disposal, and water re-cycling that lasts on the Moon for centuries, even millennia. Once we settle the Moon, it will be our home for a long time. We must begin rightly, with an ethos of conservation (Chap. 14, Norman & Reiss), which lies at the foundation of every commercial and settlement mission.

Many important things have happened in the 50 years since the last humans landed on the Moon (e.g., Figs. 1.2 and 1.3, NASA 1971b). The environmental issues that troubled people in the 1960s and 1970s were new for many. Our view is that a contemporary, twenty-first century, lunar environmental ethos must draw in everyone—scientists, construction workers, companies heading to the asteroids to mine precious metals, celebrities, politicians, tourists, and yes, even the military (Chap. 11, Stewart & Rappaport). That environmental ethos will be created for the

Fig. 1.2 Astronaut James B. Irwin, lunar module pilot, uses a scoop to make a trench in the lunar soil during Apollo 15 EVA, in 1971. Mount Hadley rises above the plain in the background. Photo Credit: NASA; David Scott

benefit of all of Earth's societies, and no doubt transferred, in part, to Martian cultures and societies when they develop. That ethos is an obligation we have to the future. Without it, the rationale for humanity's expansion beyond Earth, and use of space resources, falls short.

1.2   Themes that Cross-Cut Chapters

Five themes arise from among the details and ancillary themes in the chapters of this volume. One was planned—lunar environmental concern. We planned a light to moderate emphasis, but it emerges, at times, stronger than expected in importance

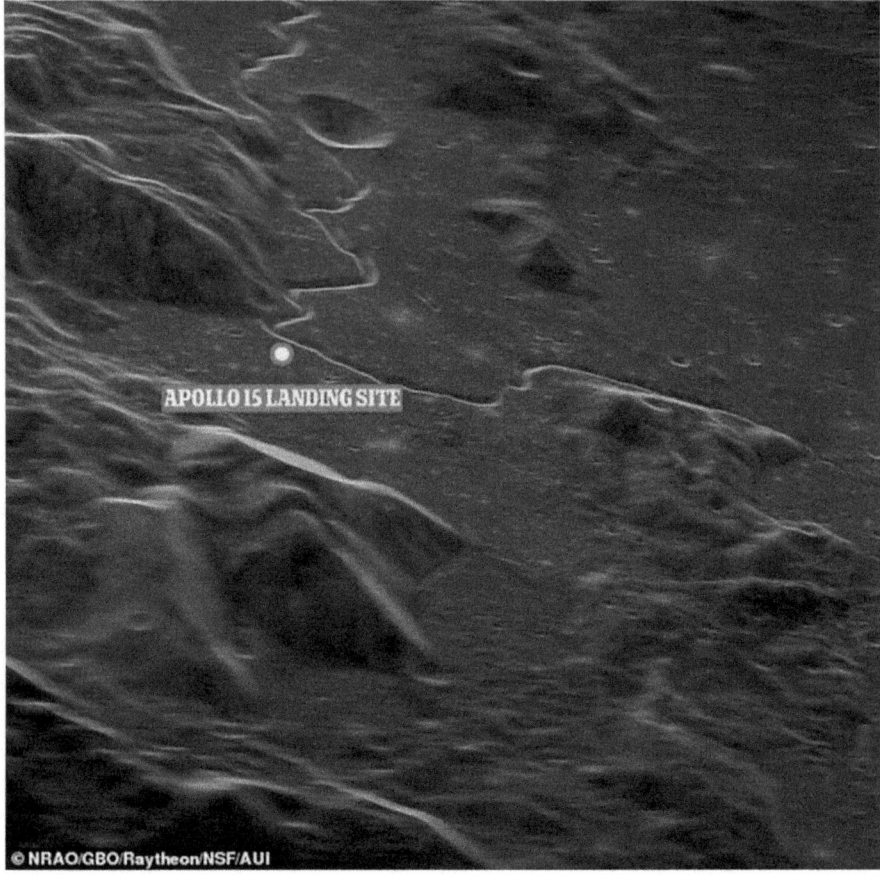

Fig. 1.3 2021 radar image of the 1971 Apollo 15 Landing Region. Credit: Sophia Dagnello, NRAO/GBO/Raytheon/AUI/NSF/USGS

and broader in scope to expand outward and include the duo of the Earth and Moon in "cis-lunar space." That was new, along with the other themes that emerged de novo. Indeed, cis-lunar space could become a dumping ground and a waste disposal problem. Space junk is already a problem for Earth, and it will expand to the space around the Moon, unless we plan very well and decide collectively on regulating mechanisms.

From the most theoretical to the most practical, the five cross-cutting themes include: (1) the expansion of research models in lunar research, (2) the strong multi-disciplinary nature of space research and its applications; (3) the siting of Moon bases according to human requirements, (4) the unexpected importance of religious expression, and again, (5) the most practical—environmental concern, from a light (e.g., Chap. 11) to an encompassing level (e.g., Chap. 14). Individual chapters of this book address these cross-cutting issues in different ways, for different human functions, and using different languages, disciplines, and suggested research models.

1.2.1 Expansion of Research Models

The expansion of research models is already hinted above. Something happens to scientists—medical researchers, sociologists, and agronomists—who realize there is another known category of human: The off-Earth human. Their first thought is to compare humans on Earth to humans on the Moon, and in myriad ways. Humanity will have humans who have settled some place other than Earth, and all of our perspectives will change accordingly. "They are the same as earthlings, aren't they?" we will ask. Or, we might wonder if they are somehow different in ways both real and imagined. The human imagination will ask these and another several million questions.

Primarily, what do we call off-Earth humans? Whatever the moniker, their existence will expand the experience of human living, human health, and inevitably human evolution, and create a new category. A new place will become a new state of being. We shall compare humans on Earth with humans living on the Moon, just as we compare humans on Earth who live in different nations, different cities, and different cultures on Earth. This new category will have immediate importance and researchers will ask: Does aging occur in the same way and at the same rate among people on the Moon, as among people on Earth? In a more practical vein: Do I need workman's compensation insurance or is life simply too dangerous for workman's comp and it makes no sense on the Moon? And even, more intimately: Do I need to be married before I live with my love in a Moon dome, or are the rules of co-habitation different? Besides, who really cares *up here?* Or … is it *out here?* It would certainly save on the oxygen bill.

In fact, the very nature of "short supplies" will change. First, air to breathe, then food that will initially be shipped at great expense from Earth, or later, grown with exquisite care in the Moon's fledgling greenhouses. Finally, medicines, and fuel to get back "home" to Earth, if anyone wants to leave (Chap. 12, Schwartz). The infinity of details in human life and culture will shift, and we shall begin to ask new questions and the same questions, differently. There will be perceived divisions among us that did not exist before the settlement of the Moon. The time between the emergence of *Homo sapiens*, 300–400,000 years ago, and today has seen many changes for the human line on planet Earth. We are set to inhabit another.

In addition to new divisions, there will also be one important newly perceived commonality. Eventually, when common cause is recognized, the humans on Earth and the humans on the Moon will feel a new joint bond and say to themselves: We are in this together. We live on a duo of planetary bodies bound to each other first by gravity, but then by humanity. We live together on these two planetary bodies, together orbiting our Sun in cis-lunar space. Mars will come along soon, but there will never be another moon like our Moon.

These changes will expand research models, from "terrestrial thinking" to "extraterrestrial thinking". Some of us thought that the first extraterrestrials would be visitors from another star system, but instead, we find that humans will be the first extraterrestrials, quite literally. All our research will shift quietly to include the new category of humans on Earth's Moon, and in contrast to other humans on planetary bodies like Mars, Europa, and Enceladus. These names will soon be familiar to us all.

Humanity will shift from terrestrial to extraterrestrial thinking, and from Earth-centered to cis-lunar thinking. It will probably happen slowly, as one after another physician, political scientist, or philosopher wonders if human experience on Earth can be extrapolated to the Moon. When Mars is settled, comparisons will expand further. These shifts, themselves, will also be the subject of speculation, and eventually philosophers will write about the repercussions of models that include humans on the Moon.

Some research questions occur to us even now. Will settlers of the Moon be moved to cooperate more with each other because they are in such a challenging environment? Or, will they be more callous and less careful with other people because a single priority is omnipresent: Survival. Will the result be that settlers are more or less sensitive? Will they resemble the settlers moving west to settle North America, and how? Will they appreciate life, itself, more or less? Much will be determined by the initial selections of space crew, the criteria used by both government and commercial programs, as well as the *self*-selection factors—some obvious and others more opaque. What kind of human wants to go to the Moon to visit? To settle? Or to die there? Who would be content to never see Earth again, except as a blue, green, and white vision in the pitch-black lunar sky? Some humans will be so content, and they are probably already born.

There will be a continuing opportunity to test notions of cooperation and the evolution of value systems. Bioethics should find much of interest because all human pursuits will be impacted by the human body's and mind's reactions to isolation, enclosure, dryness, moon dust, radiation, and all, at 1/6 of Earth's gravity. It will be a different sort of physical and medical existence, and just how different, we are just beginning to anticipate (Chaps. 6–9).

1.2.2   The Multi-Disciplinary and Pragmatic Nature of Space Research

Research on future space missions does not belong to any single science or discipline. If there is one encompassing umbrella, it is "futures research", although that has problems involving trend analysis, forecasting, and widening errors and uncertainties with the length of time into the future. Triangulation of results using multiple methods (qualitative and quantitative) is helping futures research achieve new goals.

The theories and research findings of individual disciplines can help to anticipate an array of issues and potential solutions, and the humanities (philosophy, the arts, education, and theology) can help us interpret the meaning of future scenarios for humans. Operations research has been doing this for decades—helping humans to make better decisions. In a sense, that is what traditional humanities have always helped humans to do, as well. Therefore, we are quite happy to have both the hard sciences and the softer, social sciences represented, and the humanities, as well. We have two, doctoral-level astronomers as contributors, who are a wealth of information helping to back-stop our vision, but they both chose to write on other subjects in which they have some expertise (Impey and Corbally, in Chaps. 16 and 20). The two of them represent the multi-disciplinary nature of this book and each chapter in it.

The more we know about the Earth's Moon, the better able we are to anticipate the future, although none of us ventures to "predict". Casting our vision in terms of scenarios and possibilities based on today's knowledge base is a worthwhile exercise because we are all trained in different disciplines, methods, and styles of argumentation and discourse. Eventually, all our scenarios (or others yet unimagined) will be approached carefully, dissected, and then—engineered!

The engineer has the job of translating knowledge of humans into spaceship hulls, propulsion systems, environmental controls, and comfortable accommodations. The engineer stands between scenarios of humans and real missions to the Moon. Four engineers contribute to this book (Chap. 3, Haviland; Chap. 4, Lumbreras & Perez Grande; Chap. 7, Shelhamer, a biomedical engineer), and an industrial agriculturalist (a biologist by training)—serves in that role, too (Chap. 5, Cazalis). Physiologist Braddock writes in Chaps. 6 and 9 on the opportunities and the health risks of all agriculture, industry, and construction using the Moon's regolith. In surveying the chapters in this book, we conclude that, for humans in space, nothing is pure, except the beauty of the vivid scenes of the Cosmos outside our portholes. Every discipline that can help to keep humans alive in space and on the Moon must be consulted, and this includes physicians (Newberg, Chap. 8),

The social sciences, to date, have lagged behind the physical sciences and engineering in contributing to spaces studies. Philosophers have contributed, but fewer trained social scientists have done so. We anticipate that social scientists will soon begin to contribute much more substantially than in the past. "Settlement" is, after all, a human social enterprise and it always has been. Humans live in communities, whether the components are distantly spaced (as in outback Alaska) or tightly packed (as in conjoined lunar domes). It is true that social isolation and anomie are important issues on the Moon (Chap. 19, Campa), but equally important will be social research on human cooperation, while infrastructure is built on the Moon, transportation grids are laid out, and increasingly large greenhouses are fitted with illumination of various wavelengths, waterworks, and electronics to monitor output. These projects will take human social cooperation in a configuration little known until now: Relatively small crews handling large complex machinery. Indeed, much lunar construction is scheduled to eventually be done by robotic manufacturing and 3D printers (Corbally & Rappaport, Chap. 20). It is entirely possible that most of the early settlers (and their spouses) will be engineers—even the physicians! A moon settlement is, in part, a socially organized assembly based first and foremost on engineering expertise—and good monitoring by qualified medical staff and social scientists.

Multi-disciplinarity, while seen as desirable in academia today, runs headlong into almost overwhelming tendencies to "stovepipe" knowledge. Complaints continue that disciplinary experts talk "only to each other." For that reason, we have tried to include an especially diverse group of contributors. Input from different fields is important in the study of future human exploration of space (Schwartz 2020). However, a multi-disciplinary perspective is only one side of our methodological commitment.

The other is pragmatism, and it is a natural offshoot of multi-disciplinarity. Different disciplines suggest obvious problems to each other when their experts work together in teams. It is easy to ignore obvious failures in logic or approach when conversing only with people in one's own discipline. From our perspective, even philosophical analysis (especially ethics) has a pragmatic context. It must do so, because the level of human risk and the amount of investment at stake in space missions is so very great.

As a result, this volume avoids purely philosophical speculations, but appreciates the value of applied ethics (Chaps. 12–15, Schwartz; Szocik; Norman & Reiss; Kendal), which offer useful arguments about the human rights to explore and exploit, but at the same time, they espouse the concepts of justice, equity, and responsibility in space missions. We as a species and a group of nations cannot run rapacious through our Solar System, without thinking about the future of those who follow. What we do makes a difference, while we set the standards for those who follow.

1.2.3   The Siting of Moon Bases According to Human Requirements

Site selection on the Moon will be an accommodation between human needs, the Moon's available resources, and capabilities that humans bring with them. The process begins by understanding its surface, where and how it can change, and where ice may lie under the surface, ready for human consumption (Haviland, Chap. 3). As this book goes to press, it is understood that NASA's Artemis Program launch will occur in October 2024, and Artemis III will land an astronaut team and set up base camp near the moon's south pole, where satellites confirm water ice exists (NASA 2021). See Fig. 1.4 for the latest equipment in ice detection—NASA's Volatiles Investigating Polar Exploration Rover.

Clearly, the location of the Artemis base follows standard theory in geography. Human communities locate in places where food and water resources are available. Water on the Moon will always be a precious commodity because there is no free-flowing water. The best we can hope is that we re-cycle carefully and find water beneath the regolith in shaded areas.

Fig. 1.4   NASA's VIPER lunar rover. Credit: NASA Ames/Daniel Rutter

Other chapters point to different principles in operation when a base site selection is made, now and in the future. Chap. 11, on the military on the Moon, points to strategic principles and relationships behind the location of moon bases, as well as for transportation and communications access. In the future, the Lagrange points, where vehicles can "park" in space with a minimal expenditure of fuel to keep themselves in a stable orbit, will be the location of giant construction projects. Access to Lagrange points in near-Moon space will be convenient and important, and so control of these locations cannot be assumed by any single nation-state on Earth.

Similarly, another principle from geography—break-in-bulk—suggests that the Moon may later house giant facilities for warehousing, shipping, transportation, and re-fueling. The Moon may be the location where giant ships who are returning to near-Earth space from the asteroids "park" and "unload", without having to risk a landing on Earth or the use of enormous energy to lift new cargo to orbit. Warehousing fuel on the Moon will be easier in the very cold temperatures, and in the absence of ambient oxygen. Safety protocols will be easier to follow in colder temperatures, and, where every person has been taught to handle sparks and open flames carefully, if at all. Many of these break-in-bulk principles have guided the locations of towns and cities on Earth, and they will simply become operational in three dimensions on the Moon and in cis-lunar space.

The location of moon domes for human living will follow principles that may well be like the preferences operating on Earth, and in addition, there may be new principles operating. On Earth, humans tend to locate on rivers and bodies of water, for beauty and transportation access, but there will be no bodies of water on the Moon at least until very large moon domes are constructed that can encompass entire, small ecosystems. Humans also like to live or pursue recreational activities in high places where the views are grand. There are mountains on the Moon, and we anticipate they may well be used for recreational pursuits like hiking and eventually spelunking in the Moon's caves (Chap. 20, Corbally & Rappaport).

Humans tend to settle in "beautiful" places. This tendency is not to be dismissed as foolish because it is important for "quality of life," which has good health metrics. Each time a new land is settled on Earth, there are those who come along later to complain, "All the good spots have been taken by the first people!" We ask then, what is a "good spot" on the Moon? It appears to depend on access to resources and the physical beauty of the area or the view, and the features that call to mind "beauty" can vary. On the Moon, beauty will be newly defined. A good spot on a low mountain, with a perfect view of spaceships coming and going from a landing field may qualify as "beautiful." The one, singular environmental feature that will be instantly beautiful is a view of the orb that is humanity's home planet—Earth.

Beauty may not come to operate initially in the selection of moon dome sites. In the first phases of settlement, the proximity to supplies from Earth will be especially important. If there is transport via a small train system, then access to that will be important, or to a system of self-driving vehicles on a network of tracks. Access to vehicles that are the descendants of the rovers first operating on Mars may determine which locations are prime. There will be competition for locations because human settlement always involves competition for the "best spots."

1.2.4   The Importance of Religious Expression Emerges

Quite unexpectedly, this book came to house three short chapters on religious experience, expression, and faith on the Moon. The chapters are all very different from each other. One is by an astronomer who has done lengthy periods of field research among Buddhist monks (Chap. 16, Impey). Another is by a social scientist who is also a Catholic friar, which discusses the importance of religious experience in new and challenging environments (Chap. 17, Oviedo). The third is by a cultural anthropologist who describes the factors operating in religion on the Moon, from a functional social systems perspective (Chap. 18, Murray). We emphasize that the appearance of these three chapters was not planned. Yet, three doctoral-level specialists with different backgrounds chose to write on this subject, and their proposals were accepted.

We Editors were surprised. We knew that NASA, for example, had called upon psychologists to understand the experience of isolation by humans on long space voyages. However, we did not know whether government and commercial space companies were expecting their crew to eventually involve themselves in religious experience, especially on lengthy missions. Had they thought this through and made provisions? Had they thought of how to accommodate both "religious" and "non-religious" crew members, perhaps in similar activities, such as a non-denominational type of meditation?

These and many other questions arise when it is fully acknowledged that some crew on some missions to the Moon and Mars will engage in religious practices—whether specific mission sponsors foresee this or not, like it or not, and whether it can be "used" in a positive way, during the development of off-world settlements. The Human Factor Series includes different types of chapters that anticipate religious experience among crew, both in the earlier book on Mars (Szocik 2019), and the book on Human Enhancements (Szocik 2020). It will be useful for planners to consider how mission sponsors should approach this probable eventuality, and we recommend that these three chapters be read at a single time, to gain a variety of viewpoints.

What exactly will be needed on the voyage to, or on arrival at the Moon? Will there be celebrations that bring together all the crew in religious or quasi-religious "services"? Will people of different religious backgrounds clash? Will religious observances disappear? The answer to the latter question is probably, "No." We anticipate that humans will behave like humans in all cultures known from a variety of literatures. Humans have usually engaged in something called "religion", and we believe mission planners should discuss in detail how to handle the emergence of this type of human behavior, in spite of trends in secularization in many modern industrial societies. We doubt that religious expression will simply "go away."

It is also worth paying some attention to the possible trajectories of the development of human religious experience in space, and in the settlement of the Moon and Mars. It is ardently hoped that mission planners will encourage crew and settlers (when the time comes) to journal their experiences. That corpus of stories, thoughts, and feelings could well be used to understand the religious tendencies that humans have while settling new planets far from Earth. Eventually, there may be specialized groups of settlers who espouse a specific set of religious beliefs. It is hoped that mission planners will not wait until that happens, to begin to consider how to manage the widespread desire of many humans to engage in religious observations, prayer, and religious ritual. It could be a source of serious misunderstanding between people, and the forceful or energetic pursuit of some religious practices could up-end the best of plans.

Furthermore, there is another important context for the study of religious behavior in space missions to the Moon. Religious ritual and prayer are widely used by humans in coping with stress and, in becoming motivated to engage in non-religious activities. Religious beliefs and expression could have a unique potential for the application of important principles of social organization in extremely stressful and demanding environments, as found on space missions. If religious belief systems have the potential to influence the behavior of crew or eventually settlers, then this issue must be well explored by mission planners, in both government-funded and commercial space missions to the Moon and elsewhere.

Above all, the crew must be allowed some type of minimally disruptive religious activities, perhaps in private or in small groups. On the other hand, religious activities cannot be forced upon crew if they choose not to participate. As a voluntary and non-intrusive type of activity, religious expression should probably be considered, condoned, documented, and explored verbally before liftoff of any mission to the Moon. History teaches that religious beliefs and practices have an enormous potential to soothe and to motivate, but also a strong disruptive capacity. Therefore, the issue is non-trivial, no matter how much and for how long discussion of it is sidelined.

1.2.5 Environmental Issues in Settlement of Earth's Moon and Use of Cis-Lunar Space

The lunar environmental context is very different from Earth's. The surface of the Moon accounts for only 7.4% of the total Earth area (37.9 million square kilometers of Moon surface, versus 510 million square kilometers on Earth). In terms of the site selection criteria now being used, there are not many attractive targets on the Moon. The tight real estate situation on the Moon has important consequences for environmental planning, so that all who come to the Moon can interact peaceably and not despoil the surface or the space environment surrounding the Moon.

A phenomenon named "the tragedy of the commons" has now been known for almost two centuries. It describes an eventual course of resource depletion when individuals use available resources in a manner unhampered by social structures and mutually agreeable rules of access and use. In that type of situation, self-interest runs rampant and the "common good" is not protected (Lloyd 1833). In some ways, this describes the political and economic situation on the lunar surface now. However, political ambitions may well begin soon to "carve up" the Moon's surface and create boundaries, as on Earth's surface. Even if this happens, cooperation must be pursued diplomatically to ensure environmental regulation and protection of the Moon's environment. Chap. 10 (Jurado Ripoll) reviews lunar treaties, the products of past diplomacy, and in Chap. 12, Schwartz examines the right to return to Earth, an important aspect of settlement: return migration.

The risks of lunar contamination are extremely high because large machinery, multiple types of liquid and gaseous fuels, different work parties pursuing concurrent construction projects, and human exploration, itself, will all meet and blend in an airless and dusty environment. Protecting the Moon's surface means taking extra care and working harder to clean up accidents and the normal production of by-products. Human creation and industry are messy and necessarily produce by-products, sometimes in large quantities. We see this clearly back into human prehistory in refuse middens. Protocols for cleanup on the Moon are not out of the question, but simply common sense. Some fear a type of "super-exploitation" of space that would out-pace and out-pollute even the despoiling of Earth (Elvis & Milligan 2019).

Regardless of the rationale for any space mission, even the seemingly least invasive and most environmentally friendly, there will be transformation of the lunar environment, some pollution, and some resource consumption—especially of the regolith, which is now discussed as the matrix for 3D printers (Chap. 20, Corbally & Rappaport), and use in agriculture and other pursuits (Chaps. 6 and 9, Braddock). Environmental ethics are important in their own right, but they do not exist in a vacuum without political context. As that context changes over the coming decades, common cause must be pursued and agreement on methods and measures to protect the fragile lunar surface. On Earth, the water and weather appear to eventually wear down and decompose everything (although that is not altogether true). On the Moon, that water and weather will not be operating to clean up spillage and waste. Therefore, we are on course to find where our ethical concerns and environmental goals overlap in spite of political differences. Plans for environmental protection of Earth's Moon must soon turn to a political discussion, and the sponsors of Moon missions, both governmental and commercial, must confront now-unspoken environmental policy issues openly.

1.3 Delving into the Details: An Adventure

Our introductory chapter began with the goal of "practical planning." There is no doubt that the space enterprise has had both its practical and visionary elements from the beginning. It remains both, but in this chapter we have purposefully tried to introduce pragmatism—power, food, water, housing, construction, conservation, and some politics and defense. That is the stuff of which all human settlements are made, fashioned anew to suit each generation.

Humans are going to settle their Moon and it will be an important location for a very long time, even after other planets, asteroids, and the outer moons are settled by new human populations that are sent even farther away from their home planet. We invite the reader on a journey through interesting details now being contemplated in preparation for this exodus—the exodus of a few at first, and then more and more.

By themselves, the details of industrial greenhouse agriculture, Moon treaties among Earth's nation-states, the anticipated physiology of the human body in a different gravity (what little is known), and the social organization of housing, transportation, government, and commerce—all these details are now being anticipated. A broad outline of the types of details will probably remain the same over the coming years, but the details are sure to change. We hope that these chapters will interest, entertain, and challenge everyone to stay involved in the settlement of Earth's Moon.

References

Elvis, M., & Milligan, T. (2019). How much of the solar system should we leave as wilderness? *Acta Astronautica, 162*, 574–580.

Lloyd, W.F. (1833). *Two lectures on the checks to population*. Oxford University. JSTOR 1972412. OL 23458465M.

National Aeronautics and Space Administration—NASA. (2021). NASA's Artemis base camp on the moon will need light, water, elevation. Retrieved January 27, from https://www.nasa.gov/specials/artemis/, https://www.nasa.gov/artemisprogram.

NASA. (1971a). Apollo 14 Heads Home (image).

NASA. (1971b). Apollo 15 Astronaut James Irwin before Mt. Headley (image).

National Science Foundation. (2021). New radar image of the Apollo 15 landing site (image). National Science Foundation, Greenbank Observatory. Credit: Sophia Dagnello NRAO/GBO/Raytheon/AUI/NSF/USGS.

Schwartz, J. S. J. (2020). *The value of science in space exploration*. Oxford University Press.

Szocik, K. (Ed.). (2019) *The human factor in a mission to Mars; an interdisciplinary approach*. Springer.

Szocik, K. (Ed.). (2020). *Human enhancements for space missions; Lunar, Martian, and future missions to the outer planets*. Springer.

Part I
Perspectives from Culture History

Chapter 2
Moon Traditions: An Overview of Changing Beliefs About Earth's Moon

Gerald F. Murray

Abstract This chapter, written in the context of impending lunar settlements, gives a brief historical overview of beliefs and practices that human groups have developed, through the ages and across the culturally diverse globe, about the origin of the Moon and its impact on human life. We can only speculate about unwritten prehistoric beliefs. However, the invention of writing some 5,000 years ago generated numerous texts documenting folk theories about the origin and functions of the Sun, Moon, and stars. The chapter begins with a discussion of accounts of cosmogenesis around the world, including the creation of the Moon, and the anthropogenesis of the first humans. Such accounts are often embedded in religious texts. The analysis moves to a discussion of beliefs, most of them folkloric rather than religious, about the impact of the Moon on human minds and bodies, particularly during a full moon. The chapter ends with a discussion of the traditional function of lunar phases as markers of time in human cultures. We examine the origins of days, weeks, months, and years, which will be fundamentally disconnected from the astronomical reality confronting humans living on Earth's Moon.

2.1 Introduction

This chapter will focus on the evolution of human beliefs about the Moon. It is written in the context of the impending construction of lunar bases. Human presence on the Moon presupposes a precise knowledge of the lunar environment that has never before been needed in human history. Our cognitive involvement with the Moon will be radically deepened and transformed. As a prelude to these transformations, it will be interesting and appropriate to first review the earlier ideas—some prescientific and some based on astronomic investigations—that humans have had about the Moon. In these pages I will attempt to organize, categorize, and summarize some major patterns.

G. F. Murray (✉)
Department of Anthropology, University of Florida, Gainesville, FL, USA
e-mail: murray@ufl.edu

The chapter will be organized by theme, rather than by chronology or by geographical region. The themes to be touched on are (a) folk-theories concerning the origin of the Moon; (b) the deities—male or female, major or minor, helpful or harmful—who are associated with the Moon; (c) the impacts which the Moon is believed to have, both beneficial and harmful, on human minds and bodies and on social and economic life; and (d) the major impact of solar and lunar cycles on human organization of time. With respect to the latter, the phenomenon of "phases of the Moon", an optical illusion that has played such a major part in many cultures on the organization of time, is totally irrelevant for organizing time on a lunar settlement. The same is true of the 12 h gap between terrestrial "sunrise" and "sunset". The organization of lunar time will have to be based on criteria independent of lunar phases, lunar sunrise, and lunar sunset.

My coverage of traditional lunar beliefs will be illustrative rather than exhaustive. Since lunar traditions are often embedded in religious systems, many of the examples drawn will be from major world religions whose collective membership covers a heavy percentage of human groups. Anthropologists have documented fascinating beliefs in cultural groups with only a few hundred surviving members. They will not be covered here for space reasons, not because their beliefs about the Sun and Moon are less interesting.

2.2   The Origin of the Moon and Accounts of Cosmogenesis

A corpus of literature usually called "origin myths" collects pre-scientific accounts of the origin of the Sun, Moon, stars, Earth, and different forms of life on Earth, including humans. Since the word "myth" hints, at least in colloquial English, at the meanings "lie" or "falsehood", and since the dominant origin myth of the West, which is contained in the opening chapters of Genesis, and with which several billion Jews, Christians, and even Muslims are at least familiar, is militantly believed by many to be divinely revealed historical truth, the use of the term "origin account" may be more neutral than "origin myth". It builds in no implicit judgment concerning the veracity of the account and thus avoids potential polemics tangential to the issue at hand. There are two distinguishable types of origin accounts which we can call *cosmogenesis* and *anthropogenesis*. The former presents accounts about the origin of the universe, including celestial bodies, and of the earth itself. Folk theories of anthropogenesis, in contrast, focus on the origin of humans and often begin with the sky, the earth, and the oceans as pre-existing givens. Some, but not all, origin accounts cover both.

2.2.1   The Ancient Near East: The Moon as a Divinely Created Nocturnal Lamp

By far the most widely cited traditional account around the world of the origin of the Moon is the six-day creation account followed by a day in which God rested, in the first chapter of Genesis, which is itself the first book in the Tanakh (the Hebrew Bible) and the Christian Old Testament. This account appears in modified form, in the Islamic Qur'an's account of creation. Though Jews account for less than two-tenths of one percent of the world's population, Christians and Muslims together account for close to 60% (Pew Research Center 2018). Some religious readers may consider the biblical account to be historical fact. (I have had some friends warn me with concern that "the theory of evolution" is a plot of Satan.) People on the other side of the aisle may intensely dislike "religious dogma" and have absolutely no interest in even hearing about the biblical account. Each extreme reaction has its individual logic. However, whatever one's ideological stance, the Genesis origin account is empirically and statistically the dominant pre-evolutionary origin account around the world. It must therefore be ranked high in any scale-of-importance applied to traditional accounts of the origin of the Moon.

Let us therefore examine the text. In the first chapter of Genesis, creation of everything, including the Moon, was done by Elohim, one of the many names of the Hebrew God. The Elohim of this first chapter created by *speaking—vayomer Elohim yehe* …. "And God said, let there be …." And one by one things came into existence. In this chapter he created everything, including the Sun and the Moon, only by the power of his spoken word.[1]

The posited sequence of events leading up to the creation of the Moon sets modern readers aback, even some religious readers, at least if they examine the text carefully. On Day 1 light ('or) is created and separated from darkness (khoshekh). Next comes a solid dome (raqi > a) on Day 2 placed above the earth to separate two bodies of water: the upper waters (which are presumably the source of rain), and the lower waters— the Seas (yamim). Evapotranspiration from the seas and forests as the source of rain was of course unknown in antiquity. The author thus posits two separate bodies of water separated by a dome, often translated into English as "firmament". On Day 3 the lower waters, the seas, which covered the entire earth, are gathered into one place and separated from dry land (yabashah) as a preliminary to the creation, on that same Day 3, of vegetation—specifically grass ('esev) and fruit trees ('ets pri). Then and then only, on Day 4, does Genesis report the creation of the Sun, the Moon

---

[1] Emphasis here is on "this chapter." The following chapters of Genesis contain a different account of anthropogenesis, in which the name Elohim is supplemented by YHWH, another name for the Hebrew God. The sequence of events in these chapters is so different from that of Chap. 1 that critical scripture scholars, including many religious scholars, assume that it was composed (or compiled from oral tradition) at a different time by different individuals. (We will examine specific discrepancies later in the chapter.) The seven-day account of Chap. 1 is generally viewed as being written later by a priestly elite and inserted as Chap. 1 when the first five books of the Hebrew Bible were eventually compiled into the single scroll that is today used in synagogues.

and the stars, which by implication are all attached to the solid dome and rotate with the dome around the flat, stable, immovable earth.

Modern readers encounter serious problems with the Genesis model of cosmogenesis, not the least of them being vegetation without the Sun. The entire account is radically out of kilter with modern understandings of the history of the cosmos, of plant and animal life, and particularly of Homo sapiens. Our purpose is to describe, not to critique or defend, origin accounts. However, it must be quickly pointed out that the authors or compilers of the seven-day account were involved in a polemic defense of the Israelite version of the generalized seven day week tradition that by then dominated the Ancient Near East. The Israelites had a seven-day week like other cultures in the region. Their goal was to provide religious justification for the unusual strictness of the Israelite mandate to spend the seventh day resting from labor, *and the obligation to free from labor servants, slaves, and hired hands,* who then performed much or most of the agricultural labor in this region where agriculture first emerged in human history. They created this religious justification by depicting Elohim as working six days creating the world and then resting, setting an example for the Israelites. To engineer this scenario, they ingeniously organized the creation of the cosmos into six days of divine "work" followed by a seventh day of divine rest. One can choose whether to be horrified at the scientific ignorance of the account or to be impressed by its literary creativity and its concern for those who performed backbreaking agricultural labor. (The issue of whether the text was "divinely inspired" or not is above an anthropologist's pay grade.)

Redirecting our focus to the origin of the Sun and the Moon, in the Genesis account the sun was created, not as a source of *life* but of *light*, of supplementary illumination. It was to supplement the mysterious "light" created on Day 1. One need only look around on a cloudy day. There is light that appears to be independent of that round burning celestial torch hidden by the clouds. The function of the Sun in Genesis was supplemental light. The Moon was to be a source of softer light at night, when total darkness would otherwise prevail. The Moon of Genesis, it should be noted, had its own light. The Hebrew words for Sun (shemesh) and Moon (yareakh) are in fact not even used in this chapter of Genesis. In this Hebrew passage they are both simply called *me'orot*—"lights" or "lamps" whose function was to give light. Elohim made a "big lamp" in the sky to shine during the day and a "small lamp" to shine at night. Each lamp would exercise its rule (memshelet) over the day or night respectively. The awareness that the moon has no inner light, that moonlight is simply a reflection of sunlight, would come later in human history. And it would first be formulated in Greek, Latin, and Arabic, not in Hebrew or Aramaic.

Besides its function as a luminary, Genesis attributed another function to the Moon—marking the onset of Israelite festivals. Such a calendrical function of the phases of Moon is by no means restricted to Genesis. The calendrical function of the Moon has been so widespread around the world in other cultures that it might qualify as a cultural universal. The cycle of "phases of the Moon", however, is an optical illusion that will of course be completely irrelevant to humans living on the Moon. We will discuss the implications of that paradox toward the end of this chapter.

2.2.2 China: The Moon as the Eye of a Culture Hero's Corpse

For the moment we are focusing simply on cultural theories of the Moon's origin. There is another subset of cultural theories in which the Sun, the Moon, and the elements of nature originated as different parts of the dead body of some ancestral spirit or long departed culture hero. Let's rapidly switch continents and examine this model of cosmogenesis, which dominates origin accounts in ancient China. Though there is no universally recognized single authoritative text, analogous to the Hebrew Bible, accounts of cosmogenesis and anthropogenesis were formulated in ancient Chinese writings.

There are two major figures: an anthropomorphic male giant named Pangu responsible for cosmogenesis and a hybrid female human/serpentine figure named Nüwa responsible for anthropogenesis. The stories about the female Nüwa are believed by many to have predated texts about the male Pangu. Patriarchal Confucianism and its religious companion Daoism eventually brought males and male deities into positions of greater power than females. Summaries of Chinese origin accounts often begin with the Pangu account, because it deals with the creation of the Earth, Sky, Sun, and Moon, whereas the Nüwa account, though perhaps chronologically earlier, begins with the creation of humans after the world, in some unspecified manner, had come into existence.

We can follow that order. Cosmogenesis begins with a cosmic egg in which all matter and life was potentially present but was mingled. Inside the cosmic egg was a humanoid figure named Pangu, who was asleep and slowly growing in the egg for 18,000 years. Inside this undifferentiated blob, Yin and Yang (the Chinese "female" and "male" principles which generated Earth and Sky respectively) co-mingled in cosmic chaos along with Pangu himself. Finally, the egg burst open and Pangu sprang into action. He swung his axe and separated Yin (the dark, passive female Earth) from Yang (the bright, active Sky). The two separated. The sky floated upward, and the earth emerged as a flat, stable separate entity below.

Thrilled to be alive, Pangu was determined not to be reduced again to a blob. To prevent a collapse and reversion, Pangu stood on the Yin Earth, raised his arms, and pushed upwards against the Yang Sky to keep it from collapsing again into the Earth. His pushing successfully increased the distance between the Sky and the Earth, and Pangu himself increased every day in size—a process which took another 18,000 years. By then Pangu had become a hairy anthropomorphic giant with arms, legs, horns, eyes, ears, and abundant body hair. The cosmos at that moment consisted only of Sky, Earth, and Pangu. But Pangu, exhausted from his 18,000 years of laborious exertions, died. It was from Pangu's body parts that the rest of creation emerged. His right eye became the Moon, his left eye became the Sun, his facial hair the stars. His breath became the wind, his voice thunder, his blood rivers, his sweat rain and so on.

We see here that, as in the Genesis account, the Moon emerged simply as one element in a process entailing emergence of the entire cosmos. Unlike the Genesis account, however, there is no Supreme Creator deity. Chinese Daoism, Buddhism, and village religions venerate hundreds of spirit beings, but they are all lesser deities with limited spheres of action. The existence of the original Cosmic Egg and the subsequent quasi "Big Bang" explosion are therefore left unexplained.

Unlike the Genesis account, there is no Chinese group of believers insisting that the Pangu account is word-by-word divinely revealed historical truth that must be believed under pain of post-mortem punishment. I have discussed both the Pangu and Genesis accounts in anthropology courses in Shanghai, Chengdu, and Nanjing. Chinese students are familiar with the Pangu account but laugh when I ask them if anyone insists that they "must believe" that the account is historically true. It is, in their eyes, venerable Chinese folklore, not actual history. This respectful skepticism long antedates, and has nothing to do with, the Maoist anti-religious Cultural Revolution (which I learned can now be openly criticized in Chinese classrooms). Chinese converts to Evangelical Protestant Christianity may obey their Western-trained Chinese pastors who insist that the Genesis six-day account must be literally believed—or else. (I have found Chinese Catholic priests and laity to be more laid back about humanity's evolutionary origins.) But I have heard nobody in China, nor have I encountered any Chinese text, insisting on the historicity of the Pangu account.

2.2.3 Repair of the Moon: The Chinese Female Deity

If the male Pangu was China's source of the cosmogenesis that produced the Moon, it was a female deity who was responsible for anthropogenesis, the creation of human beings. She also ensured their subsequent protection by repairing damage done to the natural world, including to the Moon and the Sun, by two evil male antagonists in their battle with each other. Unlike the text of Genesis, however, in which distinct accounts were (according to many scholars) later pasted together into a single scroll, the two accounts occur in distinct Chinese texts unrelated to each other.

In the beginning, a solitary female deity named Nüwa, with the upper body of a woman and a serpentine lower body, led a lonely existence in a world where she was the only living being. She looked into a river and was thrilled so see a woman just like herself. Her joy was dampened, and she had to laugh, when she realized it was simply her own reflection in the water. But the reflection inspired her with the idea of creating clay figurines that looked like her. She took clay and fashioned it into lifeless humanoid figurines, male and female, with faces, and with limbs—legs as well as arms. (Some texts explicitly describe the clay as yellow, a predictable Asian preference over the white of Caucasians or the black of Africans.) To her great joy, as soon as she placed them on the ground, the lifeless figurines came alive, talking, calling her "mama", and dancing.

The Earth was a flat, level structure supported by four mountains. Nüwa was so thrilled with her "children" that she feverishly created large groups, both male and female, until her hands grew tired. She then resorted to quicker creative measures entailing the rapid splattering of mud. (The humans she created with her own hands were the Chinese nobility. The others that emerged from splattered mud were commoners. Origin accounts often rationalize existing social stratification.) The newly created humans reproduced and populated the Earth beneath the peaceful Sun and the Moon.

But two destructive male warrior gods began fighting with each other for control of the world. The loser was so angry he deliberately banged his head against one of the four pillars supporting the flat earth, knocking it over and causing the Earth and Sky to rupture. If he could not rule the world, nobody would.

The results were catastrophic. The Sun and the Moon were dislodged from their place. The Earth tilted precipitously. Floods inundated the Earth; fires broke out everywhere. Of most concern to Nüwa was the unleashing of wild animals and birds of prey that suddenly brought death to the humans which she had created. She sprang into action, repairing the sky with colored stones and restoring the pillars of the earth with the legs of a cooperative Great Turtle. (The earth continued to tilt a bit, which is why western China is higher than the eastern coast and the rivers in China all flow eastward to the sea.) Thanks to Nüwa's interventions, the earth was restored and the humans who survived lived in peace. We thus see in China an account of anthropogenesis in which a loving and merciful female not only creates humans but also counteracts the cosmically damaging effect of the behavior of destructive males, in the process restoring the Sun and the Moon to their proper places in the sky.

2.2.4 Anthropogenesis in the Ancient Near East: The Destructive Female

A quite different gender scenario occurs back in the Ancient Near East. In that world region, death is brought to the world by the behavior of a female. For that we must briefly return to Genesis, but begin in Chap. 2.

The creation of humans by God's command had been mentioned briefly at the tail end of Day 6. But the details of anthropogenesis are contained in a quite different account that begins in Chap. 2. This account opens with an existing world in which there was only a barren earth with no vegetation yet. A mist ('ed) suddenly rose and moistened the dust ('afar) of the barren earth. From this moistened clay-like dust, YHWH-Elohim molded[2] a human male, breathed life into his nostrils, and the man came alive. Only then did Elohim plant vegetation and create a garden. God placed the man in the garden to take care of it and use fruit as food—except for one fruit that was forbidden to touch. The man was lonely, however, amidst all this vegetation, so God again used the ground to shape animal life forms—specifically "animals of the

---

[2] The verb is *yatsar,* which means to mold with the hands, like a potter molds clay. It is a different image of the creative act from that of Chap. 1, in which the Creator simply speaks ('amar) and creates from nothing (bar'a). Later in this second account, YHWH-Elohim plants a garden and later walks in the garden and asks Adam where he is when Adam and Eve hide themselves. The YHWH-Elohim of this account is depicted as much more human—molding clay, breathing life from his nostrils, planting a garden, walking, searching—, than the majestic, disembodied Elohim of the seven-day account.

field" and birds—as companions to the man. The man—he is named Adam, a play on the Hebrew word *Adamah*, which refers to the "soil" from which he came—gave names to all the animals. But he was still lonely. So, God put the man into a deep sleep, removed one of his ribs, and shaped it into a woman, who was named Eve (cognate with the Hebrew word for "life").

The man is delighted with his new companion but warns her about the fruit that they are not allowed to eat. (Though created as adults, they each spoke fluent Hebrew, the first human language according to later rabbinic sources.) Eve takes her first walk in the garden, passes the forbidden tree, and is cajoled by a talking snake into eating the forbidden fruit.[3] Unlike the Chinese Nüwa account, in which Nüwa restores the Sun and Moon to their place, this Ancient Near Eastern account of anthropogenesis makes no mention of the sky or its components. The Sun and Moon figure as elements only in the preceding seven-day account.[4]

A major difference between the Chinese origin accounts and the Ancient Near East account is the role of women. In China, the first woman is a creator/protector/and restorer of human life. In the Ancient Near East account, at least the one in Hebrew, humans were supposed to live forever. But it was the first woman, Eve, who brought death into the world by eating the forbidden fruit and coaxing Adam to do the same. The beneficent female role of Nüwa's is often attributed to its origin in a pre-Confucian, pre-Daoist phase of Chinese social organization that was matrilineal in its descent rules, where property would therefore have passed down through females, and where females consequently had a degree of economic and social power.

No such matrilineal phase has been documented for the Ancient Near East. The Hebrew God is referred to exclusively with masculine pronouns and with masculine verb forms. The four angels mentioned in the Hebrew scriptures (Michael, Gabriel, Rafael, and Uriel) are all male, and have positive roles. The only brief reference to a female spirit is to Lilith (Isaiah 34:14), and she is demonic, not angelic. One Hebrew bible translates her name as "night-monster". The female spirit Lilith is associated in the text with thorns, wild dogs, jackals, and snakes. The depiction of female spirits is quite different in ancient China and the Ancient Near East.

---

[3] Despite "Adam's apple" stories, Eve could not have eaten an apple. Apples did not exist yet. The fruit that Eve ate was a special fruit, unavailable to us moderns, that gave "knowledge of good and evil." And despite common Christian exegetic assertions, it was not Satan that possessed the serpent and tempted Eve. According to the text, the serpent was simply the "most clever of all animals", so clever it could talk. It was simply a garden-variety talking snake; there is nothing about Satan in the text. When God doled out punishments afterwards, he punished the snake, not Satan. The snake lost its legs and was doomed to crawl on its belly.

[4] The differences between the two origin accounts in Genesis are substantial. To partially summarize: In Chap. 1 God created vegetation on Tuesday, birds on Thursday, land animals early on Friday, and a man and a woman together on Friday afternoon, as the final act of creation. In this second account, Adam was created first, then vegetation, then birds and land animals, and finally Eve. The major shared element in both accounts is their insistence on a monotheistic framework: everything was created by Elohim or YHWH-Elohim. It functioned as a polemic denial of the polytheistic battles among multiple lesser deities found in other origin accounts. We shall examine some of these.

2.3 Hinduism, Buddhism, and the Moon

We have examined accounts from China and the Near East with respect to the origin of the Moon. Other world regions and religions have different accounts, too numerous to deal with here in detail.

Hinduism presents an interesting contrast. In the first place, it is the one major world religion that has no "founder". By one count, about 15% of the world's population currently practice Hinduism, mostly in India where 80% of the population identify as Hindu. The Vedas, the major scriptures of Hinduism, give no specific account of the creation of the Moon, but there are Hindu accounts of the creation of the world. One account involves Vishnu and Brahma. In the beginning before time, there was only an expanse of water. A giant cobra was coiled on the surface of the water. Vishnu, a high deity, slept in the protective coils of the cobra. The silence was gradually replaced by the mystical, pulsating, and increasingly loud sound "OM" that came from beneath the waters. For the first time, dawn began to break. Vishnu awoke. With him was his servant Brahma, enveloped in a lotus flower. Vishnu told Brahma "The time has come!" and commanded him to create the world. A fierce wind began blowing. Vishnu and the cobra disappeared, leaving Brahma, his lotus flower, and the wind. Brahma split the lotus flower into three parts. One became the high heavens, another became the earth, and the third became the sky in between. Brahma then covered the barren earth with vegetation, land animals and insects, birds in the air, and fish in the seas (wwgschools.org n. d.; Hindu Creation 2011).

Hinduism is a diverse religious tradition with equally diverse texts. Another account from the Rig Veda also begins with primeval waters. A golden egg emerged in the waters. Eventually Prajapati ("Lord creator") emerged from the egg. After a year of silence, he finally spoke, and the earth was created. He spoke again and the sky came into existence, with its seasons. Prajapati, however, felt lonely and had himself subdivided into a husband-and-wife couple. It was this couple who created the other gods, the elements, and humans.

Neither of these Hindu accounts deals explicitly with the creation of the Moon. There is, however, one prominent Hindu deity—Chandra—who is seen as a personification of the Moon. He is depicted in some accounts as a bellicose male rascal given to misbehavior. He fell in love with Tara, the wife of Brihaspati (the planet Jupiter). The feelings were mutual. Jupiter was, after all, an old man; the Moon was young, like Tara. With her consent Chandra kidnapped Tara. They produced a child Budha, who became the planet Mercury. Brihaspati (Jupiter), the jilted husband, was furious and prepared for war. Both he and Chandra marshalled their own planetary allies. Through the intervention of Brahma (the earlier mentioned creator deity), however, Tara eventually ended up back on Jupiter (while still pregnant with Budha) and war between the Moon and the planets was averted. On another occasion, Chandra offended the elephant deity Ganesha. Ganesha broke off one of his tusks and threw it at the Moon, cursing Chandra and in the process damaging the surface of the Moon. This is the origin of the waxing and waning of the Moon. The impact of Ganesha's tusk can still be seen in a large crater on the lunar surface. (The conflict is long past. Lunar settlers need build no shelters against terrestrial elephant tusks.)

2.3.1 Buddhism and the Moon

According to one credible calculation (Pew Research Center 2018) 7.1% of the world's population is Buddhist, which would yield a world Buddhist population of about 550 million. Nearly half of them are in China (though they account for less than 19% of the Chinese population). There are several varieties of Buddhism around the world. But with respect to the Moon, one can point out that Buddhist texts, with their emphasis on reincarnation, and on endless cycles of birth, death, and rebirth until enlightenment is reached, have no clearly delineated "creation account." For classic Buddhist scholars, the world had no beginning. The goal of life is enlightenment and liberation, not insights into the details of cosmic history. The Moon, however, does fit easily into the Buddhist cyclical vision of being by its eternally repeated cycle of reoccurring phases. It is a visible celestial reminder of the cycles of life, death, and new life.

However, for Buddhism the Moon is more than just a visual symbol. There are beliefs about the differential spiritual impact of the Moon on humans at different lunar phases and there are rituals connected with these phases. Even before the rise and spread of Buddhism, religious practices in ancient India already included a recognition of the power of the Moon to influence humans. The impact of the full moon in particular was seen as dangerous. Religious leaders encouraged people to dedicate full moon periods, nights as well as days, to an intensification of protective spiritual practices. Full moons became periods of relaxation, partial withdrawal from daily obligations, and visits to religious centers to listen to the words, and imbibe the wisdom, of spiritual leaders.

Though lunar issues were not an original concern of Gautama Buddha himself, he accepted the advice of a local king to take advantage of the gatherings in full moon periods to spread his own teachings. Lunar beliefs thus slowly made their way into the core of Buddhist teaching and practice. Two full moon practices became particularly prominent. One was the recitation by monks at full moon of texts enumerating required monastic practices. At the full moon Buddhist monks and nuns not only recite texts that embody the rules, but also publicly confess any transgression on their part of monastic disciplinary rules. The full moon thus becomes a monthly occasion for purification and spiritual intensification, a practice referred to as Patimokkha. Furthermore, for many Buddhist laity, full moon days are occasions for overnight retreats at monasteries or other spiritual centers to increase their knowledge of, and intensify their involvement in, the practices taught by the Buddha. Among the practices recommended in these retreats are meditation sessions, chanting, reading, attendance at classes, group discussions, and others.

Some online promoters of Buddhism for westerners may describe the full moon as a period of intensified spiritual energy emanating from the full moon. It is probably historically more accurate to posit the opposite. The full moon was seen as a period of increased spiritual danger. The intensified religious practices arose as a protection against these dangers, in effect converting the full moon into a period of intensified spirituality. But this intensified spirituality does not emanate from the full moon itself. It emanates from the intensified practices adopted by Buddhists to combat what was historically viewed as the negative influence of the full moon (Norrad 2017; Rathanasara 2018; Index Mundi 2020[5]).

2.4 "The Man on the Moon"

In our analysis of lunar traditions, we can leave the domain of religion, with its emphasis on spirits and rituals and spiritual authorities, and enter that of folklore, only remotely connected to spirit beings or clergy. Neil Armstrong was the first man on the moon in modern times. But the theme of the "man on the moon" is of ancient origin. Throughout history members of different cultures, when viewing the Full Moon, have perceived the image of a human face peering down on Earth from above. These are referred to as "pareidolia", images that are created when an observer (incorrectly or creatively) projects a known image onto a surface with features that vaguely resemble what is known.

The images of humans or animals on the Moon are created by the juxtaposition of large dark lunar maria (lava "seas" in Latin) and brighter elevated regions. The formation of pareidolic images, though once considered a possible sign of mental illness, is more likely a normal and possibly pan-human phenomenon. It happens frequently with clouds. By this same process, two horizontal lines next to each other in the middle of a circle with a centrally positioned third line slightly below toward the bottom of the circle will rapidly be interpreted as a human face with eyes and a mouth—even without the nose or ears.

All human cultures are exposed to the full moon, and many of them have projected images onto the lunar surface and incorporated these projections into their views of the Moon. But though the process of projection may be based on universally shared neurological underpinnings, the images seen, and the interpretation of these images take widely different cultural forms.

Let us examine several pre-modern European beliefs in this regard. Forged in the context of Christianity, Europe has had few religious concerns with the impact of the Moon on humans. Unlike Greco-Roman, Egyptian, Hindu, or Chinese traditions, there are no lunar deities in Christian theology. In the fourth century Nicene Creed, at the Annunciation Jesus "descendit de caelo"—came down from heaven—but not from the Moon. He subsequently "ascendit in caelum"—ascended into heaven—but, again, not to the Moon. Nor does Christian theology posit the existence of lesser angelic or demonic forces specifically associated with the Moon. The Moon and its

---

[5] Excellent one page overview of world religions.

phases are simply not a matter of religious concern in any variant of Christianity, be it European Catholicism or European Protestantism.

However, European societies did develop beliefs about the Moon. These lunar traditions, however, operated tangentially to European Catholicism and Protestantism and are best analytically assigned to the realm of folklore. One aspect of this folklore concerns the above-mentioned pareidolic images that emerge in the human imagination when seeing the variegated lunar environment from afar (Baring-Gold 1876).

What is common to most of the stories is that the Moon is viewed as a place of exile to which a person is sent in punishment for some transgression. One story from Germany is that of a man who was found violating the Sabbath by gathering firewood. He was "doomed to reside in the Moon" till the end of time. At full moon he can be seen seated with a bundle of firewood on his back. Another version of the story describes the man as a giant who is assigned to hard labor pouring water from the Moon to the Earth to produce high tide. He is allowed several daily hours of rest, at which point the tide ebbs. Soon after his unending labor cycle recommences. A similar story of a man banished to the Moon is even found in Shakespeare's "Midsummer Night's Dream". In general, the European versions of the man in the moon depict him as a thief or a Sabbath-violator being sent to the moon in punishment. One violator was offered the choice of burning in the Sun or shivering on the cold Moon. Without hesitation he chose the Moon. It is not as painful as the eternal fires of Christian or Islamic Hell, or even the temporary flames of Catholic Purgatory or Jewish Gehinnom (which according to the Rabbis lasts at most a year.) On the Moon, a transgressor is beyond the reach of Death, so he stays fully alive for his punishment in lunar exile. It is perpetual, unpleasant, but much less agonizing than the eternal flames suffered by those in Dante's Inferno.

In recounting these images of the man in the moon it is important to repeat that, unlike the Genesis account, which many Jews and Christians view as divinely revealed factual world history, these accounts are presented as simple folklore, much as the Pangu and Nüwa stories in China, rather than as objects of obligatory belief. This folklore should be of at least passing interest to would-be 21st-century lunar settlers. They should realize that their lunar destination has never been viewed, at least in Western folklore, as a longed-for Shangri-La. In European folklore it is rather a hardship post to which people are sent as punishment, a type of lunar penal colony.

But folkloric accounts of lunar exile seemed to have assumed the availability of air, water, and food on the Moon. It may be a bit chillier than most places on Earth, but apparently tolerable. We now know that that is not the case. This entire volume is dedicated to analyzing and planning for survival on a barren floating rock hostile to human survival without technologically advanced measures. It is quite probable that lunar settlers will not want to be consigned to Luna, like the man in the moon, for the rest of their lives. Crammed into subterranean (or sublunar) close quarters, they will want to complete their assignments as quickly as possible and scoot back to the beaches, rivers, forests, mountains, parks, and gardens of planet Earth. Unlike the man-on-the-moon and other permanently banished culprits, the settlers will have well maintained spacecraft that can transport them back home.

2.5 Lunar Eclipses

There has been a particularly prominent level of cultural concern with the possible negative impact of a lunar eclipse. The literature produced by several ancient cultures mentions a widespread belief that a "blood moon"—a lunar eclipse during the full moon—is caused by an animal devouring the Moon. The redness came from the bleeding of the Moon following the attack. The ancient Egyptians identified the animal as a pig. Among the Central American Mayan and the South American Inca, the predator was a jaguar.

Ancient Mesopotamian and traditional Chinese beliefs about eclipses, independently from each other as far as we know, gave lunar eclipses a political interpretation. Lunar eclipses could lead to, or at least predict and symbolize, the death of the king. In Mesopotamia, the attackers during a lunar eclipse were not animals but demons. In attacking the Moon during an eclipse, people thought that the demons were going after the King. Since the civilizations of the Ancient Near East had skilled astronomers (the famous Matthaean Magi who followed a star to Bethlehem are examples), they had already learned how to predict lunar eclipses, which in turn were thought to predict the death of royalty. It is reported that, to protect the king, the palace would appoint a temporary surrogate king. The fate of the surrogate was unclear. His death may have been publicly engineered to deceive the demons and to discourage follow-up against the real king. At any rate there are no historical reports of competition for the honor of temporary regal appointment.

A similar cluster of beliefs is found in China. There the lunar eclipse was caused by a dragon swallowing the Moon. Unlike in the West, the dragon in China is usually a benevolent figure, responsible for bringing the right amount of rain at the right time. The dragon in fact symbolized benevolent regal power. The clothing of emperors often had embroidered dragon figures. But during an eclipse, the dragon's attack on the Moon symbolized the impending death of the monarch. During the eclipse, the masses would beat gongs and drums to scare the dragon. They would beat their dogs to make the dogs bark and whine and participate in the noise to intimidate the dragon.

It should be mentioned that lunar eclipses are the only form of lunar misbehavior recognized by human cultures. The weather constantly misbehaves, and does the earth with earthquakes, and mountains with volcanoes, and even the Sun when it sends blistering, prolonged heat. But except during eclipses, which even the ancients learned how to predict, the Moon always behaves with gentle regularity. Religions are filled with rituals to draw rain and prevent floods. There are few if any rituals to prevent the Moon from misbehaving, which it rarely does. Even Judaism, which has a monthly New Moon ritual, does not invoke the Moon or even ask God to make the Moon behave. The Moon never misbehaves. The New Moon ritual simply asks God for blessings for the coming month, a request which the Moon itself is in no position to grant or deny.

2.6   Beliefs About a Lunar Impact on the Human Body and Mind

We have seen that the "man on the moon" folklore deals with the imaginary function of the Moon as a place of punishment for humans who transgress some rule. In this body of folk wisdom, the Moon is portrayed as a type of penal colony. But the Moon itself, though it behaves regularly and predictably can cause harm to individuals on Earth. There is a widely distributed view about negative impacts of the Moon on the bodies and minds of human beings. In these narratives the human victims are not guilty of any transgression, nor are they transported to the Moon. They suffer from lunar emanations "in situ" on Earth, itself. Nor are the victims the targets of hostile lunar spirit beings. The negative forces emanating from the Moon are natural, impersonal forces.

Furthermore, these noxious impacts are not stable features of the Moon. Rather, they fluctuate with lunar phases. Paradoxically, however, whereas modern media and singers tend to romanticize the Full Moon as a period of beauty, peace, and love, in many cultural traditions it is precisely the Full Moon that emerges as the most dangerous and harmful phase of the month. Full moon days require special protective measures.

The following is a partial list of some of the negative effects of the full moon reported in different cultures (Keral.com 2012; Lee 2014). In the absence of scientific evidence on most of these issues, we should classify them for the moment as cultural beliefs rather than as proven scientific facts. The full moon is believed to have the following effects.

- It precipitates mental and behavioral disorders or exacerbates those tendencies in people inclined to such disorders. (It is often pointed out that the word "lunatic" is derived from *luna,* the Latin word for moon.)
- It exacerbates the impact of epilepsy, asthma, bronchitis, and certain skin diseases.
- It reduces the efficacy of certain medicines. (Physicians in antiquity reported this effect.)
- It increases the likelihood that snake bites will be lethal.
- It increases bleeding during surgery.
- It reduces the restfulness of sleep.
- It exerts an impact on the female menstrual cycle.

2.6.1   The Moon and Menstruation: Scientific Evidence

Some of these beliefs have been subjected to empirical validation or refutation. Among these are the beliefs regarding menstruation and the beliefs about the negative impact of the Moon on sleep. From ancient times there was a belief that lunar phases influenced the female menstrual cycle. The word "menses" itself, as well as the derivatives "menstrual" and "menstruation", all come from the Latin word *mensis,* the word for "month", and they are also more remotely related to the Greek word *mene,* which means "moon". This direct lexical association between the menstrual flow and "month" simply reflects that the lunar month and the *average* female menstrual cycle are of the same length—about 29.5 days. Whether the lunar and menstrual cycles are synchronized in any way is a separate question.

One widespread belief, at least in the West, has been that women are more likely to ovulate during the full moon and menstruate during days of the new moon (Squier 2016). As far back in 1708, a physician impressed by the research of Isaac Newton, explained the correlation by positing that the gravitational impact of the Moon affected not only the tides but also the fluids in the human body, including menstrual flows. The impact of the moon on female monthly secretions, he argued, was stronger nearer the Equator. Lunar gravitational impact was also thought to exacerbate epilepsy and even kidney stones (Royal Museums Greenwich 2019). More recent commentators have posited possible non-gravitational impacts of the Moon. Among them are the impact of moonlight itself, and lunar electromagnetic radiation (Clue 2019).

Numerous research projects have been carried out in the last 70 years to determine empirically whether there is in fact a statistical correlation between lunar phases and onset of menses (Clue 2019). A minority of studies have detected some non-random co-occurrences. But the overwhelming thrust of the studies is that the onset of menses occurs randomly through the month, independently of lunar phases. The article thus labels the conventional wisdom, with its posited association between menses and the Moon, as a "myth", in the colloquial English sense of "falsehood", not a myth in the more academic sense of a sacred story symbolically communicating hidden truths.

2.6.2   The Negative Impact of Lunar Phases on Sleep

Scientific research has also been carried out on the impact of lunar phases on sleep. The most common proposition is that human sleep is negatively affected by the full moon. A "common sense" explanation for this pattern would be interference with sleep by the more powerful illumination of a full moon in comparison with other phases (Cajochen et al. 2013). Researchers have gathered experimental support for the general proposition, in a manner that excluded the confounding variable of exposure to lunar illumination by those subjects whose sleep was being monitored. They first point out that "endogenous rhythms of circalunar periodicity" have indeed been documented in several marine species. Certain aspects of their behavior are indeed influenced by lunar phases, and there is an "underlying molecular and genetic basis" for this lunar impact. Is the same true in humans, specifically with regard to sleep?

They found that around the full moon, EEG measures showed a 30% decrease in NREM deep sleep, a 5 min increase in the time it took to fall asleep, and a decrease of 20 min in the duration of total sleep time. These objective indicators were buttressed by subjective retrospective assessments by research participants concerning less restful sleep during the full moon. (The subjects, it should be noted, were not told that the impact of lunar phases was being analyzed. Only the researchers knew that lunar phases were being analyzed. Subjects were simply asked how well they had slept the previous night.)

Their findings also correspond with what another research project found about reduced sleep time during full moon. In this other study, subjects were simply asked how long they had slept. However, theirs was the first research to test the hypotheses using EEG measures of cortical activity. The researchers pointed out several possible confounding factors in the research and proposed even tighter designs for future research. However, unlike the support for the null hypothesis in the above-mentioned menstruation studies, this study did find a basis in the long-standing and widespread cultural belief concerning the negative impact of the full moon on sleep. The specific causal mechanisms remain a matter of speculation, but to all appearances there is a bona-fide causal linkage between full moon and lest restful sleep.

There are numerous other studies that attempt to verify or falsify traditional beliefs and propositions about the impact of the phases of the moon, particularly the full moon, on human wellbeing.

2.6.3   The Full Moon in China

With respect to full moons, however, reports of a traditional Chinese reaction indicate a much more positive attitude than the negative one found in other cultures toward a full moon. There was a frog in ancient China referred to as the *jinchan,* the "golden frog". If a jinchan appeared near your house during the full moon, it meant that wealth was about to arrive. If no golden frog appeared in the full moon, people would place a statuette of a golden frog in their homes, sitting on a pile of coins and holding a coin in its mouth. But Feng-Shui ("wind–water" geomancy) specialists would warn: Never place the golden frog in front of the main door facing outward. (Wealth would presumably drain away from the home.) Have it face inward. And people were warned not to place it in the bathroom, kitchen, bedroom, or dining room. Modern businesses in China may also strategically place the statue of a money frog. (It is best if it points toward the cash register.) But these are all faute-de-mieux secondary measures. The best guarantee of prosperity is when a live golden three-legged frog appears near your house during the full moon.

2.6.4   Phases of the Moon and Agricultural Productivity

There is also an impressive body of literature about the relevance of awareness of lunar phases for guiding agricultural practices. The basic determinant of agricultural success hinges on planting and harvesting in the proper annual season, which is governed by the position of the Earth with respect to the Sun, not the Moon. However, within the planting, weeding, and harvesting seasons, there is also a vast body of folk wisdom about which procedures are best carried out on which crops at which phase of the lunar cycle. For reasons of space, we will only refer readers to the well-known and heavily utilized Farmer's Almanac, published annually since 1818, as well as other links (McLeod 2021; Farmers' Almanac n. d.; Lunar Farming n. d.; Anderson n. d.).

2.7   The Moon as a Marker of Time: Irrelevance for a Lunar Colony

In this final section we will address a question directly affecting lunar colonies: the manner in which traditional units of time on planet Earth, determined by interactions among the Sun, the Moon, and the Earth, are incompatible with the reality of time on the Moon, where the Moon is no longer an alien body in the sky, but the planet on which you live. The biological and cultural rhythms which evolved on planet Earth are radically out of kilter with the reality of lunar and solar cycles that settlers on the Moon will encounter. We will proceed briefly but logically through four universal time units that govern human life on earth: days, weeks, months, and years. We will discuss the relevance—or irrelevance—of these traditional units for human life on a lunar settlement.

2.7.1   The Day

All mammalian species have evolved biologically embedded time clocks of roughly 24 h as a response to the timing of the rotation of the Earth on its axis. Having evolved on planet Earth, the human organism now requires roughly eight hours of sleep in any given 24 h period. It usually occurs at night, though there is flexibility. But the required 24 h oscillation between activity and sleep has non-negotiable neurological underpinnings with only minor leeway for culture-specific deviations. Sunrise or sunset on the Moon, in contrast, occurs once every 27 days. There is no way that lunar settlers (unless they are robots) could adjust to the lunar "day", working continuously during one lunar day (which lasts 16 or 17 Earth days), and then spend nine or ten earth days sleeping during the lunar night. The "day", as defined by the time between one sunrise and the next, becomes irrelevant to human life on the Moon. This presents a dilemma on a lunar colony. The lunar day becomes irrelevant. The human neurological system does not. Other arrangements will have to be instituted to accommodate human biological clocks. (They have already been instituted on the International Space Station, where "sunrise" occurs about once every 92 min.)

2.7.2 The Year

We will momentarily bypass the "week" and the "month" and skip to the "year". The terrestrial "year" also becomes irrelevant on the Moon. On earth the life of many plant and animal species is governed biologically by cycles of "years" with their seasons, produced by the axial tilt of the Earth as it circles the Sun. In non-human mammals, annual mating and other biological rhythms are seasonally governed. Humans override annual mating cycles and are ready to reproduce all year long. However, human life is still governed economically by yearly cycles. The major livelihood pursuits of pre-urban human societies—hunting and farming being the most prominent—were governed by the biologically determined annual cycles of plants and other animals. Humans could not hunt animals whose annual migratory cycles brought them elsewhere, nor could they plant wheat in the snow. In short, humans share with fellow mammals, and even with plants, the universal cycles of the "day" and the "year", both of which are determined by the interaction between the Sun and the planet they have live on.

These seasonal oscillations will be irrelevant to life in a lunar settlement. Whereas the 24 h earth day, with its biological clock, will still govern human life on the Moon, the oscillations of the lunar year will not. The lunar "year" in fact lasts only about 12 lunar days—i.e., the Moon will have rotated on its axis only about 12 times before it reaches the same position relative to the Sun. (The nearby Earth will have made 365 rotations on its axis in the same time.) The year as a unit of time thus becomes biologically irrelevant on the Moon. It also becomes, at least in principle, behaviorally irrelevant. That is, lunar settlers will not have to time their activity to seasonal migration of animals or a seasonally determined planting and harvesting season or seasonal variations in rainfall or temperature. Every day in the lunar year can in principle be spent in the exact same activities, with no climatically determined seasonal variations. Will it be boring? That is not the issue here. We are simply analyzing the irrelevance of terrestrial time units to life on the Moon.

2.7.3 The 30 Day Month

We can turn in that vein to the other universal time units governing human life on planet Earth: the "week" and the "month"? What will happen to the "week" and the "month" on the Moon?

As for the "month", its lunar origin is clear. The universal optical illusion of "lunar phases" has engendered in human cultures the phenomenon of the "month", with the "new Moon" generally chosen as the onset of a new month.[6] And the cultural norm around the world has been a cycle of 12 named months per year to roughly correspond

---

[6] The word "month" itself is cognate, in the Germanic branch of Indo-European languages, with the word for moon. The English lexical couple moon /month is paralleled in German with Mond / Monat, in Dutch with maan / maand, in Swedish mane/månad, in Norwegian måne / måned. There

with the seasons of the solar year. The correspondence, however, is rough, and this has created a dilemma. There are 29.5 days between one new Moon and the next, which generates 354 days per 12 month "lunar year". The time between one solar equinox and the next, however, which is one measure of the solar year, is slightly over 365 days. The lunar year is about 11 days shorter than the solar year. This will mean that over time a given lunar month (all of which usually have their own names) can occur in the spring or summer as well as in the fall or winter of the solar year. The solution adopted in the Jewish world, whose religious calendar is still based on new moons and full moons, is a bit complicated. An additional month is inserted into the calendar seven times in every nineteen-year cycle. Thus Passover (which was originally an agropastoral festival) will always fall sometime in the northern solar spring, and Hanukkah always falls sometime in the solar winter.[7]

A much simpler solution has emerged that eliminates the Moon from consideration. Following the example of the Roman Julian calendar and its more recent sixteenth century variant, the Gregorian calendar, the world at large eventually adopted a simpler solution: Get rid of the link between the "month" and the Moon, simply divide the 365 day year into twelve solar months of roughly equal length, and every four years simply add an extra day to the shortest month. Thus, January is always in the northern winter, July in the summer, and so forth. This is now the dominant arrangement in the globalized world. Even countries that maintain a lunar calendar for religious festivals follow the Gregorian solar calendar in ordinary daily life.

But is the "month" even a necessary unit of time on the Moon? The answer is a tentative no, barring convincing arguments to the contrary. When standing in the Moon its phases do not exist and it cannot signal time periods. This is not to argue for its elimination. Counting months will cause no harm to lunar life, and lunar settlers will arrive with pre-existing habitual yearly and monthly rhythms that they will want to protect. How can one celebrate a birthday without a birth month? And what about wedding anniversaries? It is a challenge imagining traditional human social life without months, no matter how alien the unit is to the objective planetary conditions on the Moon.

---

is a similar association in Mandarin Chinese: "moon" and "month" are both "yuè". January is "1st-moon", February "2nd-moon", etc.). This association of the words for "moon" and "month" is by no means universal. In the Romance languages each comes from a different root, as is true in the Slavic and Semitic languages. In Hebrew, for example, "month", *khodesh*, is related to the word "new", *khadash*, not the Hebrew word "moon", *yareakh*.

[7] In the Muslim world, whose religious calendar is also lunar based, this discrepancy is not viewed as problematic. The 12 lunar months are simply counted in order with no adjustments being made to keep them lined up with the seasons of the solar year. The sacred month of Ramadan can thus fall during any season of the solar year. Christians have opted for a solar year; Christmas, for example, always falls on Dec. 25th, whatever the phase of the Moon. There is one minor surviving lunar intrusion in the Christian calendar. Easter comes on the first Sunday after the first full moon after the Spring equinox (March 21.) Other Easter-related festivals like Palm Sunday and Ash Wednesday are determined by counting backwards from that Sunday.

2.7.4   The Seven-Day Week

That brings us to our final unit, the mysterious, artificial seven-day week that has now become the norm on a globalized planet. It is mysterious and artificial in that it has no apparent basis in any solar, terrestrial, or lunar cycles. Where did it come from? And is it a useful time-unit to export to the Moon? Can lunar settlers dispense with the seven-day week on the Moon, e.g., voting to switch to a five-day or ten-day week, on which the final day, in either case, would be a day of relief from work. Can humans on the Moon free themselves from the repetitive seven-day cycle that now frames human life in almost the entire Earth?

To answer that, it will be useful first to address question #1: Where did this seven-day week come from? Here we tread into a controversy briefly discussed earlier in these pages. Those familiar with the Judeo-Christian Bible and/or the Qur'an may assume that the seven-day week is "based on the bible", that it is a divinely instituted division of time that has come to dominate calendrical systems around the world. As I will point out in Chap. 18, judging from research done on astronauts, many lunar settlers will come with pre-existing religious beliefs and practices in which the seven-day week with its concluding sabbath is assumed to be of supernatural origin, specifically from the first Chapter of Genesis.

They will hopefully continue their sabbath observance but are encouraged to re-examine their view of the origin of the seven-day weekly cycle. Research in the Ancient Near East, specifically Sumeria in Mesopotamia, and later Babylonia, has documented the practice of the seven day week several centuries—perhaps a millennium—preceding the traditional date for the composition of Genesis by Moses in the desert (the 1500's BCE) Many Scriptural scholars date the composition of the text 1,000 years later, during the Babylonian Captivity.

An anthropological addendum to those historical insights would be that the seven-day week arose precisely in that part of the world where agriculture first emerged to eventually replace the hunting of wild animals and the gathering of wild vegetation. Anthropologists have documented that the spread of farming, though it increased human food production, introduced into human life unprecedented new patterns of back-breaking, time-consuming labor that were never experienced before. The Genesis punishment by which Adam would henceforth earn bread with the sweat of his brow reflected an objective feature of farming. This gives rise to the hypothesis, which can be stated but not tested here, that the emergence of back breaking agricultural field labor led to the emergence of a new biological requirement for a rest period beyond the 8 h sleep requirement that had already been hard-wired into the human genome.

The seven-day cycle, with six days of hard labor and one day of freedom from labor, was a cultural response to this new biological need. The victims of merciless, incessant field labor during certain phases of the agricultural cycle were those of the lower echelon of the increasingly stratified societies of the Ancient Near East—peasants, servants, serfs, slaves, and landless day-laborers who had to sell their labors to landowners. It is no anthropological accident that the seven-day week with its day of rest *arose precisely in the same region where agriculture first emerged.*

The first chapter of the Genesis creation account, with its seven-day framework, came long after the spread of the seven-day week. It functioned to give religious justification to the mandate to cease labor on the seventh day and *to desist from making one's servants, slaves, or hired labor to work on that day.* The priestly authors did this by creatively organizing the entire creation into six days of work by God, having him rest on the seventh day, and inserting this text as the first chapter into an already-existing creation account whose following chapters present a different order of creation unrelated to the sabbath rest. There is nothing new in these propositions; the consensus of many scripture scholars posits separate authorship and a later date for the composition of Chap. 1. What is being proposed here in addition is an explicit linkage between this process and the observation that the seven-day week first appeared precisely in that world region where anthropologists have documented the earliest emergence of agriculture. Agriculture, it is here proposed, created a new biological urgency for additional rest, beyond nocturnal sleep, for the social classes that had to perform the field labor. The Genesis seven-day account embedded this humanly important requirement into a religious framework.

2.7.5 Time at a Lunar Colony

Now, back to the lunar colony. Is the seven-day week an appropriate division of time—six days of work, one day of rest—to export to a lunar colony? The answer is a tentative yes. Some oscillation between workdays and leisure days will continue to be necessary on the Moon. It has continued to be necessary on Earth even when most people no longer engage in back-breaking agriculture, and even when the six days of work have been reduced to five. The seven-day week divides the month into four viable and roughly equal periods for rest from labors.

The human organism does not require a seven-day work/rest cycle. It could easily tolerate a four-day work week followed by a fifth day of rest, or even an eight-day work week followed by a ninth day of rest. However, there are at least three good reasons for adhering to the seven-day week on the Moon. (1) Lunar settlers would already arrive with that custom. (2) Some of the Jewish and Christian settlers will continue to practice their religion, in which either Saturday or Sunday is a religiously mandated day of rest. This will be discussed in Chap. 18. (3) A lunar base will be in frequent, perhaps daily, contact with a terrestrial base of operations—Houston, in the case of an American lunar base. It would make sense for that settlement not only to calibrate clocks on the Moon with Houston time on Earth, but also to maintain the same seven-day weekly schedule that governs the lives of Houston colleagues. The same utility of the seven-day week would also be true, mutatis mutandis, of a Chinese, Saudi, or Israeli lunar settlement. As a rule in dealing with terrestrial time schedules, "If it ain't broke, don't fix it."

This chapter began with a rapid historical journey through different cultural beliefs and practices with respect to the Moon. The following chapters will now turn to discussing the practical challenges of physically traveling to the Moon and establishing a permanent base there.

References

Anderson, M. (n. d.). Hunting by the moon phase. *Moon connection*. Retrieved from https://www.moonconnection.com/moon_phase_hunting.phtml.

Baring-Gold, S. (1876). Curious myths of the middle ages. *Internet archive*. Retrieved from https://archive.org/details/curiousmythsofmi00bariuoft/page/190.

Cajochen, Ch., Altanay-Ekici, S., Münch, M., Frey, S., Knoblauch, V., & Wirz-Justice, A. (2013). Evidence that the lunar cycle influences human sleep. *Current Biology, 23*(15), 1485–1488.

Clue. (2019). The myth of moon phases and menstruation. Retrieved from https://helloclue.com/articles/cycle-a-z/myth-moon-phases-menstruation.

Farmers' Almanac. (n. d.). Wikipedia. Retrieved from https://en.wikipedia.org/wiki/Farmers%27_Almanac.

Hindu Creation. (2011). From the Satapatha Brahmana, first millenium B.C., Rig Veda, The Big Myth, Distant Train, inc. Retrieved from http://www.bigmyth.com/download/HINDU_CREATION.pdf.

Index Mundi. (2020). China religions. Retrieved from https://www.indexmundi.com/china/religions.html (Excellent one page overview of world religions).

Keral.com. (2012). the three-legged money frog: Some dos and donts. Retrieved from https://web.archive.org/web/20121128051445, http://www.keral.com/Realestate/The-Three-Legged-Money-Frog-Some-Dos-and-Donts.html

Lee, J.J. (2014). Lunar eclipse myths from around the world. *National geographic*. Retrieved from https://www.nationalgeographic.com/science/article/140413-total-lunar-eclipse-myths-space-culture-science.

Lunar Farming. (n. d.). Moon made farms. Retrieved from https://www.moonmadefarms.com/lunar-farming.

McLeod, J. (2021). Why do we garden by the moon? Farmers' Almanac. Retrieved from https://www.farmersalmanac.com/why-garden-by-the-moon-20824.

Norrad, S. (2017). How to celebrate the full moon like a Buddhist. *Elephant Journal*. Retrieved from https://www.elephantjournal.com/2017/03/how-to-celebrate-the-full-moon-like-a-buddhist.

Pew Research Center. (2018). The global religious landscape. Retrieved from https://www.pewforum.org/2012/12/18/global-religious-landscape-exec/.

Rathanasara, K. (2018). The significance of the full moon. Dhammakami Buddhist Society. Retrieved from https://dhammakami.org/2018/08/26/the-significance-of-the-full-moon/.

Royal Museums Greenwich. (2019). Can the moon really affect our health? Retrieved from https://www.rmg.co.uk/stories/topics/can-moon-affect-our-health-behaviour.

Squier, Ch. (2016). This is what the moon actually does to your period. Grazia. Retrieved from https://graziadaily.co.uk/life/real-life/moon-period-cycle-menstruation-lunar/.

wwgschools.org. (n. d.). Retrieved from https://www.wwgschools.org/ClassDocuments/Hindu%t20Creation%20Story.pdf.

Part II
Engineering, Industrial, and Agricultural Perspectives

Chapter 3
The Changing Lunar Surface Environment: Hazards and Resources

Heidi Fuqua Haviland

Abstract The changing lunar surface environment reviews the current state of knowledge characterizing the lunar surface including observed and simulated conclusions. This chapter discusses potential hazards and possible mitigation for humans and machines at the lunar surface including the near surface plasma, dust, and high energy radiation. The implications of these processes for a human settlement at or within the lunar surface, are discussed for nominal and extreme solar conditions. This chapter also discusses the driving processes that contribute to the dynamic environment and changing surface including the movement of the Moon through the Earth's magnetosphere each month as well as the varying solar inputs over each solar cycle. The lunar water cycle and location of volatiles is discussed. Lastly, the potential resources are considered, including water, iron, and titanium, at and near the lunar surface and challenges for a self-sustaining lunar settlement.

3.1 Introduction

The Moon, Earth's closest celestial neighbor, is a familiar face in the night sky. Due to the Earth's gravitational tides, the nearside lunar hemisphere always faces towards the Earth and the farside points away. A lunar day in the mid latitudes consists of approximately fourteen 24-h days of sunlight followed by fourteen 24-h days of darkness. The lack of a global magnetic field and atmosphere leave the solid surface of the Moon to interact directly with its space environment. This includes a constant stream of plasma emanating from the solar corona, bombardment from meteorites, and high energy radiation. During one lunar orbit around the Earth, the Moon traverses into and out of the Earth's magnetosphere experiencing distinct plasma regimes. The highly cratered lunar surface experiences near constant bombardment from micrometeorites, and extreme temperatures, especially at the poles. Temperatures at the surface can reach 125 °C (260 °F, 400 K), during the day and to below −220 °C (−370 °F, 50 K), at night depending on latitude and the thermophysical properties of the

H. F. Haviland (✉)
NASA Marshall Space Flight Center, Alabama, USA
e-mail: heidi.haviland@nasa.gov

© The Author(s), under exclusive license to Springer Nature Switzerland AG 2021
M. B. Rappaport and K. Szocik (eds.), *The Human Factor in the Settlement of the Moon*,
Space and Society, https://doi.org/10.1007/978-3-030-81388-8_3

local regolith (Williams et al., 2017). Within permanently shadowed craters, temperatures are predicted to be even colder (Sefton-Nash et al., 2019). The lunar surface experiences regular small scale seismic and periodic large scale shaking (Nunn et al., 2020) that needs to be considered in future habitat designs. In this chapter, we review the dust, atmosphere, and plasma at the lunar surface under natural conditions. This includes understanding the driving solar and earth mechanisms which contribute to the lunar surface forming processes. We discuss the hazards of this natural environment to human and robotic operations and potential mitigation plans. Finally, we review the possible resources identified at the lunar surface and will provide a summary and closing thoughts for a future human settlement in terms of dealing with and thriving in the challenging environment of the lunar surface.

3.2 Background: The Dynamic Lunar Surface

The solid surface of the Moon is made up of a highly fractured layer of silicate, oxygen-rich, material consisting of rocks, soil, and dust. We use the term "regolith" and "soil" here interchangeably. Regolith is produced by the mechanical breakdown of rock by impacts and by the melting and fusing of rock, mineral grains, or glass fragments welded together with interstitial vesicular glass, which is known as agglutinates (Heiken et al., 1991). These regolith surface grains are irregular in shape varying from spherical to highly angular. In general, the average regolith thickness of the highlands is 10–15 m and 3–4 m in the mare regions. Approximately 20% (by mass) of the regolith particles are smaller than 20 μm (McKay et al., 1991) with electrostatic forces being able to loft or levitate the grains smaller than 10 μm (i.e., Colwell et al., 2007). Charged dust grain lofting at the lunar terminator has been proposed to explain the horizon glow observed by Surveyor (Rennilson & Criswell, 1974) as well as the Apollo dust experiments (O'Brien & Hollick, 2015). This has been proposed to be the result of changing electrostatic processes which occur with the passing of illumination into darkness. These dust grains have been observed being levitated at <1–10 m above the surface, with lofted particles being proposed to reach altitudes up to 100 km (Stubbs et al., 2007). This model shows the highest altitudes the levitated and lofted dust grains reached peaked just after the dawn terminator inversely proportional to grain size. The smaller particles (~0.01 μm) reached the highest altitudes. While recent spacecraft observations did not detect the existence of this population of nano-particles over the terminator (Szalay & Horányi, 2015), recent lab experiments have shown dust grains to loft to comparable heights when considering the inclusion of micro-cavities (Farrell et al., in press; Wang et al., 2016).

Due to the lack of a global magnetic field and atmosphere, the lunar surface interacts directly with a thin layer of loosely bound particles in a collisionless atmosphere, or a surface-bounded exosphere. Plasma (charged particles exhibiting collective neutral behavior) exists at the lunar surface originating primarily from the solar wind as well as are re-emitted from the surface. Recent studies have found that these three phenomena, the dust, plasma and atmosphere of the Moon, are all tightly

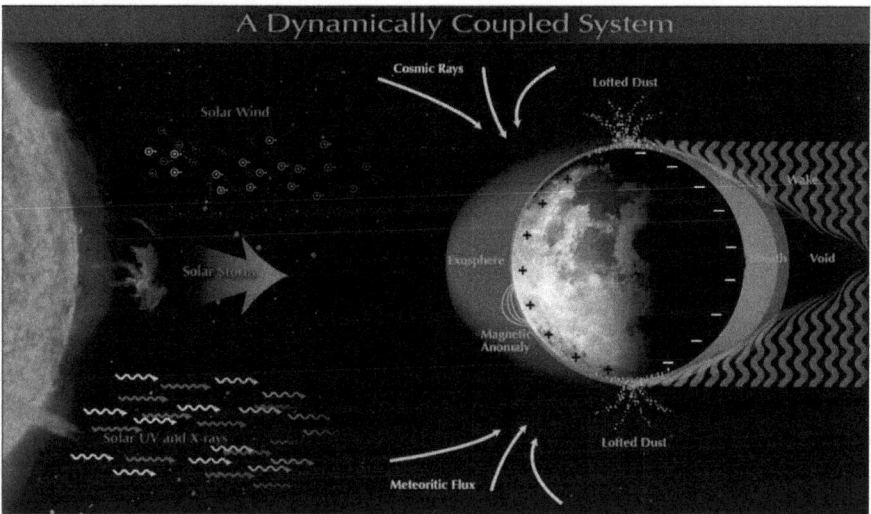

Fig. 3.1 An overview of the interconnected lunar surface dust, atmosphere, and plasma environments with solar drivers (Farrell et al., 2012b)

interconnected (Farrell et al., in press; Halekas et al., 2012). Moreover, the Moon is enshrouded by a dusty low-density plasma which comes from the impact of meteorites and space plasma directly onto the surface, Fig. 3.1. These phenomena are not unique to the Moon, and have been observed at all airless bodies in the solar system.

This is not a stagnant environment, but one that is very rich and dynamic at the molecular level. The external processes that are constantly at work shaping the regolith provide surface weathering, sputtering and exospheric cycling on the diurnal (28.5 days), annual, and solar cycle (11 years) levels. The day and nightside hemispheres have different plasma environments. The dayside photoelectron population increases the plasma density and charge dissipation. Trailing the Moon is a plasma wake cavity that forms due to the of balance of plasma current systems formed by the absorption of solar wind plasma on the dayside leaving a near vacuum region on the nightside. The increase in meteoroid flux to the lunar surface has been observed following traversing of the Leonids, Geminids, and Quadrantids meteor showers (Colaprete et al., 2016). Solar cycles variance includes the increase in solar flux, and associated decrease in galactic cosmic rays. In addition, high energy solar storms (including Coronal Mass Ejections, Solar Event Particles, and Solar Flares) will impinge on the Moon increasing surface radiation, and plasma processes including surface charging.

Several other factors influence the on-going processes in addition to the temporal changes of the lunar space environment. Localized regions of enhanced crustal magnetization stand off the solar wind plasma forming mini-magnetospheres (i.e., Halekas et al., 2008a) which are also associated with preserved dark and light albedo features visible in the regolith known as swirls (e.g., Deca et al., 2020). It is thought that space weathering around the crustal magnetic fields is the cause of these swirl features (Kramer et al., 2011).

Latitude dependent phenomena has also been observed including the detection of water ice within permanently shadowed craters at the poles (Colaprete et al., 2012). The LCROSS mission observed $5.6 \pm 2.9\%$ by mass concentration of water ice in addition to other volatiles including light hydrocarbons, sulfur-bearing compounds, and carbon dioxide (Colaprete et al., 2010). Recently water molecules have also been observed within the dayside regolith (Honniball et al., 2020) and within non-polar shadowed regions suggesting the trapping of water molecules within micro-cold traps (Hayne et al., 2020). Moreover, we see a hydrogen and water cycles at the surface which vary diurnally (Hendrix et al., 2019), suggesting the recombination of solar wind hydrogen with silicate molecules of the lunar regolith can produce water molecules which may then in turn sputter towards a cold trap where these volatiles are being concentrated. Micrometeoroid impacts are thought to be the primary driver of water motion with observed intensified concentration during meteor storms (Benna et al., 2019). This surface frost has been suggested to accumulate 0.0125 ± 0.0022 wt% water within the upper ~1 m of regolith during each lunar day, or 190 ± 30 ml of water per cubic meter of regolith (Livengood et al., 2015), providing a small but renewable resource for lunar settlements.

3.3   Earth and Solar Drivers to Lunar Surface Processes

The interaction of the solid surface with the near-surface space environment encompasses processes that move atoms or dust particles from one form to the other. This includes: photon-stimulated desorption, thermal desorption, impact vaporization, reflection of solar wind protons, and the bombardment of the solar wind (sputtering, surface charging, and surface chemistry) (Farrell et al., in press). There are three main drivers of these processes: solar radiation, meteoroid impact flux, and the impingement of solar wind plasma particles. The thermal environment and space weathering processes are impacted by the visible portion of solar radiation at the lunar surface (e.g., Domingue et al., 2014). The lunar surface is being constantly bombarded by high-energy meteoroids, thus, processes such as *regolith gardening* describe the redistribution of lunar surface material. Meteoroid impacts also dislodge a large amount of neutral atoms which contribute to exosphere dynamics. The solar wind is a near constant flow of protons and electrons emanating from the sun. The solar wind normally consists of mainly hydrogen protons (95%), helium (2–4%), and metals (Farrell et al., in press). Nominally, at the Moon's location, the solar wind travels at approximately 400 km/s and carries the interplanetary magnetic field with it (~5 nT). It has an 11-year cycle corresponding to the solar cycle and can vary between fast and slow wind conditions. With the passing of a Coronal Mass Ejections, solar wind speeds can double with a plasma density increase of a factor of 5–10 (Farrell et al., 2012a). When discussing dynamic processes at the lunar surface, it is important to understand their context including driving conditions from the Sun and Earth.

3.4   Hazards: The Challenges of the Harsh Natural Lunar Surface Environment to Human and Robotic Operations

There are several known risks of living and operating humans and robots at the lunar surface. Here we focus on three hazards most relevant to the natural environment: mechanical abrasion from dust grains, radiation from high energy particles, and the potential to charge and discharge surfaces within the near vacuum space environment. These hazards continue outwards to orbiting satellites, vehicles, or platforms operating within the cis-lunar environment; however, we focus only on surface hazards in this chapter.

3.4.1   Dust

Following the Apollo missions, it is now known that the abrasive nature of lunar surface dust grains pose several challenges for human and robotic systems operating in the presence of this dust. Moreover, system engineers acknowledge the potentially hazardous nature of the dust and are designing requirements to mitigate against and remove dust from human habitats. Dust particles less than 10 μm diameter pose a threat to crew inhalation, with grains smaller than 2.5 μm having the greatest likelihood for deep respiration (NASA, 2020). In addition, the mechanical nature of the agglutinate particles' irregular shapes, also provide adhesion of these particles onto objects at the surface. Rover wheels and moving mechanisms can be inhibited by these dust particles. Due to the small particle size, lunar dust grains are subject to electrostatic charging and inter-particulate forces (Carroll et al., 2020) which can result in lofting, levitating, and adhesion to surface objects. Suffice it to say, dust, and its transport and mitigation, will continue to be a part of future lunar settlements.

3.4.2   Radiation

For the majority of the Moon's orbit, it is outside of Earth's magnetic field. Moreover, the protective shielding of the Earth's atmosphere is non-existent at the Moon. Thus, human and electronic systems operating at the surface of the Moon need to take into consideration the radiation environment of space. The ionizing radiation experienced at the Moon comes from galactic cosmic rays and solar energetic particles similar to the exposure in free space, however, the lunar sphere blocks half of the free space radiation. These provide the total ionizing dose effects and single event effects (NASA, 2020). In addition, these high energy particles interact with lunar regolith and produce secondary neutrons. The solar cycle does influence the amount of galactic cosmic radiation experienced at the Moon (Adams et al., 2007). The radiation experienced by residents will need to be constantly monitored by a lunar settlement as well as creative solutions to minimize the radiation experienced within daily life. Some have proposed burrowing into the lunar regolith or building habitats within caves and lava tubes. Regardless of the exact solution, living permanently outside of the Earth's atmosphere will require constant attention to ionizing radiation for human health and safety, and electronics that can be susceptible to degradation during an experienced high energy particle or neutron.

3.4.3 Surface Charging

One of the primary concerns for a human and robotic operation on the surface of the Moon is spacecraft charging, which involves electrical charge build up on the lunar surface regolith, habitats, vehicles, or space suits. Spacecraft charging is the balance of currents between the space environment and spacecraft. It is largely dependent on the energy and density of the plasma, photoemission, and secondary emissions. Surface-plasma interactions increase with certain material properties such as dielectrics or mechanical designs like extended booms. The low density of the ambient plasma at the Moon and high secondary emissions also contribute to high rates of surface charging. The dayside surface plasma includes a population of photoelectrons which increases the plasma density. In general in the nominal solar wind, an object in full sunlight will charge to a low positive electric potential (~5–10 V, Stubbs et al., 2014). This is due to photoemission dominating over plasma currents. However, in shadow where no photoemission occurs, this object will charge to large negative potentials (−100 V, and up to −1 kV in the plasma sheet) (Farrell et al., 2008; Halekas et al., 2008b). The worst case observed was due to a solar energetic particle (SEP) event, which caused the surface to charge to −4.5 kV on the lunar nightside within the central wake cavity (Halekas et al., 2007, 2009).

For objects moving on the lunar surface such as rover wheels or astronaut's boots, tribocharging is important to fully understand and mitigate against. Tribocharging is the process where charge gets transferred from the surface regolith to a moving object (Jackson et al., 2015). Recent simulations suggest the lack of dissipation within a shadowed region could result in large buildup of electrical charge especially for moving systems (Zimmerman et al., 2012). This effect is similar to the buildup of static electricity on Earth which might result in an occasional shock during discharge for a small amount of charge, however, the same phenomena causes the dissipation of a large amount of charge results in a lightning strike from cloud to the ground. This illustration helps us to understand the importance of fully characterizing tribocharging and charge build up on the Moon. In addition, the lunar regolith is highly resistive (>1e−9 S/m) which increases the charge build up between objects and the surface. This can cause damage to materials and electronics. Dissipation of charge needs to be included in all aspects of life and operation within the lunar surface space environment.

Fig. 3.2 Vertical obstructions to horizontal flowing solar wind generate local plasma environments which pose a challenge to operations such as roving and drilling (Rhodes et al., 2020) (*Image credit* Jay Friedlander, NASA GSFC)

At the terminator including the poles, the solar wind flow is horizontal to the surface. Large objects, including depressed crater walls or extruding mountains, pose an obstruction to the flowing solar wind causing a plasma void and wake structure on the downwind side, suggesting local plasma effects that will change the surface charging regime (Farrell et al., 2010; Zimmerman et al., 2011, 2012). See Fig. 3.2.

3.5 Resources at the Lunar Surface

There are several chemical and mineralogical resources at the lunar surface. The most interesting is the existence of water, both in the form of hydrated regolith and water ice within permanently shadowed regions and micro-cold traps. We note the water cycle on the Moon has a very minimal naturally occurring replenishment, as previously discussed. Water can be used for life support systems and propellant when it is broken down into hydrogen and oxygen components. However, the extraction technology development is in work and will depend on the precise composition, including purity, state, availability, and global distribution of water molecules (Anand et al., 2012). In addition to water, lunar ices include molecular hydrogen, mercury and other metals, carbon and hydrocarbons, ammonia, methane, and other chemicals all of which can be used as resources (Colaprete et al., 2010).

Surface regolith is a plentiful resource globally. Regolith may be able to be used for the construction of cement structures, habitats, and infrastructure including roads and rocket landing/launch pads. Moreover, oxygen extraction direct from the abundant lunar regolith is an option. Ilmenite reduction products include Fe, and Ti, with either O_2 or H_2O (Anand et al., 2012). Sources for raw metals like Fe and Ti can be reused for 3D printed spacecraft or habitat parts. Ilmenite is found in the Porcellarum KREEP (Potassium, Rare Earth Elements and Phosphorus) Terrane (PKT) enriched in the equatorial nearside Mare region. The PKT is known to be enriched in Uranium and Thorium as well as Rare Earth Elements, compared to non-PKT soils. However, compared to their terrestrial counterparts, the lunar concentrations of U and Th are much less. Aluminum is plentiful within highland regolith. Silicon is another plentiful molecular component within the lunar regolith at large. All of which may be of interest to future lunar settlements. The solar wind implants volatiles on surface regolith which includes the following: H, N, C and He, F, Cl (Anand et al., 2012; Crawford, 2015). While these exist in small quantities (<125 ppm), they are thought to be readily accessible. In addition, water molecules have been observed within what is thought to be volcanic glass deposits in non-polar regions (Honniball et al., 2020).

Solar power is readily available and can be harnessed for power needs; however, the long duration and extreme cold temperature experienced during the lunar night pose a challenge for current solar array and battery systems. At the equator, the lunar night can last up to ~14 days. While the opposite is true for locations at the poles, where solar illumination can reach up to 90% of the lunar orbit. Radioisotope heater and power systems exist and may provide solutions to these power and thermal control problems.

3.6   Summary

The lunar surface hosts several dynamic processes between the dust, atmosphere and plasma. The sun, including solar wind and solar radiation, earth, including magnetic field, and meteoroid impact flux drive lunar surface processes in this realm. Hazards for human and robotic systems living and operating at the lunar surface include mechanical abrasion by dust particles, ionizing radiation, and surface charging. There are several resources within this natural environment that can be tapped into by a lunar settlement. These include water, regolith, oxygen, metals, silicon, rare earth elements, and solar power. These need to offset the challenges of surviving and operating within the extreme cold of the lunar night. While much has been learned about the surface of the Moon over the past half century in particular, challenges remain. Future settlers will provide creative engineering, prospecting, and research needed to overcome these current hurdles.

References

Adams, J. H., Bhattacharya, M., Lin, Z. W., Pendleton, G., & Watts, J. W. (2007). The ionizing radiation environment on the moon. *Advances in Space Research, 40*(3), 338–341. https://doi.org/10.1016/j.asr.2007.05.032

Anand, M., Crawford, I. A., Balat-Pichelin, M., Abanades, S., Van Westrenen, W., Péraudeau, G., et al. (2012). A brief review of chemical and mineralogical resources on the Moon and likely initial in situ resource utilization (ISRU) applications. *Planetary and Space Science, 74*(1), 42–48. https://doi.org/10.1016/j.pss.2012.08.012

Benna, M., Hurley, D. M., Stubbs, T. J., Mahaffy, P. R., & Elphic, R. C. (2019). Lunar soil hydration constrained by exospheric water liberated by meteoroid impacts. *Nature Geoscience, 12*(5), 333–338. https://doi.org/10.1038/s41561-019-0345-3

Carroll, A., Hood, N., Mike, R., Wang, X., Hsu, H.-W., & Horányi, M. (2020). Laboratory measurements of initial launch velocities of electrostatically lofted dust on airless planetary bodies. *Icarus, 352,* 113972. https://doi.org/10.1016/j.icarus.2020.113972

Colaprete, A., Schultz, P., Heldmann, J., Wooden, D., Shirley, M., Ennico, K., et al. (2010). Detection of water in the LCROSS ejecta plume. *Science, 330*(6003), 463–468. https://doi.org/10.1126/science.1186986

Colaprete, A., Elphic, R. C., Heldmann, J., & Ennico, K. (2012). An overview of the lunar crater observation and sensing satellite (LCROSS). *Space Science Reviews, 167*(1), 3–22. https://doi.org/10.1007/s11214-012-9880-6

Colaprete, A., Sarantos, M., Wooden, D. H., Stubbs, T. J., Cook, A. M., & Shirley, M. (2016). How surface composition and meteoroid impacts mediate sodium and potassium in the lunar exosphere. *Science, 351*(6270), 249–252. https://doi.org/10.1126/science.aad2380

Colwell, J. E., Batiste, S., Horányi, M., Robertson, S., & Sture, S. (2007). Lunar surface: Dust dynamics and regolith mechanics. *Reviews of Geophysics, 45*(2), 1–26. https://doi.org/10.1029/2005RG000184

Crawford, I. A. (2015). Lunar resources: A review. *Progress in Physical Geography, 39*(2), 137–167. https://doi.org/10.1177/0309133314567585

Deca, J., Hemingway, D. J., Divin, A., Lue, C., Poppe, A. R., Garrick-Bethell, I., et al. (2020). Simulating the Reiner gamma swirl: The long-term effect of solar wind standoff. *JGR-Planets, 125,* e2019JE006219. https://doi.org/10.1029/2019JE006219

Domingue, D. L., Chapman, C. R., Killen, R. M., Zurbuchen, T. H., Gilbert, J. A., Sarantos, M., et al. (2014). Mercury's weather-beaten surface: Understanding mercury in the context of lunar and asteroidal space weathering studies. *Space Science Reviews, 181*(1–4), 121–214. https://doi.org/10.1007/s11214-014-0039-5

Farrell, W. M., Halekas, J. S., Horanyi, M., Killen, R. M., Grava, C., Szalay, J. R., et al. (in press). The dust, atmosphere, and plasma at the Moon. In *New views of the Moon 2* (pp. 1–92).

Farrell, W. M., Stubbs, T. J., Delory, G. T., Vondrak, R. R., Collier, M. R., Halekas, J. S., & Lin, R. P. (2008). Concerning the dissipation of electrically charged objects in the shadowed lunar polar regions. *Geophysical Research Letters, 35*(19), 1–5. https://doi.org/10.1029/2008GL034785

Farrell, W. M., Stubbs, T. J., Halekas, J. S., Killen, R. M., Delory, G. T., Collier, M. R., & Vondrak, R. R. (2010). Anticipated electrical environment within permanently shadowed lunar craters. *Journal of Geophysical Research, 115*(E3), 1–14. https://doi.org/10.1029/2009JE003464

Farrell, W. M., Halekas, J. S., Killen, R. M., Delory, G. T., Gross, N., Bleacher, L. V., et al. (2012a). Solar-storm/lunar atmosphere model (SSLAM): An overview of the effort and description of the driving storm environment. *Journal of Geophysical Research E: Planets, 117*(10), 1–11. https://doi.org/10.1029/2012JE004070

Farrell, W. M., et al. (2012b). Dynamic response of the environment at the Moon (DREAM) years 1–3 executive summary. NASA Lunar Science Institute. https://lunarscience.nasa.gov/wp-content/uploads/2012/03/DREAM_Summary.pdf

Halekas, J. S., Delory, G. T., Brain, D. A., Lin, R. P., Fillingim, M. O., Lee, C. O., et al. (2007). Extreme lunar surface charging during solar energetic particle events. *Geophysical Research Letters, 34*(2), 1–5. https://doi.org/10.1029/2006GL028517

Halekas, J. S., Delory, G. T., & Brain, D. A. (2008a). Density cavity observed over a strong lunar crustal magnetic anomaly in the solar wind: A mini-magnetosphere? *Planetary and Space ..., 56*, 941–946. https://doi.org/10.1016/j.pss.2008.01.008

Halekas, J. S., Delory, G. T., Lin, R. P., Stubbs, T. J., & Farrell, W. M. (2008b). Lunar Prospector observations of the electrostatic potential of the lunar surface and its response to incident currents. *Journal of Geophysical Research: Space Physics, 113*(A9), n/a–n/a. https://doi.org/10.1029/200 8JA013194.

Halekas, J. S., Delory, G. T., Lin, R. P., Stubbs, T. J., & Farrell, W. M. (2009). Lunar surface charging during solar energetic particle events: Measurement and prediction. *Journal of Geophysical Research: Space Physics, 114*(A5), n/a-n/a. https://doi.org/10.1029/2009JA014113.

Halekas, J. S., Poppe, A. R., Delory, G. T., Sarantos, M., Farrell, W. M., Angelopoulos, V., & McFadden, J. P. (2012). Lunar pickup ions observed by ARTEMIS: Spatial and temporal distribution and constraints on species and source locations. *Journal of Geophysical Research, 117*(E6), E06006. https://doi.org/10.1029/2012JE004107

Hayne, P. O., Aharonson, O., & Schörghofer, N. (2020). Micro cold traps on the Moon. *Nature Astronomy*. https://doi.org/10.1038/s41550-020-1198-9

Heiken, G. H., Vaniman, D. T., & French, B. M. (Eds.). (1991). *Lunar sourcebook, a user's guide to the Moon*. Cambridge University Press.

Hendrix, A. R., Hurley, D. M., Farrell, W. M., Greenhagen, B. T., Hayne, P. O., Retherford, K. D., et al. (2019). Diurnally migrating lunar water: Evidence from ultraviolet data. *Geophysical Research Letters, 46*(5), 2417–2424. https://doi.org/10.1029/2018GL081821

Honniball, C. I., Lucey, P. G., Li, S., Shenoy, S., Orlando, T. M., Hibbitts, C. A., et al. (2020). Molecular water detected on the sunlit Moon by SOFIA. *Nature Astronomy*. https://doi.org/10. 1038/s41550-020-01222-x

Jackson, T. L., Farrell, W. M., & Zimmerman, M. I. (2015). Rover wheel charging on the lunar surface. *Advances in Space Research, 55*(6), 1710–1720. https://doi.org/10.1016/j.asr.2014. 12.027

Kramer, G. Y., Combe, J. P., Harnett, E. M., Hawke, B. R., Noble, S. K., Blewett, D. T., et al. (2011). Characterization of lunar swirls at Mare Ingenii: A model for space weathering at magnetic anomalies. *Journal of Geophysical Research E: Planets, 116*(4), 1–18. https://doi.org/10.1029/ 2010JE003669

Livengood, T. A., Chin, G., Sagdeev, R. Z., Mitrofanov, I. G., Boynton, W. V., Evans, L. G., et al. (2015). Moonshine: Diurnally varying hydration through natural distillation on the Moon, detected by the lunar exploration neutron detector (LEND). *Icarus, 255*, 100–115. https://doi. org/10.1016/j.icarus.2015.04.004

McKay, D. S., Heiken, G., Basu, A., Blanford, G., Simon, S., Reedy, R., et al. (1991). The lunar regolith. In *Lunar sourcebook* (Vol. 7, pp. 285–356). Citeseer.

NASA. (2020). SLS-SPEC-159 cross-program design specification for natural environments (DSNE) Rev H.

Nunn, C., Garcia, R. F., Nakamura, Y., Marusiak, A. G., Kawamura, T., Sun, D., et al. (2020). Lunar seismology: A data and instrumentation review. *Space Science Reviews, 216*(5). https://doi.org/ 10.1007/s11214-020-00709-3

O'Brien, B. J., & Hollick, M. (2015). Sunrise-driven movements of dust on the Moon: Apollo 12 Ground-truth measurements. *Planetary and Space Science, 119*, 194–199. https://doi.org/10. 1016/j.pss.2015.09.018

Rennilson, J. J., & Criswell, D. R. (1974). Surveyor observations of lunar horizon-glow. *The Moon, 10*, 121–142.

Rhodes, D. J., Farrell, W. M., & McLain, J. L. (2020). Tribocharging and electrical grounding of a drill in shadowed regions of the Moon. *Advances in Space Research, 66*(4), 753–759. https:// doi.org/10.1016/j.asr.2020.05.005

Sefton-Nash, E., Williams, J.-P., Greenhagen, B. T., Warren, T. J., Bandfield, J. L., Aye, K.-M., et al. (2019). Evidence for ultra-cold traps and surface water ice in the lunar south polar crater Amundsen. *Icarus, 332*, 1–13. https://doi.org/10.1016/j.icarus.2019.06.002

Stubbs, T. J., Halekas, J. S., Farrell, W. M., & Vondrak, R. R. (2007). Lunar surface charging- A global perspective using lunar prospector data. In *Dust in planetary systems, ESA SP-643* (pp. 1–4).

Stubbs, T. J., Farrell, W. M., Halekas, J. S., Burchill, J. K., Collier, M. R., Zimmerman, M. I., et al. (2014). Dependence of lunar surface charging on solar wind plasma conditions and solar irradiation. *Planetary and Space Science, 90*, 10–27. https://doi.org/10.1016/j.pss.2013.07.008

Szalay, J. R., & Horányi, M. (2015). The search for electrostatically lofted grains above the Moon with the Lunar dust experiment. *Geophysical Research Letters, 42*(13), 5141–5146. https://doi.org/10.1002/2015GL064324

Wang, X., Schwan, J., Hsu, H. W., Grün, E., & Horányi, M. (2016). Dust charging and transport on airless planetary bodies. *Geophysical Research Letters, 43*(12), 6103–6110. https://doi.org/10.1002/2016GL069491

Williams, J.-P., Paige, D. A., Greenhagen, B. T., & Sefton-Nash, E. (2017). The global surface temperatures of the Moon as measured by the diviner lunar radiometer experiment. *Icarus, 283*, 300–325. https://doi.org/10.1016/j.icarus.2016.08.012

Zimmerman, M. I., Farrell, W. M., Stubbs, T. J., Halekas, J. S., & Jackson, T. L. (2011). Solar wind access to lunar polar craters: Feedback between surface charging and plasma expansion. *Geophysical Research Letters, 38*(19), 3–7. https://doi.org/10.1029/2011GL048880

Zimmerman, M. I., Jackson, T. L., Farrell, W. M., & Stubbs, T. J. (2012). Plasma wake simulations and object charging in a shadowed lunar crater during a solar storm. *Journal of Geophysical Research E: Planets, 117*(8), 1–11. https://doi.org/10.1029/2012JE004094

Chapter 4
Power System Concepts for a Lunar Base

S. Lumbreras and Daniel Pérez Grande

Abstract The design of the power system for a Lunar base is full of interesting challenges. The system must include infrastructure for generating, storing, and distributing energy. All these systems must be optimized for weight, given that the single largest cost component is the transportation of materials. The characteristics of the Moon mean that solar power is the most attractive alternative, either by installing panels on the surface (currently the cheapest alternative), or by setting them in orbit. Nuclear power can be an interesting complement to solar given its constant availability. The transmission of energy from an in-orbit solar generator could be achieved by Microwave or Laser Distance Power Transmission. Then, the power would likely be distributed by a DC network built on underground aluminium cables. The development of a Lunar base could fuel the development of key technologies both for Earth and deep-space applications, in addition to fostering a new era in space exploration and being a stepping stone on our way to Mars. This chapter presents an introductory discussion on the topic, together with references to in depth analyses.

4.1 The Interest of Lunar Power Systems, Challenges, and Opportunities

4.1.1 Power Systems on the Moon: Essential Infrastructure for the Lunar Base

The generation and distribution of power has been recognised as an essential working infrastructure since the beginnings of the space race, acknowledging that space stations present considerably different conditions than "on-the-ground" bases, and

S. Lumbreras (✉)
Institute for Research in Technology, Universidad Pontificia Comillas, Madrid, Spain
e-mail: slumbreras@comillas.edu

D. Pérez Grande
ienai SPACE, Madrid, Spain
e-mail: danielpg@ienai.space

that the landscape of the technically feasible and economically efficient changes rapidly as technologies emerge and develop. There is a continuity of works that explore the needs and opportunities for energy generation and transportation in space, updating their costs as they evolve on Earth (Khan et al., 2006; Lior, 2001; Snyder, 1961). However, some technologies may benefit of accelerated development by being deployed in space first: options such as Space-based Solar Power (SBSP) or Microwave Distance Power Transmission (MDPT), which will be discussed in more detail in the following sections, might have in a lunar base an ideal starting project that might result in these technologies being developed enough for their efficient use on Earth. The technological spillovers (i.e., technological advances or cost reductions that appear in a different setting from the one where the main effort is being conducted) that can be anticipated in the space sector are indeed interesting (London Economics, 2018), especially interesting in the context of energy, where they could in the distance future be essential to cover humanity's needs for power generation.

4.1.2   Sending Materials to the Moon: Current Perspectives

The main challenge that makes lunar power systems difficult is, of course, the extremely high cost of sending materials to the surface of the Moon or to the cislunar space (the sphere surrounding the Moon where objects orbit it). Commercial payload transportation services to the lunar surface do not yet exist, but NASA's 2018 Commercial Lunar Payload Services (CLPS) program provides a good reference for future capabilities by a diverse cast of companies over the coming decades. Small lunar landers tasked with delivering payloads under 100Kg (Astrobotic's Peregrine, Masten Space Systems' XL-1, Intuitive Machines NOVA-C, Draper Labs' Artemis-7) for fixed contracts to launch and deliver missions (Potter, 2018) have costs around $100 M. This implies a prohibitive present-day cost of around $1 M/kg for bringing materials to the surface; indeed, Astrobotic had previously stated a $1.2 M/kg cost for integration, communications and delivery to lunar surface (Astrobotic, 2013). Future larger landers such as Blue Origin's Blue Moon or SpaceX Starship Moon variant may be able to largely reduce said costs by delivering payloads in the order of a few to hundreds of metric tons, respectively, if sufficient demand for recurring missions is available; nonetheless, payload cost delivery estimates are not publicly available for these landers.

In search for a more affordable alternative, an option which may prove fruitful is to locate solar power production infrastructure not directly on the surface but in lunar orbit. This infrastructure would then supply an on-ground receiver through wireless power transmission, a concept first proposed in the 1941 short story "Reason", by Isaac Asimov and known today as SBSP (SBSP, Glaser, 1992); the first ever mission aiming to demonstrate SBSP was flown on USAF's X-37B space plane and tested successfully in 2020 (Rodenbeck et al., 2020), opening the possibility for this technology to be deployed on a much larger scale in the future. Focusing solely on transport cost, as of 2020, the current capacity to deliver payloads to trans-lunar injection orbits is reserved for heavy-lift launch vehicles such as the Chinese Long March 5 (9,400 Kg payloads, Pingqi et al., 2019), ULA's Delta-IV Heavy (11,290 Kg payloads, Hart, 1998) or SpaceX's Falcon Heavy (in the range of 16,800 Kg—Mars Injection—to 26,700 Kg—GEO—payloads, undisclosed). The publicised launch cost of these vehicles is between $10 k/kg and $30 k/kg which likely puts companies wishing to deliver turnkey solutions for ferrying payloads to cislunar space at $100 k/kg, at most. These figures represent a ten-fold to one-hundred-fold decrease in cost compared to payload delivered to the surface, making SBSP an attractive option to supply power to a lunar base in cost-constrained scenarios. It is worth noting, however, that SBSP does not fully eliminate the need for on-ground infrastructure such as the receiver and distribution elements or the storage element (which, nonetheless, would be reduced compared to grounded infrastructure). These elements must be factored into the final cost comparison, together with risk and complexity analyses.

Although attractive, SBSP introduces complexities from the perspective of the optimal location for this space-based infrastructure. On Earth, geostationary orbits have been proposed for these applications since they maximize the coverage window and limit the alignment and pointing requirements between the orbiting platform and on-ground receiver. On the Moon however, stable lunar-stationary (or seleno-stationary) orbits do not exist, since their radius is larger than the Hill radius for the moon (the distance at which objects enter the sphere of gravitational influence of the Moon): the sphere of influence of the Moon is around 58,000 km (Chebotarev, 1964), whereas the distance to an orbit with an orbital period equal to the rotation period of the Moon is around 88,000 km, as per Kepler's law. Thus, the distance between the SBSP stations and the lunar base becomes the critical parameter in this analysis: larger orbits increase the transmission windows and reduce the need for storage elements both in orbit and on ground, but typically increase the size of the orbital collector and receiver elements, whereas smaller orbits imply the opposite. The analysis of particular orbital and system configurations and sizing is beyond the scope of this chapter but existing solutions in the literature have been proposed around "frozen orbits" (those orbits where the effects of the Moon-Earth-satellite three-body problem, solar radiation pressure and the non-spherical gravity field of the Moon cancel out, eliminating or largely reducing the need for station keeping manoeuvres), near sun-synchronous polar orbits or Near-Rectilinear-Halo-Orbits (NRHO, Fig. 4.1) which can maximize coverage of lunar polar regions and solar irradiance (Gillespie et al., 2020; Criswell, 2000). For bases located outside the polar regions, SBSP infrastructure situated in the stable L4, L5 Lagrange points (regions in space where the gravitational forces of two bodies, such as Earth and the Moon, cancel out; Fig. 4.2) could provide almost continuous coverage to a lunar base and to large areas of the moon, at the expense of, perhaps unfeasibly, large distances equal to the Earth-Moon separation note that the NRHO also orbit the Lagrange points, but in this case the unstable L1, L2 positions (Soto & Summerer, 2008). All of the aforementioned locations may be attractive if SBSP is considered, as we will discuss below; it is also worth noting than the cost of sending materials to the near-moon Lagrange points

 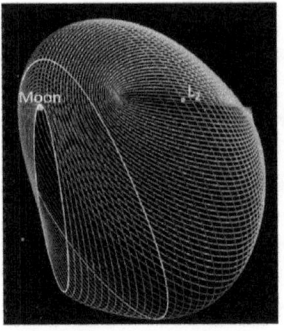 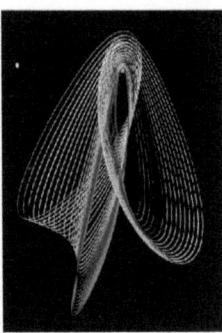

(a) The L_1 and L_2 southern halo families of orbits in configuration space.

(b) Zoomed-in view of the L_2 halo family delineating the bounds of the NRHOs in white.

(c) Zoomed-in view of the L_1 and L_2 NRHOs.

Fig. 4.1 Earth-Moon NRHO orbits. From: Zimovan et al. (2017), with permission by the authors

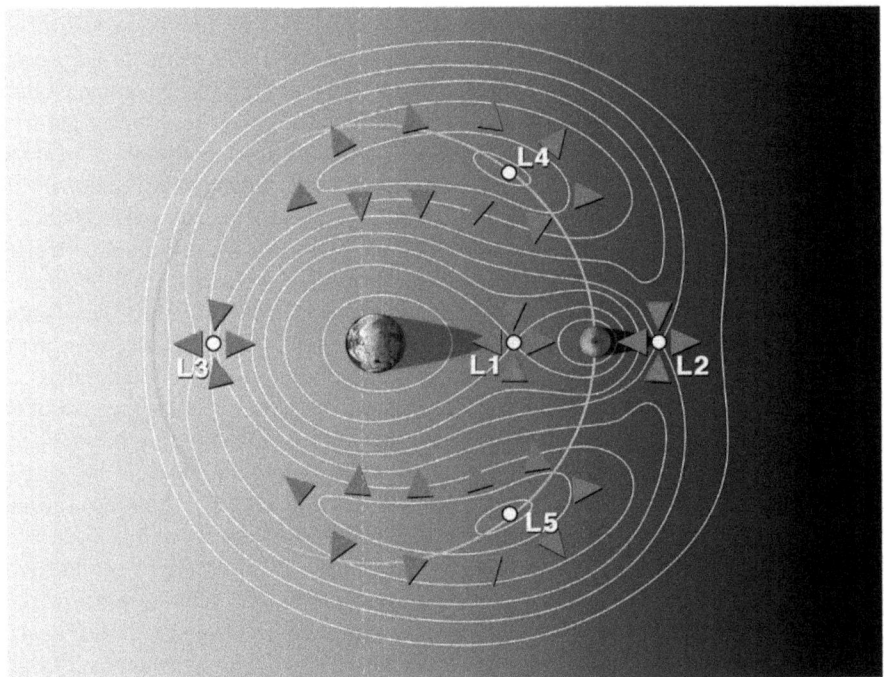

Fig. 4.2 The Lagrange points for the Earth-moon system. *Credit* David A. Kring, LPI-JSC Center for Lunar Science and Exploration

can be much lower than into lunar orbit itself (Hamera et al., 2008), especially if low energy transfer manoeuvres are utilized (Koon et al., 2001).

Several options are available for powering the lunar base, including solar and nuclear alternatives. Their relative suitability depends on factors such as the size of the base (which we will assume to be of a few tens of astronauts) and the activities that will be performed, from research to mining.

When dealing with electricity generation, one issue that does not appear on Earth are the differences with respect to heat evacuation. Thermodynamic power generation cycles generate work (which can be used to produce electricity) by transferring heat (for instance, coming from a nuclear reactor) and work in and out of a system where temperature and pressure vary. There are a multitude of different types of cycles (Rankine cycles, as in the first steam engine, Otto cycles as in gasoline automobiles, Diesel cycles for Diesel engines, etc. The most relevant cycles for power production are Ericsson, Stirling and Rankine). These cycles differ in how temperature and pressure vary, but they have in common the fact that there must be a heat source and a cold source for the cycle to work. The largest the temperature difference, and the highest the ability to exchange heat between them, the higher the efficiency of the cycle. Fuels are used to generate the hot source. On Earth, the cold source is commonly built from cooling water. In space, the temperature is much lower than on Earth (around 3 K), but heat cannot be evacuated efficiently due to the vacuum present in space. Because there is no possibility for conduction (heat transfer between objects that are touching) or convection (heat transfer to the fluid that surrounds an object), the only physical mechanism available for heat dissipation is radiation (heat transfer though electromagnetic radiation). This is the most inefficient way of heat transfer, and depends on the surface of the radiator, which means that when thermodynamic cycles are used in space, large radiators are needed, presenting further challenges in terms of their applicability to solutions which are heavily mass constrained, such as those requiring delivery to the lunar surface Nevertheless, if we assume that the radiators could be built, we would reach efficiencies in the energy conversion that would be much higher than on Earth (for instance, increasing it by 44% in the case of the Rankine cycle (Tarlecki et al., 2007; Toro & Lior, 2017). On the lunar surface, the atmosphere is too thin to be useful for cooling purposes, and the conductivity of the soil is much reduced, so there would be a challenge in finding a cold source. However, if the Lunar base was installed close to water or ice deposits, there would be a possibility in using them for this aim, which would also be useful as water would need to be melted for its use in human habitation, either as drinking water, as a source of oxygen, or to produce rocket fuel.

This takes us to one of the most relevant considerations for the design of a Lunar power system. The location of the outpost is key: it determines what resources would be available, namely the hours of irradiation and whether water ice is accessible. It is desirable to have as many hours of continuous irradiation as possible, as this will allow to use solar panels reducing the need for storage in batteries or otherwise. This strongly depends on location, and the Moon contains areas where there is almost continuous sunlight (the poles during their polar summers, SotoSoto & Summerer, 2008) or darkness (certain craters or areas of the poles during their polar winters) or cycles of ~15 Earth-days of daylight and ~15 of darkness, with temperatures from $-127\,°C$ to $+173\,°C$). In two of the five Lagrange points, L4 and L5, there is continuous solar irradiation, so SBSP installed there could have negligible storage needs. This, nonetheless, may be offset by other systems requirements non-negligible, and perhaps unpractical, losses due to large distances. Alternatively, other orbits such as HALO can maximize exposure to solar irradiation at lower distances to the surface, at the cost of smaller coverage windows; both effects would lead to considerable requirements for on-board power storage. In addition, locations close to the Poles and in particular the Shackleton crater in the South Pole would have the double advantage of having almost continue sunlight during polar summer and access to water ice, so they would be especially attractive as a location for the base (Khan et al., 2006). In addition, the use of pit craters, which are believed to be the entrances to caves by analogy to the Earth (Hong et al., 2014), mean that cooler, stable temperatures can be guaranteed for the base while keeping access to continuous sunlight for extended periods. In what follows, we will assume that this has been the case and that either the Lunar base is established close to the Shakleton crater or an area that offers a similar level of solar irradiation and access to water ice. The selection of a polar location may limit the extension of a given mission to a few months if the location becomes shadowed in the pole winter season or, conversely, add requirements for SBSP or non-solar power production such as nuclear power.

The rest of this chapter covers the fundaments of energy generation, storage, and distribution on the Moon, as well as some basic ideas about telecommunications on the Lunar surface. Although there is a focus on the economic viability of the alternatives, it should be noted that the Lunar base should not be solely driven or even limited by a too strong focus on cost, as the development of new technology is one of the main results expected from this unprecedented project.

4.2   Generating Power on the Moon

Generating power is the first step in the design of the Lunar power system. The minimum power demand that is assumed for a lunar base is around 100 kW (Criswell, 2000; Soto & Summerer, 2008; Duke et al., 1989), a figure which should drive any preliminary concept design for power generation.

Three main options are available to serve this demand: solar photovoltaic (either on surface or in orbit) and nuclear; depending on the parameters of the mission, at least two of these may be required in tandem. Solar thermal options may also be available but have been discarded due to their general low efficiency and requirements for both collectors and radiators. As explained above, although designs on Earth are based on LCOE, the main metric for the design is specific power, that is, the amount of power that the infrastructure can generate per kg of mass.

4.2.1   Solar Panels on the Surface

Solar panels on the surface are the most straightforward alternative for electricity generation (Fig. 4.3) and have been used in large-scale space applications such as the International Space Station (ISS), which has 84–120 kW average power and 240 kW peak power (Wright, 2017). Geostationary commercial satellites have demonstrated up to 20–30 kW power production from solar arrays in platforms such as the ViaSat, Boeing's 702HP bus, Echostar or Maxar's SSL1300. Solar power generation has also been demonstrated on many historic missions to the lunar surface: NASA's Surveyor landers were all solar powered and the Apollo 11 Early Apollo Scientific Experiment Package (EASEP) was also powered by solar cells (although the lander itself was powered by fuel cells), the Soviet Union's Luna 17, 21 and 24 missions also demonstrated solar power and, more recently, China's Chang'e 3 and 5 missions were both powered by on-the-ground solar panels. Traditionally, photovoltaic cells have faced a theoretical efficiency limit of 34%, expressed as the percentage of solar irradiation that is actually converted to electricity, which is derived from the physics of the photoelectric effect that is their foundation. However, recent breakthroughs mean that we can produce multi-junction photovoltaic cells, which use a combination of different semiconductor materials (indium gallium phosphide, gallium arsenide

Fig. 4.3 Solar arrays for energy generation, greenhouses for food production and habitats shielded with regolith. *Credit* @ESA—P. Carril

and germanium) to harvest more energy from the solar spectrum and to beat this limit, currently exceeding 47.1% (Vossier et al., 2017).

In addition, the absence of an atmosphere means that solar panels on the Moon are much more efficient than on Earth. On Earth, about 55–60% of solar energy gets either reflected or absorbed before it reaches the panel through clouds, gases, and dust. This means that the total energy generated from the panel on the Moon will double the one that would be obtained on Earth. It should be noted that this situation is considerably different from the one on Mars, where the distance means that solar irradiance is around half the one on Earth (or the surface of the Moon), so that twice as many panels would need to be installed for the same capacity. In addition, enormous sandstorms cover the red planet once every 35–70 Martian days (Appelbaum & Flood, 1990) and solar irradiance gets reduced by more than 60% of its original value. This means that larger backup capacities or storage must be installed, which together with the higher transportation costs to Mars mean that solar power is much more attractive on the Moon that on Mars. Actually, if we were shown the final designs for the power system of a Lunar and a Martian base, we would be sure to identify which one is which by looking at the reliance on solar power: The Moon base has a much larger potential for solar that should be reflected in the installed capacity. An additional advantage is that there have been designs for much thinner cells in the past few years with an MIT team recently announcing they had created a one-nanometer solar cell, so thin that could sit on a soap bubble without bursting it (Choi, 2016).

All this means that readily available alternatives can offer already specific powers over 5 kW/kg (Warmann et al., 2020), which is above the 1 KW/kg need that we considered as the absolute minimum for Lunar applications. Solar panels on the Lunar surface would be therefore a viable alternative for powering the base if a location with constant or almost-constant irradiation is selected. This would be preferred mainly because of storage needs, which would require increasingly large batteries (or other storage means) for growing periods of darkness. For this reason, if the base was not located at the South Pole and was instead established in an area with 14 Earth-days of light versus 14 days of dark, solar panels could be unviable. A solution for this would be produce solar power from locations in cis-lunar space and beam it to the Moon using SBSP.

4.2.2 Solar Panels in Orbit

SBSP is the concept of collecting solar power in outer space and sending it back to Earth (or, in this case, the Moon) for its consumption. Space-based solar power benefits from the greater efficiencies that arise from higher irradiation and the absence of atmosphere, although the benefits may be offset by the coverage windows of the receiving station from the orbiting infrastructure and losses associated to energy transmission.

4.2.3 Generating Energy in Space

SBSP has been theorized as a future large-scale solution for renewable global energy consumption since the 1970s (Glaser et al., 1974). However, no current proposal is economically viable for Earth: in order to reach sufficient efficiencies, it is estimated that costs would need to drop by more than two orders of magnitude (Warmann et al., 2020), although many studies have been carried out by NASA and other institutions related to the concept (Fig. 4.4). However, this could change if solar panels could be built in space, or on the Moon. Indeed this application is flagged as the first potential "win" of in-space-manufacturing technologies currently in researched by NASA (Zheng et al., 2018; Herrik, 2019) or companies such as Made-In-Space (Patane et al., 2017). There is currently active research in SBSP, with China having commissioned a project to actively pursuit in the next decade (Rossenbaum & Susso, 2019).

Installing SBSP in cis-lunar space also offers the possibility of semi-continuous irradiation, or perhaps continuous irradiation if more than one of the Lagrange points or near-lagrange orbits are covered. However, it still bears a much higher cost due to the large scale engineering project of producing an SBSP platform, which could be

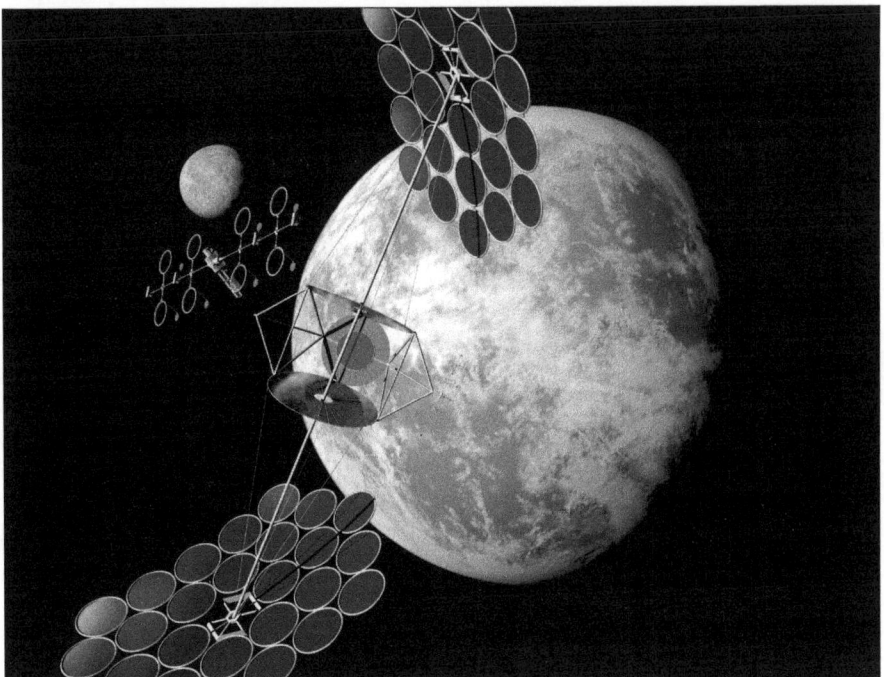

Fig. 4.4 NASA Integrated Symmetrical Concentrator SPS concept, an interesting opportunity for power generation in lunar missions. *Credit* NASA (public domain)

comparable to the cost of other remarkable orbiting missions such as the NISAR–ISRO satellite ($1.5B) or even closer to that of scientific missions such as the James Webb telescope (~$10B), which means that it would only be attractive if a permanent base or a long range of future missions to the lunar surface are planned.

4.2.4 Sending Energy to the Base

In addition to the cost barrier, SBSP introduces the considerable technological challenge of deploying large solar arrays, similar in size to the ones found in the ISS (about the size of an American Football or Soccer fields), and transmitting energy to the surface for its use, which would involve some form of wireless transmission (Fig. 4.4). Laser Beam technology and Microwave Distance Power Transmission seem the most promising candidates.

Wireless energy transfer was already demonstrated in the 60s (Brown, 1984), with an experiment that sent microwaves to power a small helicopter. In this experiment there were already the main elements of microwave distance power transmission: the source of radiation, and a receiver with a rectifier (what we refer to as *rectenna*). Since then, considerable progress has been made to increase the capacity, efficiency and distance covered by the transmission. For instance, already twenty years ago a Japanese team build a system that consisted in a set of solar panels attached to a microwave transmitter and a rectenna. The system was able to send 300 W with a 75% efficiency (Matsumoto, 2002). Since then, the US has been actively involved in the development of this technology. Much of recent research has focused on producing extremely efficient rectennas and rectenna arrays, as usually more than one receiver is needed. Efficiencies are commonly over 80% now. Frequencies around 6 GHz, which minimize losses at average atmospheric conditions and need relatively small antennae—that is why they are the second most used frequency for wifi communications (Strassner & Chang, 2013). However, a Lunar device would not have to deal with atmospheric losses and could therefore benefit from higher frequencies, which could allow smaller sizes of the emitting and receiving devices, which would translate in lower costs for the system. The rectennae would probably need to be installed in arrays close to the Lunar base. This is another advantage of a Lunar SBSP, as the land use for the receiving stations on Earth is a concern that does not appear on the Moon. It should be noted that most designs propose beam energy densities that are low enough to not be harmful if human beings—or the base as a whole—were to be unintentionally exposed. This means that SBSP would not need to be any less safe than ground-based solar panels.

Laser Transmission uses a laser beam sent to a photovoltaic receiver or to a thermal conversion system. This involves the challenge of tracking the position of the receiver so that the beam can be accurately directed towards it. In addition, the multiple energy conversions would mean that there would be only around 28% of efficiency using current designs (Raible et al., 2011). This means that microwave transmission may initially seem like the most viable alternative for sending the main energy generated by SBSP to the Lunar base; however, due to the large distances between the Lagrange points and the lunar surface, the transmission losses of RF systems may be prohibitive, leaving laser transmission as the only possibility, albeit at its limited efficiencies. Analyses in the literature can be found to compare pros and cons for both systems (Soto & Summerer, 2008).

Therefore, solar power seems the most promising alternative to powering a lunar base. Solar power deployed on the ground is presumably much more cost efficient than SBSP, due to the large development costs of the platform, even considering the exorbitant costs of landing payloads on the lunar surface. This is especially true if deployment can be achieved by the astronauts themselves without incurring in additional costs, such as those of deploying the infrastructure through robotic precursor missions. Nevertheless, SBSP might be attractive under some conditions, could provide continuous power to a permanent base and would make it a good candidate project to develop the technology for future uses on Earth.

4.3 Nuclear Power

If solar irradiation is not available, there are other technologies available. Namely, nuclear fission has the advantage of being able to generate electricity for a long period of time using a very limited mass amount of fuel. In particular, radioisotope thermoelectric generators (RTG, RITEG) are the most interesting technology for the Lunar base. RTGs are a type of nuclear battery that uses an array of thermocouples (a thermocouple is a pair of wires made of different metals which are joined to take advantage of the thermoelectric effect) to convert the heat released by the decay of a radioactive material into electricity. This conversion is made possible by the Seebeck effect, which means that an electric potential appears when there is a temperature gradient. The most interesting feature of this type of generators is that it has no moving parts, so its maintenance is much easier than other types of fission reactors. As such, RTGs have been proven in spaceflight historical missions from Apollo, to Voyager, Viking and all of the outer planets missions: Galileo, Cassini and New Horizons; on the ground, they have been incorporated in both the Curiosity and Perseverance Mars rovers, and even some remote terrestrial locations such as the lighthouses built by the Soviet Union in the Arctic (Office of Technology Assessment Congress of The United States & United States Congress Office of Technology Assessment, 1995). RTGs are a convenient solution when there is no access to solar power and maintenance is difficult or impossible. It is necessary to remember that excess heat needs to be removed to be able to maintain the temperature gradient, so that radiators must be built in the necessary proportions; achieving the temperature gradient in the vacuum of space can prove challenging from an engineering perspective. It would also be possible to use excess heat for heating purposes of the base if needed.

Their single most relevant drawback is the environmental concerns associated to their use: even long after their useful life has been completed, they present a radioactive hazard that must be managed. For the most common fuel, Pu-238, there is a half-life of 88 years. In addition, Russia is the main procurer of Pu-238, and NASA has tried to incentivize the use of solar instead of RTGs also because of political reasons (Kramer, 2011); nevertheless, in 2016, NASA's Space Technology Mission Directorate unveiled the Kilopower project (Fig. 4.5): "a near-term technology effort to develop preliminary concepts and technologies that could be used for an affordable fission nuclear power system to enable long-duration stays on planetary surfaces" (Skelly & Wittry, 2019, Gibson et al., 2017).

The main advantage of RTGs is constant availability. This may be interesting even if most of the power was generated by solar, as power shortages can be so dangerous for a human settlement that several backup options (e.g. batteries and nuclear) would be necessary. RTGs are also very robust with respect to extreme temperature variations and high radiation fields, which is particularly interesting for Lunar applications.

In addition, RTGs have a good specific capacity, with 4–5 W/kg already available in existing small applications. A 2006 NASA study found out that the 21 kW that were necessary for a manned mission with a handful of astronauts could be provided wither with a 18,000 kg solar system or with a 9,000 kg nuclear reactor (Rucker et al., 2016). It should be noted that these numbers should not be compared (Surampudi, 2011) to the specific capacities that are available for larger applications, where, as explained above solar can exceed 5 kW/kg in ultra-thin designs. RTGs are inherently small units that would be used in modular designs, so it is reasonable to expect that their specific capacity would stay relatively constant for larger capacities. This means that solar exceeds RTGs in specific power for several orders of magnitude, and therefore should constitute the foundation of the Lunar power system if the placement of the base makes it possible. Which of the two main solar options (on-surface or in

Fig. 4.5 NASA kilopower concept for lunar and Mars human exploration missions. *Credit* NASA (public domain)

orbit), would depend not only on the location but also on other global aspects of the design, such a redundancy, maximum mass to the surface, etc.

4.4 Energy Storage

Energy storage would be key for two main purposes. First, it is necessary to guarantee the continuity of energy supply in the case of an incident in the power generation system. The need for storage would be calculated based on the emergency consumption of the base, estimated the number of days that would be necessary to get the generation system to work again safely. In addition, energy storage can balance power generation with its consumption. This is especially necessary in the case of the Lunar base. One of the key energy needs will be heating and cooling. As stated above, temperatures vary in the range −157 to 127 °C, so the energy necessary to stabilize temperature will vary widely with the Lunar day. If the base is located, as speculated above, in a cave close to the South Pole, temperature oscillations will be much more benign, and the peak demand for energy (its maximum value) will be lower with respect to the base (the minimum one). When evaluating storage options for the Lunar base, the main two parameters to consider are specific energy (how much electricity can be stored in the device per kg of weight) and specific power (how quickly we can extract this energy).

4.4.1 Conventional Batteries

There are several technological options available for storage. The most straightforward alternative are conventional electrical batteries. Among the very long list of existing technologies, Nickel-Hydrogen ($Ni\text{-}H_2$) and Lithium-Ion and (Li-Ion) are the most promising in terms of specific energy and power. The ISS recently upgraded its $Ni\text{-}H_2$ batteries with Li-Ion. For the same volume, Li-ion batteries can handle twice the charge, so only half as many lithium-ion batteries are needed during replacement. In addition, although Li-Ion batteries typically have shorter lifetimes than $Ni\text{-}H_2$ batteries, the specific design that has been installed in the ISS has a much longer useful time than the previously installed batteries. The current parameters for the batteries in the ISS; which could be improved for the Lunar based, can reach 265 W h/kg in terms of specific energy, and 340 W/kg in terms of specific capacity. A complete account on their characteristics can be found in (Harding, 2017).

4.4.2   Fuel Cells

Weight would not be as important if the storage system could be built on the Lunar surface rather than transported from Earth. This opens the possibility of storing energy not in the form of electricity but in a chemically. This could be achieved by building fuel cells using materials available on the Lunar surface. A fuel cell is an electrochemical cell that converts the chemical energy of a fuel (often hydrogen) and an oxidizing agent (often oxygen) into electricity through a pair of redox chemical reactions. If the Lunar base is located close to the South Pole, it would be feasible to build the cells on the Lunar surface using water, as long as other necessary components (catalysts and the structure for the cell) are transported via rocketry. Although the efficiency of batteries can triple that of fuel cells (Peng & Chen, 2009), fuel cells could still be interesting as a complement to electrical batteries.

4.4.3   Other Options

Other forms of storage that could be built on the Lunar surface include mechanical storage such as flywheels. Flywheels store rotational energy, which can be then converted to electricity when needed. The best feature of these devices is that they can extract energy much faster than any other storage method. The NASA G2 flywheel for spacecraft energy storage, was intended not for use in space but for testing equipment in a laboratory setting. Given that flywheels could be built using materials available on the surface, they might pose an interesting alternative to meet very high, short peak loads if they emerge in the distant future of the base, maybe from research or from activities such as mining.

4.5   Energy Transmission and Distribution

The energy transmission and distribution system ensure that the power from the generation system reaches demand. We traditionally identify transmission with long distances and high power, while distribution takes care of the short distances to demand and lower powers.

4.5.1   Wireless Transmission

The single most important component of a transmission system would be the transmission of the energy generated in orbit if SBSP is used, which includes the microwave distance power transmission system with its emitter and the necessary rectennae on the surface. If SBSP is used, it would be feasible to install rectennae close to every facility that needs some power supply. This means that it would be unnecessary to build an electrical power network linking the different facilities. Even if SBSP is not used, microwave distance power transmission can be an interesting alternative to power facilities that are distant from the main location of the base when there is no constant radiation or nuclear power is not desired or is preferred to be placed at a safe distance in case of a possible accident or to minimize radiation effects on the base.

If SBSP is not used, it would probably not be realistic to imagine a long-range transmission system based on cables given the weight of transmission lines. If there are several distant locations of the Lunar base, each would need to be supplied independently.

4.5.2 Distribution Through Cables

Within each main location, the energy distribution system carries out the electricity to the final demands. The system could be designed to use Alternate Current (AC), or Direct Current (DC). AC is the dominant system on Earth given its ease for energy transformation in terms of changing voltage through transformers or to and from mechanical energy through engines and generators. However, technology has advanced tremendously on the DC side, except on circuit-breakers for high currents, which are technologically very challenging given that there is no moment with zero current. This means that building large, meshed networks in DC is difficult with current technology. However, the size of the base will be far smaller than this, with total powers in the range 100 kW–1 MW. In addition, the architecture of the network will likely be radial or radially operated: there will likely only be a single generation node from which power will be sent to the points with demand for electricity. Although there can be other links built for the purposes of reliability, their circuits will be open unless there is a contingency, so the network will be operated radially.

In addition, when there are weight constraints, AC systems of higher frequencies are used. For instance, the standard in avionics is not the usual 50–60 Hz but 400 Hz. This increases the losses due to the "skin effect", where the alternating current gets distributed more densely on the surface of the conductor and this density decays exponentially toward the centre. In addition, voltage drops are much larger at higher frequencies (because reactance increases). This is not a problem in avionics because distances are very small, but could be an issue in the longer distances in the Lunar base. For these reasons, DC is a very attractive alternative for the Lunar network.

As far as materials are concerned, we need cables that are as lightweight as possible. Aluminium has half the mass of an equivalent copper conductor and is used in most terrestrial applications; furthermore, large quantities of aluminium are available on the Moon, in the form of anorthite, which has been proposed as a prime candidate for being exploited through in-situ resource utilization (Sanders, 2018, Ellery, 2020, Faierson & Logan, 2010). NASA and others are developing the first examples of the technologies needed for these future endeavours (Fig. 4.6). On Earth, the most cost-effective design for lines is overhead, where they are mounted on towers that isolate them from the ground. However, building towers on the Moon is not feasible, so the cables would need to be insulated and laid directly on the surface. In addition, the extreme temperature variations on the Moon mean that cables would need to be buried to avoid their deterioration in the long term.

Fig. 4.6 NASA's Regolith Advanced Surface Systems Operations Robot (RASSOR) excavating simulated extraterrestrial soils. In-situ-resource-utilization strategies may be crucial for prolonged habitation on the lunar surface. *Credit* NASA

Fig. 4.7 Connecting the Earth with the Moon through ESA's Project Moonlight: a lunar communications and navigation service. *Credit* @ESA

If fuel cells are used, there would be a possibility for energy distribution using hydrogen and oxygen rather than electricity, which could be interesting as a complement to power distribution.

4.5.3   Communications on the Moon

The SBSP infrastructure has been proposed in the literature as a prime candidate for doubling as a communications platform, since the Lagrange points or near-lagrange orbits should provide almost constant coverage of both the lunar base and the Earth (Soto & Summerer, 2008). If necessary, smaller satellites may be employed as to form a communication relay networks (Babuscia et al., 2016; ESA, 2020); ESA is perhaps the most advanced agency in this aspect, having proposed Project Moonlight as a communications infrastructure for the Moon (Fig. 4.7).

4.6   Conclusions

The Lunar base is an interesting challenge full of opportunities. The location of the base is, perhaps, the single most important factor for the design of its power system. Assuming locations close to the South Pole, where more stable temperatures and access to water ice can be granted, solar irradiance would be relatively constant during polar summers, which would allow for the installation of surface panels. SBSP could be deployed to compensate the lack of direct sunlight during the polar winters. SBSP would be much more attractive than surface panels if a location without constant irradiation is chosen. The space-based infrastructure could be installed in around the Earth-Moon Lagrange points or in orbits around Lagrange points, providing almost constant coverage or constant coverage if more than one point is used. The cost of development, deployment and operation of this infrastructure may result to be prohibitive, even when factoring the cost reductions of payload deliveries to the surface versus lunar orbit or even of lunar orbit versus Lagrange orbits. In any case, the core of the generation system would be solar power given the ideal conditions of irradiance and lack of atmosphere. Nuclear power (RTGs) could be interesting as a complement or perhaps necessary as a backup for the lunar base.

If SBSP is used, microwave distance power transmission would be used to radiate the energy back to an array of surface rectennae, so that multiple, distant locations could be supplied; however, microwave transmission may be restricted by the large distances between the space-based infrastructure and the lunar surface, in which case laser transmission may be the only available alternative. In the shorter distances within the base, a DC network built from underground aluminium cables, sourced through in-situ resource utilization strategies, seems the most reasonable alternative.

The development of a Lunar base could fuel the development of key technologies both for Earth and deep-space applications, in addition to fostering a new era in space exploration and being a stepping stone on our way to Mars. This is even more true in the case of its power system, which could be the first step towards dramatically innovative technologies such as SBSP or Microwave Distance Power Transmission, which could fuel the Earth in a perhaps-not-so-distant future.

References

Appelbaum, J., & Flood, D. J. (1990). Solar radiation on mars. *Solar Energy, 45*(6), 353–363.

Astrobotic. (2013). *Astrobotic unveils lower lunar delivery pricing.* https://www.astrobotic.com/2013/07/08/astrobotic-unveils-lower-lunar-delivery-pricing/.

Babuscia, A., Divsalar, D., Cheung, K. M., & Lee, C. (2016, March). CDMA communication system performance for a constellation of CubeSats around the Moon. In *2016 IEEE Aerospace Conference* (pp. 1–15). IEEE.

Brown, W. C. (1984). The history of power transmission by radio waves. *IEEE Transactions on Microwave Theory and Techniques, 32*(9), 1230–1242.

Chebotarev, G. A. (1964). Gravitational spheres of the major planets, moon and sun. *Soviet Astronomy, 7*, 618.

Choi, C. (2016). New ultrathin solar cells are light enough to sit on a soap bubble. *Live Science.*

Criswell, D. R. (2000). Lunar solar power system: Review of the technology base of an operational LSP system. *Acta Astronautica, 46*(8), 531–540.

Duke, M. B., Mendell, W. W., & Roberts, B. B. (1989). Strategies for a permanent lunar base. *Lunar Base Agriculture: Soils for Plant Growth*, 23–35.

Ellery, A. (2020). Sustainable in-situ resource utilization on the moon. *Planetary and Space Science, 184*, 104870.

ESA. (2020). *Lunar satellites.* https://www.esa.int/Applications/Telecommunications_Integr ated_Applications/Lunar_satellites.

Faierson, E. J., & Logan, K. V. (2010). Geothermite reactions for in situ resource utilization on the moon and beyond. In *Earth and space 2010: Engineering, science, construction, and operations in challenging environments* (pp. 1152–1161).

Glaser, P. E., Maynard, O., Mackovciak, J., & Ralph, E. (1974). Feasibility study of a satellite solar power station.

Glaser, P. E. (1992). An overview of the solar power satellite option. *IEEE Transactions on Microwave Theory and Techniques, 40*(6), 1230–1238.

Gibson, M. A., Oleson, S. R., Poston, D. I., & McClure, P. (2017, March). NASA's kilopower reactor development and the path to higher power missions. In *2017 IEEE Aerospace Conference* (pp. 1–14). IEEE.

Gillespie, D., Wilson, A. R., Martin, D., Mitchell, G., Filippi, G., & Vasile, M. (2020, October). Comparative analysis of solar power satellite systems to support a moon base. In *71st International Astronautical Congress, IAC 2020.*

Hamera, K., Mosher, T., Gefreh, M., Paul, R., Slavkin, L., & Trojan, J. (2008, March). An evolvable lunar communication and navigation constellation concept. In *2008 IEEE Aerospace Conference* (pp. 1–20). IEEE.

Harding, P. (2017). *EVA-39: Spacewalkers complete the upgrading of ISS batteries.* https://www.nasaspaceflight.com/2017/01/spacewalkers-upgrading-iss-batteries/.

Hart, D. (1998). The Boeing company eelv/delta IV family. In AIAA Defense and Civil Space Programs Conference and Exhibit (p. 5166).

Herrik, K. (2019). *Building solar panels in space might be as easy as clicking print.* https://www.nasa.gov/feature/glenn/2019/building-solar-panels-in-space-might-be-as-easy-as-clicking-print.

Hong, I., Yi, Y., & Kim, E. (2014). Lunar pit craters presumed to be the entrances of lava caves by analogy to the earth lava tube pits. *Journal of Astronomy and Space Sciences, 31*(2), 131–140.

Khan, Z., Vranis, A., Zavoico, A., Freid, S., & Manners, B. (2006). Power system concepts for the lunar outpost: A review of the power generation, energy storage, power management and distribution (PMAD) system requirements and potential technologies for development of the lunar outpost. *AIP Conference Proceedings, 813*(1), 1083–1092.

Koon, W. S., Lo, M. W., Marsden, J. E., & Ross, S. D. (2001). Low energy transfer to the Moon. *Celestial Mechanics and Dynamical Astronomy, 81*(1–2), 63–73.

Kramer, D. (2011). Shortage of plutonium-238 jeopardizes NASA's planetary science missions. *PhT, 64*(1), 24.

Lior, N. (2001). Power from space. *Energy Conversion and Management, 42*(15–17), 1769–1805.

London Economics. (2018). *Spillovers in the space sector.*

Matsumoto, H. (2002). Research on solar power satellites and microwave power transmission in Japan. *IEEE Microwave Magazine, 3*(4), 36–45.

Office of Technology Assessment Congress of The United States & United States Congress Office of Technology Assessment. (1995). Nuclear wastes in the arctic: An analysis of arctic and other regional impacts from soviet nuclear contamination.

Patane, S., Joyce, E. R., Snyder, M. P., & Shestople, P. (2017). Archinaut: In-space manufacturing and assembly for next-generation space habitats. In *AIAA SPACE and astronautics forum and exposition* (p. 5227).

Peng, B., & Chen, J. (2009). Functional materials with high-efficiency energy storage and conversion for batteries and fuel cells. *Coordination Chemistry Reviews, 253*(23–24), 2805–2813.

Pingqi, L. I., Wei, H. E., Haosu, W. A. N. G., Yu, M. O. U., Dong, L., & Jue, W. A. N. G. (2019). Technical feature analysis of the LM-5 overall plan. 中国航天 (英文版), 18(1), 21–26.

Potter, S. (2018). *NASA announces new partnerships for commercial lunar payload delivery services.* https://www.nasa.gov/press-release/nasa-announces-new-partnerships-for-commercial-lunar-payload-delivery-services.

Raible, D. E., Dinca, D., & Nayfeh, T. H. (2011). Optical frequency optimization of a high intensity laser power beaming system utilizing VMJ photovoltaic cells. *2011 International Conference on Space Optical Systems and Applications (ICSOS)* (pp. 232–238).

Rodenbeck, C. T., Jaffe, P. I., Strassner, B. H., II., Hausgen, P. E., McSpadden, J. O., Kazemi, H., & Self, A. P. (2020). Microwave and millimeter wave power beaming. *IEEE Journal of Microwaves, 1*(1), 229–259.

Rossenbaum, E., & Susso, D. (2019). *China plans a solar power play in space that NASA abandoned decades ago.* https://www.cnbc.com/2019/03/15/china-plans-a-solar-power-play-in-space-that-nasa-abandoned-long-ago.html.

Rucker, M. A., Oleson, S. R., George, P., Landis, G., Fincannon, J., Bogner, A., et al. (2016). Solar vs. fission surface power for mars. In *AIAA SPACE 2016* (pp. 5452)

Sanders, G. B. (2018). Overview of past lunar in situ resource utilization (ISRU) development by NASA.

Skelly, C., Wittry, J. (2019). *Kilopower.* https://www.nasa.gov/directorates/spacetech/kilopower.

Snyder, N. W. (1961). *Energy conversion for space power.* American Institute of Aeronautics and Astronautics.

Soto, L. T., & Summerer, L. (2008). Power to survive the lunar night: An SPS application? In *59th International Astronautical Congress.*

Strassner, B., & Chang, K. (2013). Microwave power transmission: Historical milestones and system components. *Proceedings of the IEEE, 101*(6), 1379–1396.

Surampudi, S. (2011). Overview of the space power conversion and energy storage technologies. *NASA-jet propulsion laboratory.*

Tarlecki, J., Lior, N., & Zhang, N. (2007). Analysis of thermal cycles and working fluids for power generation in space. *Energy Conversion and Management, 48*(11), 2864–2878.

Toro, C., & Lior, N. (2017). Analysis and comparison of solar-heat driven stirling, brayton and rankine cycles for space power generation. *Energy, 120,* 549–564.

Vossier, A., Riverola, A., Chemisana, D., Dollet, A., & Gueymard, C. A. (2017). Is conversion efficiency still relevant to qualify advanced multi-junction solar cells? *Progress in Photovoltaics: Research and Applications, 25*(3), 242–254.

Warmann, E. C., Espinet-Gonzalez, P., Vaidya, N., Loke, S., Naqavi, A., Vinogradova, T., et al. (2020). An ultralight concentrator photovoltaic system for space solar power harvesting. *Acta Astronautica, 170,* 443–451.

Wright, J. (2017). *About the space station solar arrays.* https://www.nasa.gov/mission_pages/station/structure/elements/solar_arrays-about.html.

Zheng, Y., Kong, J., Huang, D., Shi, W., McMillon-Brown, L., Katz, H. E., & Taylor, A. D. (2018). Spray coating of the PCBM electron transport layer significantly improves the efficiency of pin planar perovskite solar cells. *Nanoscale, 10*(24), 11342–11348.

Zimovan, E., Howell, K., & Davis, D. (2017). Near rectilinear halo orbits and their application in cis-lunar space. https://www.researchgate.net/publication/319531960_NEAR_RECTILINEAR_HALO_ORBITS_AND_THEIR_APPLICATION_IN_CIS-LUNAR_SPACE.

Chapter 5
Plants Under the Moonlight: The Biology and Installation of Industrial Plants for Lunar Settlements

Roland Cazalis

Abstract Food security is a critical issue for future human settlement on the Moon and Mars, as NASA's Artemis Program outlines. The ongoing resupply of the International Space Station is not feasible for a permanent settlement. It is essential to set up an ecosystem on site that ensures food, water, oxygen, carbon dioxide removal, and waste recycling, so that the crew can achieve autonomy at a lower cost. Plants will play a crucial role in this ecosystem because they are involved in each of its components. In this chapter, we consider that living and working on the Moon, then Mars, is the natural continuation of a human journey that began long ago in Africa. We propose a four-step strategy to achieve food security for a lunar crew. Dietary planning will structure the approach and determine the selected plants, the method to validate candidate plants, biomass production, and manufacturing meals. In this strategy, the plants will also play an essential role in the crew members' psychological comfort, underlining the profound meaning of the ecosystem concept in the human world.

5.1 Introduction

Plants have been part of the space adventure since its beginnings with the Soviet/Russian station Salyut 1. The objective was obvious: lay the first stones of the ecosystem that was necessary for humans to live in space outside the Earth's biosphere for an extended duration. As its name suggests, an ecosystem is an integrated and self-sufficient whole, without additional elements brought from the Earth. It only uses the efficient power sources available in situ, and in which the perishable components could be self-regenerated (Medina, 2020). In other words, it is a bioregenerative life support system (BLSS). Higher plants become an essential component of the ecosystem. They play five key roles: food and bioactive molecules production, oxygen production, carbon dioxide reduction, water management, and metabolic waste recycling. Once exposed to light, green plants convert carbon dioxide

R. Cazalis (✉)
University of Namur, Namur, Belgium
e-mail: roland.cazalis@unamur.be

(CO_2) into food and oxygen (O_2) through photosynthesis. The crew members breathe oxygen and convert it back to CO_2. Plants absorb water up through their roots and release water vapor through their stomata, the openings on the leaves. The transpiration process thus can be used to convert contaminated water into drinking water. Photosynthesis and transpiration will produce food and water in the ecosystem. Therefore, plants' natural functioning helps close the habitat's different loops in the outdoor space planned as an ecosystem.

The exo-agriculture project plans to have astronauts master the ability to grow plants on the Moon, Mars and other planets, and onboard spacecrafts during long-term space missions. We limit our horizon to a region of outer space from 200 to 450 km above sea level, where space plant research has taken place, in view of replicating plants' farming in future missions to the Moon and Mars. The goal is to achieve such tasks within the next 30 years of the space adventure (Darrin & O'Leary, 2009). In preparation for this venture, this study analyzes recent discoveries in plant biology to see how they can help develop a BLSS in which plants could play their expected role. This objective addresses the first point concerning plants' behavior in relation to radiation and microgravity, which are two concomitant factors. Then, we address a second point which includes a dietary plan, plant choice and biomass production. The last point considers one of plants' ancillary roles in the ecosystem, namely, psychological comfort. The final synthesis will be the occasion to make proposals for the installation of a lunar colony.

5.2 Plants' Growth and Development in the Outer Space

5.2.1 Plants' Behavior in Ionizing Radiations

Plants' viability in the outer space zone subject to radiation[1] is the first factor to consider. In situ exposure and space radiation simulation through exposure to both x- and γ-rays assess plants' behavior under ionizing radiation. Scientists used different doses to complete genomic, proteomic, and morphogenetic analyses (Desiderio et al., 2019). Both kinds of experiments provide access to the damage of these radiations and consider the means of protection. In this respect, the strong Earth's geomagnetic field and its thick atmosphere protect our planet from radiation exposure that reaches a world average dose of 2.5 Sv per year (Naito et al., 2020). In contrast, the Moon has a weak magnetic field and a thin atmosphere that enable radiation and even micrometeorites to reach the soil (Naito et al., 2020). Typically, the Moon's surface experiences two kinds of radiation plus a third one, the chronic galactic cosmic rays (GCR) and the sporadic solar energetic particle events (SPEs). The third components result from the interaction field with the Moon's soil, leading to neutrons' release and gamma radiation (Naito et al., 2020).

---

[1] Gray (Gy) is the special name for the SI unit of absorbed dose (1 Gy = 1 J/kg), while, Sievert (Sv) is the name for the SI unit of equivalent and effective dose (1 Sv = 1 J/kg) (ICRP 2012).

Recently, China's Chang'E4 lander made the first-ever measurements of the radiation exposure on the Moon's surface. They measured an average dose equivalent of 1369 µSv per day on the Moon's surface, while in the same period, the ISS's onboard dose equivalent was 731 µSv per day (Zhang et al., 2020). A simulation of the effective dose equivalent of GCR particles at the lunar surface gives 416 mSv per year and 2190 mSv per event of SPEs (Naito et al., 2020). Comparing the putative values with what occurs on the Earth's surface helps to reevaluate the Earth's geomagnetic and atmospheric protecting effect and the necessity to protect the plants on the Moon.

With this data on the lunar situation in mind, we have to consider the conditions for installation. First of all, ionizing radiation is known to deposit energy inside living tissues causing functional damages. However, plants are more tolerant to ionizing radiation than animals (Arena et al., 2014). The effect of radiation varies significantly depending on the type of radiation, delivered dose, species, developmental stage and genetic traits (De Micco et al., 2011; Holst & Nagel, 1997). For instance, high-LET (Linear Energy Transfer) radiation (e.g. protons and heavy ions) are more dangerous than low-LET ones (e.g. X- and gamma-rays). They generate more cell death. Likewise, doses of X-rays up to 10 Gy in humans cause death within days or weeks, mostly due to infections resulting from a depletion of white blood cells. Such doses applied to mature bean plants' tissues do not induce any detrimental effects in the mutations of leaves' anatomical traits (De Micco et al., 2014) in both plant and mammalian cells (De Micco et al., 2011; Durante & Cucinotta, 2008; Wei et al., 2006).

Scientists have put forward various hypotheses to explain plants' tolerance to radiation, as shown by plants' persistence in the Chernobyl Exclusion Zone. Nonetheless, radiation affects plants' growth in important aspects (Mousseau et al., 2014; Santos et al., 2019) in this zone. The hypotheses involve features at the cytological, genetic and physiological levels. For instance, plants' radioresistance includes a thick specialized cell wall (Charlesby 1955, 107) and the accumulation of phenolic compounds with antioxidant properties (Agati et al., 2009; Graham et al., 2004). Polyploidy occurrence is another protecting factor against the harmful effect of mutations (Comai, 2005; Endo & Gill, 1996). The ability to increase the amount of antioxidants and phenolic compounds in cells and to trigger the activation of scavenger enzymes could matter. The effect is the removal of free radicals that the radiation stress generates (Arena et al., 2013; De Micco et al., 2014; Fan et al., 2014). Finally, plants are modular systems that can continue to grow after shedding damaged organs or parts of them (De Micco & Aronne, 2012).

However, radioresistance does not mean that plants are not sensitive to high doses. Indeed, the macromolecules, in particular DNA, are the critical targets of radiation. The interaction of radiation with cells can occur through direct interaction of radiation with cell components. It can occur as well through indirect damage caused by elevated radiation and reactive oxygen species (ROS) production. Among the different physiological processes affected, photosynthesis can be seriously affected by heavy ions, X- and gamma-rays. The main target of the injuries is the D1 protein of the photosystem II reaction-center. In this respect, high doses of gamma-irradiation (37.5 or 112.5 Gy) are responsible for decreasing photosynthesis, chlorophyll content and photosynthetic electron transport rate in bean and soybean plants (Stoeva & Bineva, 2001; Ursino et al., 1977).

A priori, the seeds are better protected from radiation aggression than the plants' aerial parts because of their cuticles. However, this is not always the case. For instance, *Arabidopsis thaliana* seeds were (Califar et al., 2018) exposed to Antarctica's upper polar stratosphere using scientific balloons (Smith & Sowa, 2017). This experiment emulated in-depth space exposure to ionizing radiation. The plant material received an average dose rate of 5047 Gy per hour (Ave et al., 2011; Benton, 2012). The analysis showed significantly reduced germination rates. Whole-genome sequencing revealed elevated somatic mutation rates correlated with an array of structural genome variants. However, if ionizing radiation may cause severe injuries to the cell, they also activate different response mechanisms to repair the type of damage.

Exposing plants to a very high dose rate of ionizing radiation has sometimes limited relevance in natural conditions found in space. Instead, other teams recommend using chronic and low dose rates under multigenerational exposure (Mousseau & Møller, 2020). These conditions are more in tune with plants' cultivation outside the Earth. The experiments could occur on nuclear accident sites such as Chernobyl and Fukushima, atomic bomb test sites, and other areas in the world with naturally high radiation levels. This approach helps investigate exposure to ionizing radiation sources over a long period. This is highly informative to establish parallelism with how cosmic radiation could affect plants' behavior and the importance of time in plants' adaptative process in outer space cultivation. The survey shows that, in general, plants cultivated over several generations on the indicated sites show elevated genetic damages, thus reducing the viability of pollen and seeds. Growth rates are slow and increase developmental abnormalities. The ionizing radiation effect occurs in the maternal generation pass over to the offspring (De Micco et al., 2011).

Finally plants generally exhibit a favorable biological response to very low dose radiation exposures or hormesis (Arena et al., 2014). As such, low–dose gamma-rays (1 or 2 Gy) stimulate Arabidopsis seedlings growth (Kovács & Keresztes, 2002) and accelerate photosynthesis, respiration and electron transport rate (Kim et al., 2004; Kurimoto et al., 2010). Lettuce seeds irradiated with gamma-rays up to 30 Gy enhance the content of chlorophyll *a*, chlorophyll *b* and carotenoids (Marcu et al., 2013).

These few data show that plants do not respond in the same way according to the variety, the dose and the nature of the radiation. Much work is needed to achieve, or at least begin to achieve a classification that combines the type of plant, the nature and amount of radiation and its effects.

These data suggest a practical rule. Because plants with heterogeneous abilities must be protected from the stress of ionizing radiation and from the extreme temperature variations that prevail on the Moon and meteoroids, it is necessary to bury the farm on the Moon beneath layers of lunar regolith. Such a structure can draw inspiration from the ESA project's lunar base (Fig. 5.1). We will further specify the details of the structure's internal organization. Alternatively, some zones offer natural shielding to minimize exposure time and maximize shielding. Examples of such areas could include the lunar vertical and horizontal lave tubes discovered in

Fig. 5.1 Representation of a module of a lunar village (Credit: ESA/Foster + Partners). The dome is protected from radiation, temperature variations and meteorites thanks to a layer of regolith. Outdoor light could be collected via special windows through the regolith-based protective layer. The village is a network of modules linked together by tunnels

2009 in the Moon's Marius Hills, *Mare Tranquillitatis*, and *Mare Ingenii* (Naito et al., 2020). Because of the human/plant mutualism that the lunar venture involves, we can draw the following practical rule of thumb from this study. We must simply apply the same shielding conditions to the plants and crew members to ensure their full viability in the Moon settlement.

5.2.2   Plants' Behavior in Microgravity

The second factor to consider is the impact of microgravity on plants' growth and development. Plants live under the influence of a ubiquitous force of 1 g called gravity. This weak force has a considerable effect in shaping the world as we know it. Gravity affects plants' growth and development from the plants' molecular constituents to their whole body (Vandenbrink et al., 2014). Consequently, microgravity is a novel environmental condition for plants, and the genome may lack a particular gene set to respond to it. However, plants' plasticity could have enough resources to control such unusual environmental factor (Medina, 2020). Understanding the actual contribution of gravity on plants' physiology requires to modify the magnitude of this force. It is possible to simulate microgravity via tools such as clinostat, random positioning machines (RPM) and magnetic levitation devices. These platforms frequently change gravity direction without reducing it (Kiss et al., 2019). Real weightlessness is set on spaceflight and orbiting spacecraft, such as on Salyut, Mir or ISS. Furthermore, using a centrifuge on board helps simulate reduced gravity conditions found on the Moon (0.17 g) and Mars (0.38 g), thus facilitating the study of plants' behavior in such conditions.

The continuous improvement of the spacecraft plant hardware has contributed significantly to the obtention of plants in space (Stankovic, 2018). Thus, the use of the Phyton-3 chamber on Salyut-6 *Arabidopsis thaliana* flowered in 1980. In 1982, the same plants were grown from seed to seed in space for the first time (Harland, 2004). The development of novel technologies to grow plants on the ISS has confirmed that it is possible to complete the full cycle of plants from seed to seed in space. We can mention two successful experiments that scientists achieved successively in 2001–2002, called the "Advanced AstroCulture" (ADVASC). *Arabidopsis* seeds obtained from the first experiment successfully germinated in the second one. The authors reported some phenotype changes relative to control but with a quite comparable germination rate (Link et al., 2014). Plant crops, such as soybean, were obtained in the ADVASC system with healthy seeds and a germination rate similar to ground control plants (Zhou, 2005).

Achieving healthy plants in a microgravity environment means that this cue can be compensated for, at least to some extent. On that point, experiments show that when the cue for gravitropism is present at low strength, other tropisms may be applied to direct plant growth. For instance, light can drive plant growth orientation with both a negative root phototropism and a positive shoot phototropism (Wyatt & Kiss, 2013). Spaceflight studies have reported roots positive phototropic responses to red light (Millar et al., 2010) and blue light (Vandenbrink et al., 2016). In this respect, Villacampa et al. (2021) used the European Modular Cultivation System (Brinckmann, 2005), installed in the ISS from 2008 to 2018. First, they applied different gravity levels (microgravity, 0.1 g; Moon; Mars; near-earth g-level; 1 g) to blue-light stimulated wild-type Landsberg ecotype *Arabidopsis thaliana* seedlings. They showed a replacement of gravitropism by blue-light-based phototropism signaling at microgravity level (Vandenbrink et al., 2019), and found a striking stress response at 0.1 g. Secondly, the same team studied the changes in 6-day-old *Arabidopsis* seedlings (Col-0) grown in the Seedling Growth experiments onboard the ISS (Valbuena et al., 2018) at three g-levels (microgravity, Mars gravity level and 1 g ground reference run) and under red light photostimulation. The team spent the last two days of the experiment controlling the seedlings in darkness. The results were consistent with previous reports showing that red light stimulated proliferation (Reichler et al., 2001; Valbuena et al., 2018). They concluded that red-light photostimulation might help plants to overcome some of the harmful effects of the spaceflight environment. Scientists found similar results in their experiments with blue-light photostimulation (Herranz et al., 2019).

These promising discoveries should not neglect that reduced gravity impacts plant growth and development. The changes are expressed at the anatomical and transcriptomic levels. *Inter alia*, reduced gravity environment specifically induces dysregulation of the genes involved in photosynthesis in *Arabidopsis thaliana* (Villacampa et al., 2021). Similar changes were observed in *Brassica rapa* plants grown in space (Jiao et al., 2004) and in *Oryza sativa* plants grown in simulated microgravity (Chen et al., 2013). These changes may affect the plant value. On this point, chemical analysis of "Outredgeous" red romaine lettuce grown on the ISS evidenced significant changes in elementary and antioxidant content, which may affect the nutritional value of lettuce (Khodadad et al., 2020).

From these experiments, we can conclude that plants show automorphogenesis in space, a critical aspect masked by gravity on Earth (Stankovic, 2018). Nevertheless, the expression of this automorphogenesis requires other cues, such as red and blue lights in the absence of the gravity vector. From there, it is necessary to test the reduced gravity compensation system for the plants selected for biomass production. Moreover, the observed changes in elementary and antioxidant content in the "Outredgeous" red romaine lettuce grown on the ISS require further research to assess the correlation between gravity cue compensation and plant nutritional value. The work could be undertaken in devices such as the new plant habitat facility aboard the ISS, the Advanced Plant Habitat (APH). The latter is the most extensive research facility for plant growth. It is a critical tool to progress in the study of plants' behavior under microgravity. *Arabidopsis* and dwarf wheat (cv Apogee) were successfully grown from seeds to validate the facility (Monje et al., 2020). Finally, the growing body of data demonstrates that edible plants can grow in outer space, as evidenced by three astronauts snacking salad leaves recently harvested on the ISS (Heiney, 2017). What is feasible on the ISS station can be done on the Moon Gateway of the Artemis project. It can also be reciprocated on a spaceflight to Mars, or on a larger scale, on the Moon or Mars. Both have more favorable conditions than on the orbital stations.

5.3   Dietary Plan, Selection of Plants and Biomass Production

In the perspective of a human installation on the Moon, the purpose of plant biology in space is to contribute to edible biomass production for the crew's food autonomy. This science must also provide know-how enabling plants to fulfill their crucial role in the colony's ecosystem. Accordingly, Veggie is the last generation of the crop production system launched on the ISS in early 2014. The first system was explicitly designed for biomass production rather than for plant biology under microgravity (Zabel et al., 2016).To achieve food biomass production, we propose a four-level operating scheme: dietary plan, selection of plants, production mode, and meal manufacturing, assuming safety and palatability, and ensuring variety to avoid menu fatigue.

5.3.1 Dietary Plan

In the decision chain we propose, dietary management comes first. One must ensure a balance between carbohydrate, protein and lipid, fiber and microelements, as well as dietary energy. The dietary energy intake of a crew member is about 2800 kcal (Kovalev et al., 2020). The source could be only a plant or a mixture of plant and animal. From this point of view, based on research developed on the space stations from Salyut to ISS and in loop-closed and ground-based experiments, researchers have developed various optimized diet meal plans. The latter should satisfy crews' basic nutritional needs in long-term manned missions. The dietary menu comprises biomass produced in situ with BLSS regenerated ingredients (e.g. Cooper et al., 2012; Fu et al., 2016). However, these menus remain putative because they are proposed without knowing if the ingredients available in a ground-based BLSS will also be available in space.

Balanced dietary nutrients are critical because it is not solely an intake of energy. They regulate health as a whole and are essential for the gut microbiome, and more generally, for immunity and intestinal health (He et al., 2017; Ma et al., 2017). Despite this, a consensus is still missing on space diet and nutrition requirements (Fu et al., 2019). Eating habits probably form an obstacle to a diet model that recalls that eating is, first of all, a cultural act and that ingredients and their presentation belong to the staging of such an act. However, let's suppose that the first installation on the Moon represents a real challenge by the size of the task. In light of the Artemis project, we saw that international cooperation was necessary to divide the work. The same should occur in the diet model, and certain pragmatism must take the lead for present and future missions. Indeed, not everything is possible in the spatial environment with regards to exo-agriculture. To reinforce pragmatism, when human beings are sent on a mission that surpasses them while bringing them under the spotlight, they know how to compensate for the discomforts of space life.

5.3.2 Selection of Plants

The second level concerns the list of candidate plants capable of maximizing the first level. Diet is the guiding principle of this choice, i.e., the nutritional value of what we eat, including the special needs that a space environment requires. However, we must consider other criteria, such as the plant's size and dwarf species. We must also take into account the plant's adaptability to the Moon, its short life cycle, polyploidy that gives an additional reserve of plasticity for adaptation (Medina, 2020), high harvest index, recipe elaboration, conservation, etc. (Chunxiao and Hong 2008; Wheeler, 2017; Carillo et al., 2020). The complete food-chain overview leads to a relatively broad starting base, i.e., an initially high number of candidate plants, taking into account interchangeable plants for the same nutritional qualities to avoid menu fatigue (Douglas et al., 2020).

El-Nakhel et al. (2019) describe a case of plant selection based on criteria favorable to human life support systems. They compare two differently pigmented butterhead *Lactuca sativa* L. (red and green Salanova) cultivars produced in closed soilless cultivation. They were assessed in terms of morphometric, mineral, bioactive and physiological parameters. The authors showed that red Salanova exhibited higher fresh biomass, water use efficiency, lipophilic antioxidant activity, total phenols and total ascorbic acid. Red Salanova had 37.2% less nitrate than green Salanova. According to the authors, this cultivar's contribution has yet to be assessed in air regeneration and water recycling to complete the choice of red Salanova as a new candidate cultivar for BLSSs.

With the dietary plan in mind, this example shows that the initial list of candidates must be screened through a scale to track crop testing progression for space. Such a filter will refine the list at the end of the process while increasing the selected plants' suitability. Concerning the scale to assess crops' readiness, Romeyn et al. (2019) propose a tool to track the preparation and the testing of a crop species for use in the space environment. The device includes basic horticultural testing, cultivar trials, growth and yield, space-like environment adaptation, hardware, along with the nutritional value, the acceptability and the safety aspects of the crop. They recommend a 1–9 scale. The basic level "1" refers to potential crop identification. The final stage "9" refers to crops ready for consumption by the crew in space. The ISS's crew members snacking the "Outredgeous" red romaine lettuce means this salad reached the "9th" level and joined the final list. Reaching a consensus between competing nations is not easy. Still, once an efficient process to achieve results has been identified, we avoid wasting time and energy to reinvent what has already been found to invest in the next step. Romeyn's tool, which can always be improved, is a protocol to approve candidate plants, knowing that approval occurs only at level "9".

In an attempt to advance the concept of sustainable food production in a closed environment, various crop plants have already been tested on space stations and in direct interaction with astronauts. Other ground-based experiments are taking place in a closed-loop plant growth unit. One example is NASA's Biomass Production Chamber used for over a decade (1988–1999), and in which lettuces, potatoes, radishes, rice, soybeans and wheat were all successfully grown (Stutte, 2016). Other plant production systems incorporate a crew into the design of the living and working habitat. This the case of the Soviet/Russian BIOS-2 projects in Krasnoyarsk, Siberia or more recently, the Chinese Lunar Palace 1 at the Beihang University in Beijing, where the BLSS effective closure was tested. In other words, all these experiments provide a large amount of data on plants' behavior in an integrated closed-loop system. They help validate that in situ plant production effectively plays its expected role in the ecosystem in a ground-based environment. In other words, the production should provide a balanced diet, ensure the release of enough oxygen and the absorption of carbon dioxide, and support the full water cycle. Finally, the plant production system effectively removes atmospheric contaminants from the air. In short, an ecosystem is set up and operates adequately.

The crop plants used in the different plant production programs are often the same and reflect the eating habits. Nevertheless, we wonder if the selection of these plants respects any of the specifications of a dietary plan. In this trend, the European Space Agency's project MELiSSA (MicroEcological Life-Support System Alternative) from the Higher Plant Chamber has incorporated current crops in the assays. Candidate crops include durum and bread wheat, rice, potato, soybean, lettuce, and beetroot. Lettuce is the favorite crop monitored in the MELiSSA Pilot Plant at the Universidad Autónoma de Barcelona (Peiro et al., 2020). This preference is due to much data collected on many cultivars and its short, easy and quick growth cycle. This is why the lettuce reaches Romeyn's level "9". Consequently, the remaining crop plant candidates have to enter the same process as the "Outredgeous" romaine lettuce did to obtain their level "9" on the ISS.

So far, no fruit tree candidate has been introduced into the testing and production program because of these plants' size and cycle length (Graham, 2016). Recently, the United States Department of Agriculture (USDA) obtained a dwarf plum tree (*Prunus domestica*) comparable in size to other herbaceous spaceflight candidate crops, such as sweet peppers (*Capsicum annuum*) (Graham et al., 2015). These plum trees overexpress the flowering locus T1 (FT1) gene taken from *Populus trichocarpa*, a poplar specie (Srinivasan et al., 2012, 2014). Besides disrupting the apical dominance, the FT1 gene has other effects. It accelerates the flowering and fruiting cycles and disrupts the required cold dormancy period between fruiting cycles. The plum trees then flower and fruit continuously. In addition to providing fresh fruits to the crew, plums can prevent bone density loss in ground-based rodent and human models (Schreurs et al., 2016; Wallace, 2017) induced by microgravity and ionizing radiation during long-duration spaceflight. The dwarf plum tree may be the first fruit tree candidate to be tested on the ISS to determine if it reaches a level "9". Probably, other fruit trees, aimed to balance the crew's diet or to prevent microgravity's collateral effects, can enter a breeding program and yield similar promising results. This research work falls under the first field of the proposed strategy.

5.3.3 Biomass Production

The third level of the strategy concerns space farming practices. Various ongoing experiments try to simulate crop biomass production in close systems. For instance, the EDEN ISS mobile Facility set up in Antarctica will help us analyze some critical points. The EDEN ISS Facility focuses on cultivating plants in controlled conditions and ontesting hardware, microbiology, food quality and safety, and energy aspects, such as the required crew time. Scientists used the latest technology, such as LED lighting (Zabel et al., 2020). The EDEN ISS project rationale supposes crop monitoring supported by expert backrooms on the ground, which prevents the crew from spending time to maintain them. The system is endowed with an updated phenotyping camera system so that the ground expert agronomists could instruct the staff (Zeidler et al., 2019). This dual monitoring is instrumental on the actual ISS and future spaceflight, on the Moon Gateway and even in the Moon and Mars settlements. In the near future, the phenotyping experiment dataset will contain enough examples to train a deep-learning algorithm to diagnose crop health state, monitor the plant's development via telemetry and instruct an on-site operator.

In the experiment, scientists aimed to have the selected crop species grow into fresh, pick-and-eat vegetables. One of the critical aspects of this experiment was to grow all crops simultaneously in the same space and under the same conditions. The compromise climate is more related to the near-term space greenhouses. Individualcrops were not produced under optimal conditions because of the cost and complexity of such optimized system (Zabel et al., 2020). This last aspect shows the difference between the science plant in the APH on the ISS and biomass production in real space conditions to feed the crew members.

Due to the unique Moon characteristics, we must improve plant productivity to sustain human settlements, and this is directly correlated with light quality. In this respect, quantum dots technology has been utilized in various applied fields (Kargozar et al., 2020); it appears as an alternative to improve photosynthetic efficiency. A NASA-funded agriculture study shows that CIS/ZnS quantum dots incorporated into a film passively modify the solar spectrum by down-converting UV and blue photons to orange and red photons. Applied on *Lactuca sativa* L. (cv. "Outredgeous" red romaine lettuce) within a semi-closed plant growth system, this technology increases edible dry and fresh biomass and total leaf area (Parrish et al., 2021). Quantum dots into flexible luminescent agriculture films could help enhance the photosynthetic efficiency and productivity in Moon farming.

Plant's nutrient supply is another aspect of biomass production that is useful to bear in mind in the near-future Moon settlement. The system uses must be different in a microgravity environment than on the Moon's ground base. In spaceflight to Mars, the Veggie facility and an improved version for larger volume could be used, depending on the crop turnover while preserving the cabin's ecosystem.

For Moon farming, the most interesting system is vertical farming. The latter involves plant cultivation in vertically stacked irrigation systems with nutrients recapture and recycling. The design minimizes operational costs and maximizes productivity (Benke & Tomkins, 2017). The hydroponic/aeroponic concept used in the EDEN ISS Project could be applied, notably the nutrient-film technique, which could be used as the nutrient delivery system. In this system, small rock wood plugs support the plant within the growth tray lids (Zabel et al., 2017).

Plants grown hydroponically or aeroponically represent an alternative for space-based plant growth systems. However, the appropriate system that works in the outdoor space has yet to be developed, including in the planetary system. Such a device could be inspired by the Astro Garden™ Aeroponic Plant Growth System (Moffatt et al., 2019). Indeed, many of the challenges identified with microgravity aeroponics can be addressed through parabolic flight testing with Zero-G Corporation. However, other challenges will need to be addressed through extended duration testing on the ISS. In their Plant Growth System, Moffatt et al. (2019) tested the physics of aeroponic spray delivery in microgravity through parabolic flight testing with Zero-G Corporation. They answer two important issues. First, the spray reaches the root bundle regardless of gravitational conditions. Aeroponics is then a viable nutrient delivery mechanism for a spacecraft environment. Second, forced air was an efficient method for removing free droplets from the airstream at flow velocities compatible with the root structure integrity. However, design modifications are still necessary to control the solution on the surfaces within the root zone.

Because of the time limitations of freefall periods during the parabolic flight (~15 s), they could not fully validate a hydroponic nutrient system in another test chamber. The nutrient solution fluid would move along the root structure through capillary wicking in the absence of gravity. Such experiments must be undertaken in the ISS.

The parabolic flight was also used to demonstrate the possibility to control the fluid stability in nutrient delivery to seed cartridges. The nutrient was delivered to seeds through capillarity along a wick structure and through porous plates with a felt backing. The nutrient flow could be accurately controlled with backpressure on the porous material. With these experiments, they cover plant nutrition from seed germination to the mature plant.

The hydroponic/aeroponic option avoids using regolith and traditional horticulture practice, which requires a large amount of growth substrate and generates elevated waste production over a long period. The ideal nutrient solution is aeroponic. It requires no soil with extremely limited water use. Very little nutrient use is enough, and the solution involves no pesticides and low food waste. Aeroponic cultivation is more productive than other methods, and differs according to species (Eldridge et al., 2020).

Furthermore, research in vertical farming is very active in responding to and improving urban farming's growing trends, including LED lighting and residual heat valuation (SharathKumar et al., 2020). Moon agriculture could benefit from this venture. Such a strategy requires solving water access in the lunar soil, and sufficient energy production to support vertical farming. Alternatively, we can consider using the Moon regolith as a support base for plants' growth (cf. Braddock in this volume, Chap. 6). However, some experiments tend to prove that the Moon and Mars regolith simulants used as an in situ cultivation substrate seem unable to support the growth of targeted crops without organic matter supplementation (Duri et al., 2020; Eichler et al., 2021; Wamelink et al., 2019).

With these data in mind, we can complete the facility schematic suggested in Fig. 5.1 for edible biomass production. The image of a network of modules pedagogically helps visualizing the lunar village (cf. Corbally and Rappaport in this volume, Chap. 21). The plants are selected according to the dietary plan and other criteria that deal with the ecosystem's requirements. The selected plants have first been subjected to cultivation cycles so that they are free of any pathogens that could endanger the farm. Plants are cultivated in an hydroponic/aeroponic farming, with a LED lighting system or better, the CIS/ZnS quantum dots lighting. Each module is designed to be autonomous as an ecosystem and thus has its own micro-farm in vertical culture (Fig. 5.2). In medium and long term installations, each micro-farm could specialize in a range of crops to optimize production for the village.

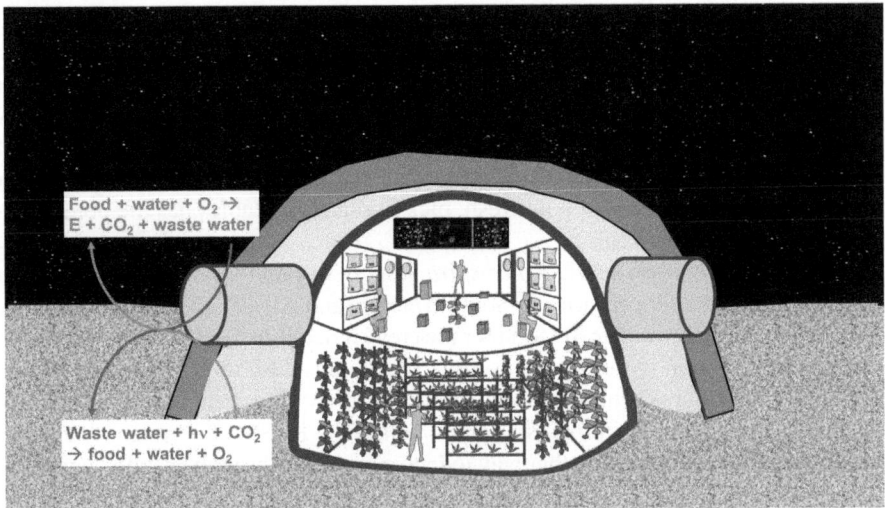

Food + water + O_2 →
E + CO_2 + waste water

Waste water + hν + CO_2
→ food + water + O_2

Fig. 5.2 Inside view of a lunar village module. Each dome is thought of as an autonomous ecosystem where the crew and the farm live in a co-dependent relationship expressed by the systems of two equations on the left (E = metabolic energy; $h\nu$ = light energy). The habitat is divided into one section for the crew and another one for the farm respecting the plants/humans proportion. The farm lighting (LED/quantum dots) could be complemented utilizing outdoor light like in the Prototype Lunar/Mars habitat (Giacomelli et al., 2012), or channeled through a fiber-optic system

5.3.4 Meal Manufacturing

The last level relates to meal manufacturing. As mentioned before, eating is not only about ingredient intake. The shape, appearance and origin of the ingredient, especially if it is recognizable, carry weight in the product's acceptability as a food. Taste is also a determining factor. If plants are the only nutritional sources, then the menu will be vegetarian and special attention is required to ensure that it is balanced. However, based on the previous loop-closed ecosystem, such as Bios-3 or Luna Palace 1, the crew expressed a need for a certain amount of animal protein (Manukovsky et al., 2005). Chinese eating habits include insects as a source of protein. As such, silkworm and yellow mealworm may be one solution in providing animal protein to astronauts on the BLSS. This alternative was successfully tested in Lunar Palace 1. Furthermore, the worms feed the inedible plant biomass contributing to the in situ recycling system (Fu et al., 2016). To this day, scientists miss data on worms' behavior in space regarding growth and sustainability. Such knowledge could then include worms in the protein diet. Nevertheless, eating worms is not immediately evident if these entities are not part of the eating habits. Crushing and mixing 'exotic ingredients' with other components could be a solution to enrich the nutritional value of the diet and save its sanitary quality.

In a similar vein, one can also opt for a completely vegetarian menu. However, since this menu is not universally accepted, using a 3D bioprinter to manufacture food would help imitate the shape and flavor similar to those of fish or meat. Plants alone or plants combined with insects could be raw ingredients to feed the bioprinter. This subterfuge lures the guest efficiently and is psychologically effective for non-vegan team members. Such a trick is possible because the human being is a complex entity. Autosuggestion belongs to the arsenal of means to adapt to the new environment, including adapting to new eating presentations. Moreover, the 3D manufactured vegan meat or fish presented with fresh, pick-and-eat vegetables has the potential to delight the crewman's soul.

Likewise, the cyanobacteria *Arthrospira platensis*, commonly known as spirulina, has been harvested for centuries in South America and Africa. This ingredient could be added to the mix to create 3D food for the crew. Indeed, spirulina could supplement the diet with protein, vitamin A and iron. Spirulina is easy to grow, and multiplies rapidly. Furthermore, it turns carbon dioxide into oxygen. It is highly resistant to outer space radiation. Spirulina is a component of the BLSS's ecosystem that is part of the MELiSSA project. Another way to incorporate it is through pseudo-seasoning. Dried powdered spirulina with 10 g sprinkled on food each day is enough to satisfy most dietary requirements (Lasseur et al., 2010).

At the logistic level, whether on the Moon's orbital station, in long-duration spaceship or in the first planetary settlement, it is necessary to rationalize biomass production and meals manufacturing by linking farm and kitchen managements.

5.4 Plants for Psychological Comfort

There are numerous studies about indoor plants' impact on human health and comfort based on at least four criteria: photosynthesis, transpiration, psychological effects, and air purification (Deng & Deng, 2018). There are beneficial impacts of plants on gases at home or in the office. Plants' critical processes, such as photosynthesis and transpiration, remove carbon dioxide from the atmosphere and release oxygen in it. These events contribute to adjusting the room temperature and humidity to some extent (Deng & Deng, 2018). They are part of the ecosystem of the Moon settlement and the long-term spaceship.

Phytoremediation is another plant property that removes indoor airborne pollutants such as particulate matter, gaseous contaminants and volatile organic compounds (VOCs) released by hardware and indoor human activity (Deng & Deng, 2018). The release can occur inside the spaceship, on the ISS and in the next Moon Gateway, inside the planetary settlement.

Fortunately, plant phytoremediation contributes to indoor air quality improvement. For instance, plants can release negative air ions (NAI) under certain conditions, and positively impact air quality. NAI can improve health at certain levels of concentration (Yue et al., 2020). We could enhance the NAI's concentration by increasing the number of plants submitted to pulse electric fields (Zhu et al., 2016). As such, safer indoor air quality would improve mood and task performance, and explain why indoor plants reduce nervousness and anxiety. However, air quality are not the whole story. Some authors attribute the positive effect to room arrangement in the workplace, and plants play their role in the mix (Kim et al., 2018).

Regardless of the elementary argument above mentioned, the reason for the positive effects on human mood remains relatively obscure. A recent study shows the possibility of applying air phytoremediation under microgravity. In this experiment, the high benzene removal plant *Chlorophytum cosmosum* shows an increase of removal efficiency under microgravity (in RPM) compared to normal gravity conditions. This supplemented efficiency was attributed to an accumulation of auxin hormones in the plant's shoot part, and maintained open plant stomata enhancing benzene phytoremediation efficiency (Treesubsuntorn et al., 2020). Scientists at Energia, the agency for the Soviet manned space progress, were impressed with the positive psychological effects of exo-gardening on cosmonauts who decided to use it to support the crew's mental health (Zimmerman, 2003). Likewise, orchids were the first plants flown on Salyut 6 to investigate the psychological effect of the crew's interaction with plants (Zabel et al., 2016).

In practice, whether in the confined environment of the spacecraft or in the Moon's settlement, plants can contribute to the team's psychological well-being in two ways. First, by exo-gardening, which is an active way of expressing emotional mutualism by taking care of the plants. Second, by the purification of the ambient air operated by the plants. On this topic, the cosmonauts Valery Ryumin and Vladimir Lyakhov relate how a mature kalanchoe plant helped them overcome loneliness and depression in 1979. In 1982, aboard Salyut 7, Valentin Lebedev noted that gardening allowed him to appease his anxiety (Zimmerman, 2003).

A further plant contribution that may be very useful in a planetary settlement is through an indoor park. Plants can bring beauty and peace inside the dome. The latter could have green walls, horizontal gardens with flowers, alleys of different levels, and indoor tree groves inspired from the Garden by the Bay in Singapore. Plants could be placed alongside the structures to emulate a tree with its branches and leaves (Fig. 5.3). Beauty setup from natural matter has a notable psychological and spiritual impact on the human mind. Furthermore, this park is another opportunity for the crew to do gardening and maintain the trees alive. This park is a place to stay, to walk and to breathe.

5.5   Synthesis and Recommendations

Experiments' results on the ISS in Veggie and APH facilities shed light on the plant's viability in reduced gravity, and more specifically in our study, in the Moon and Mars environments. Based on this, we propose a four-step approach to carry out edible biomass production to supplement the supply of food for the crew in orbital stations or for the first colony settlement on the Moon. The experience acquired here will help to better plan the Mars installation.

The first step is to design the dietary plan which then structures the other actions. We must establish a balanced diet necessary for the crew in microgravity conditions,

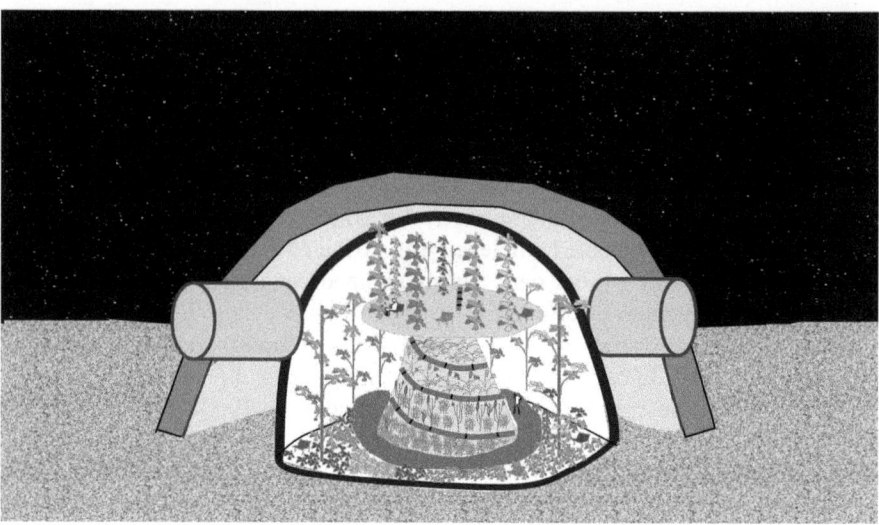

Fig. 5.3 Inside view of a concept park in the lunar village. It is a place for gardening and relaxation contributing to the psychological comfort of the villagers

and find the plants capable of providing the ingredients to make up this diet. Other sources of ingredients other than plants are available, especially in protein provisions, such as spirulina or edible insects. Finally, dwarf fruit trees, such as plums, are now among the candidates, particularly to counter spatial osteoporosis.

The second step is to validate the candidate plant. For this, we can follow the protocol proposed by Romeyn et al. 2019. ISS astronauts must validate the last step since they live in microgravity and in confinement, which will be the rule in the Moon's colony.

The third step is to plan out the biomass production from the selected plants and possibly from other ingredient sources too. The colony being installed in an underground-like structure, the plants grow under similar conditions. The ecosystem requires proximity between plants and humans. We recommend vertical farming with LED/quantum dots lighting and support adapted to hydroponic/aeroponic cultures, which is agriculture in the making. This rapidly developing technology makes it possible to considerably reduce water consumption and increase production, while avoiding soil use. An operator must conduct space farming under the supervision of a team of agronomist experts on Earth to guarantee the crew's food security. Through this interaction, space learns from the Earth, and the Earth learns from space.

Following the strategy adopted in EDEN ISS, the different species can be cultivated under the same conditions, which simplifies the process, especially during a first installation on the Moon. The last stage of the food chain is the meals' processing. We propose to use a 3D bioprinting system fed by the various raw materials obtained in the farm to produce different types of food with varied forms, aspects, and presentations that recall our food habits. Overall, the aim is to achieve a balanced diet.

Thus, it is beneficial to be in synergy with the food psychology of the crew members. This strategy implies that at least part of the food preparation must be centralized and in line with the farm's production possibilities, to always guarantee the crew's food security. Finally, plants can still play an essential role in the team's psychological well-being through micro exo-gardening in the spacecraft or in an exo-park of the settlement, owing to the pivotal role that plants play in human psychology.

Biomass production is one of the ecosystem's key elements, which means that if plants are to produce oxygen, absorb carbon dioxide and contribute to water recycling, there must be a ratio between the mass of the plant and the mass of the crew. Moreover, the rate of plants' photosynthesis varies according to the plant's cycle. Therefore, a constant percentage of young plants in the farm are necessary to guarantee the ecosystem's stability. All these details contribute to the viability of the ecosystem.

For the ecosystem to function well, it is essential to solve the issue of in situ water availability. Indeed, there is a difference between water in the lunar subsoil and water in a reservoir. Moreover, if we find traces of living organisms during the water exploration, we should decide the expedition's future course. Consequently, the water issue is a top priority for the project's viability as it is conceived. Next comes the production of sufficient energy.

Outer space settlement is the new stage of the human journey, and we must remain open to what is to come. The plants in the Moon's farms will eventually evolve in order to better adapt to their environment. The crew members will also develop, at least psychologically. The future is more open than ever.

References

Agati, A., Stefano, G., Biricolti, S., & Tattini, M. (2009). Mesophyll distribution of 'antioxidant' flavonoid glycosides in Ligustrum vulgare leaves under contrasting sunlight irradiance. *Annals of Botany, 104*, 853–861.

Arena, C., De Micco, V., Aronne, G., Pugliese, M. G., Virzo, A., & De Maio, A. (2013). Response of Phaseolus vulgaris L. plants to low-LET ionizing radiation: Growth and oxidative stress. *Acta Astronautica, 91*, 107–114.

Arena, C., De Micco, V., Macaeva, E., & Quintens, R. (2014). Space radiation effects on plant and mammalian cells. *Acta Astronautica, 104*, 419–431.

Ave, M., Boyle, P., Brannon, E., Gahbauer, F., Hermann, G., Höppner, C., Hörandel, J., Ichimura, M., Müller, D., & Obermeier, A. (2011). The TRACER instrument: A balloon-borne cosmic-ray detector. *Nuclear Instruments and Methods in Physics Research Section A: Accelerators, Spectrometers, Detectors and Associated Equipment, 654*, 140–156.

Benke, K., & Tomkins, B. (2017). Future food-production systems: Vertical farming and controlled-environment agriculture. *Sustainability: Science. Practice and Policy, 13*, 13–26.

Benton, E. (2012). Space radiation passive dosimetry. *The Health Risks of Extraterrestrial Environments*. https://three.jsc.nasa.gov/articles/BentonPasssiveDosimetry.pdf

Brinckmann, E. (2005). ESA hardware for plant research on the international space station. *Advances in Space Research, 36*, 1162–1166. https://doi.org/10.1016/j.asr.2005.02.019.

Califar, B., Tucker, R., Cromie, J., Sng, N., Schmitz, R. A., Callaham, J. A., Barbazuk, B., Paul, A.-L., & Ferl, R. J. (2018). Approaches for surveying cosmic radiation damage in large populations of arabidopsis Thaliana seeds—Antarctic balloons and particle beams. *Gravitational and Space Research, 6*(2), 54–73.

Carillo, P., Morrone, B., Fusco, G. M., De Pascale, S., & Rouphael, Y. (2020). Challenges for a sustainable food production system on board of the international space station: A technical review. *Agronomy, 10*, 687.

Charlesby, A. (1955). The degradation of cellulose by ionizing radiation. *Journal of Polymer Science, 15*, 263–270.

Chen, B., Zhang, A., & Lu, Q. (2013). Characterization of photosystem I in rice (Oryza sativa L.) seedlings upon exposure to random positioning machine. *Photosynthesis Research, 116*, 93–105.

Chunxiao, X., & Hong, L. (2008). Crop candidates for the bioregenerative life support systems in China. *Acta Astronautica, 63*, 1076–1080.

Comai, L. (2005). The advantages and disadvantages of being polyploid. *Nature Reviews Genetics, 6*, 836–846.

Cooper, M. R., Catauro, P., & Perchonok, M. (2012). Development and evaluation of bioregenerative menus for Mars habitat missions. *Acta Astronautica, 81*(2), 555–562. https://doi.org/10.1016/j.actaastro.2012.08.035

Darrin, A., & O'Leary, B. L. (Eds.). (2009). *Handbook of space engineering, archaeology, and heritage.* CRC Press.

De Micco, V., Arena, C., Pignalosa, D., & Durante, M. (2011). Effects of sparsely and densely ionizing radiation on plants. *Radiation and Environmental Biophysics, 50*, 1–19.

De Micco, V., & Aronne, G. (2012). Morpho-anatomical traits for plant adaptation to drought. In R. Aroca (Ed.), *Plant responses to drought stress: From morphological to molecular features* (pp. 37–62). Springer-Verlag, Berlin Heidelberg.

De Micco, V., Arena, C., & Aronne, G. (2014). Anatomical alterations of Phaseolus vulgaris L. mature leaves irradiated with X-rays. *Plant Biology, 16*, 187–193.

Deng, L., & Deng, Q. (2018). The basic roles of indoor plants in human health and comfort. *Environmental Science and Pollution Research, 25*, 36087–36101. https://doi.org/10.1007/s11356-018-3554-1

Desiderio, A., Salzano, A. M., Scaloni, A., Massa, S., Pimpinella, M., De Coste, V., Pioli, C., Nardi, L., Benvenuto, E., & Villani, M. E. (2019). Effects of simulated space radiations on the tomato root proteome. *Frontiers in Plant Science, 10*, 1334. https://doi.org/10.3389/fpls.2019.01334

Douglas, G. L., Zwart, S. R., & Smith, S. M. (2020). Space food for thought: Challenges and considerations for food and nutrition on exploration missions. *The Journal of Nutrition, 150*(9), 2242–2244. https://doi.org/10.1093/jn/nxaa188

Durante, M., & Cucinotta, F. A. (2008). Heavy ion carcinogenesis and human space exploration. *Nature Reviews Cancer, 8*, 465–472.

Duri, L. G., El-Nakhel, C., Caporale, A. G., Ciriello, M., Graziani, G., Pannico, A., Palladino, M., Ritieni, A., De Pascale, S., Vingiani, S., Adamo, P., & Rouphael, Y. (2020). Mars regolith simulant ameliorated by compost as in situ cultivation substrate improves lettuce growth and nutritional aspects. *Plants, 9*, 628.

Eichler, A., Hadland, N., Pickett, D., Masaitis, D., Handy, D., Perez, A., Batcheldor, D., Wheeler, B., & Palmer, A. (2021). Challenging the agricultural viability of martian regolith simulants. *Icarus, 354.* https://doi.org/10.1016/j.icarus.2020.114022

El-Nakhel, C., Giordano, M., Pannico, A., Carillo, P., Fusco, G. M., De Pascale, S., & Rouphael, Y. (2019). Cultivar-specific performance and qualitative descriptors for butterhead Salanova lettuce produced in closed soilless cultivation as a candidate salad crop for human life support in space. *Life, 9*, 61. https://doi.org/10.3390/life9030061

Eldridge, B. M., Manzoni, L. R., Graham, C. A., Rodgers, B., Farmer, J. R., & Dodd, A. N. (2020). Getting to the roots of aeroponic indoor farming. *New Phytologist, 228*, 1183–1192.

Endo, T. R., & Gill, B. S. (1996). The deletion stocks of common wheat. *Journal of Heredity, 87*, 295–307.

Fan, J., Shi, M., Huang, J.-Z., Xu, J., Wang, Z.-D., & Guo, D.-P. (2014). Regulation of photosynthetic performance and antioxidant capacity by 60Co γ-irradiation in Zizania latifolia plants. *Journal of Environmental Radioactivity, 129*, 33–42.

Fu, Y., Li, L., Xie, B., Dong, C., Wang, M., Jia, B., et al. (2016). How to establish a bioregenerative life support system for long-term crewed missions to the Moon or Mars. *Astrobiology, 16*, 925–936.

Fu, Y., Guo, R., & Liu, H. (2019). An optimized 4-day diet meal plan for 'Lunar Palace 1.' *Journal of the Science of Food and Agriculture, 99*, 696–702. https://doi.org/10.1002/jsfa.9234

Giacomelli, G. A., Furfaro, R., Kacira, M., Patterson, L., Story, D., Boscheri, G., Lobascio, C., Sadler, P., Pirolli, M., Remiddi, R., Thangavelu, M., & Catalina, M. (2012). Bio-regenerative life support system development for Lunar/Mars habitats. In *42nd International Conference on Environmental Systems 2012, ICES* 2012.

Graham, L. E., Kodner, R. B., Fisher, M. M., Graham, J. M., Wilcox, L. W., Hackney, J. M., Obst, J., Bilkey, P. C., Hanson, D. T., & Cook, M. E. (2004). Early land plant adaptations to terrestrial stress: A focus on phenolics. In A. R. Hemsley & I. Poole (Eds.), *The evolution of plant physiology* (pp. 165–168). Elsevier.

Graham, T., Scorza, R., Wheeler, R., Smith, B., Dardick, C., Dixit, A., Raines, D., Callahan, A., Srinivasan, C., Spencer, L., Richards, J., & Stutte, G. (2015). Over-expression of FT1 in plum (prunus domestica) results in phenotypes compatible with spaceflight: A potential new candidate crop for bio-regenerative life-support systems. *Gravitational and Space Research, 3*(1), 39–50.

Graham, T. (2016). Trees in space: No longer the forbidden fruit. *Environmental Scientist, 25*(1), 44–47.

Harland, D. (2004). *The story of the MIR space station.* Springer.

He, L., Han, M., Farrar, S., & Ma, X. (2017). Editorial: Impacts and regulation of dietary nutrients on gut microbiome and immunity. *Protein and Peptide Letters, 24*, 380–381.

Heiney, A. (2017). Space gardener Shane Kimbrough enjoys first of multiple harvests. *NASA Kennedy Space Center.* https://www.nasa.gov/feature/space-gardener-shane-kimbrough-enjoys-first-of-multiple-harvests

Herranz, R., Vandenbrink, J. P., Villacampa, A., Manzano, A., Poehlman, W. L., Feltus, F. A., Kiss, J. Z., & Medina, F. J. (2019). RNAseq analysis of the response of Arabidopsis thaliana to fractional gravity under blue-light stimulation during spaceflight. *Frontiers in Plant Science, 10*, 1–11.

Holst, R. W., & Nagel, D. J. (1997). (1997). Radiation effects on plants. In W. Wang, J. W. Gorsuch, & J. S. Hughes (Eds.), *Plants for environmental studies* (pp. 37–81). Lewis Publishers.

ICRP, (2012). Compendium of dose coefficients based on ICRP publication 60. ICRP Publication 119. *Annals of the ICRP, 41*(Suppl.).

Jiao, S., Hilaire, E., Paulsen, A. Q., & Guikema, J. A. (2004). Brassica *rapa* plants adapted to microgravity with reduced photosystem I and its photochemical activity. *Physiologia Plantarum*, 281–290.

Kargozar, S., Hoseini, S. J., Milan, P. B., Hooshmand, S., Kim, H.-W., & Mozafari, M. (2020). Quantum dots: A review from concept to clinic. *Biotechnology Journal, 15*, 2000117. https://doi.org/10.1002/biot.202000117

Khodadad, C. L. M., Hummerick, M. E., Spencer, L. E., Dixit, A. R., Richards, J. T., Romeyn, M. W., Smith, T. M., Wheeler, R. M., & Massa, G. D. (2020). Microbiological and nutritional analysis of lettuce crops grown on the international space station. *Frontiers in Plant Science, 11*, 199. https://doi.org/10.3389/fpls.2020.00199

Kim, J. H., Baek, M. H., Chung, B. Y., Wi, S. G., & Kim, J. S. (2004). Alterations in the photosynthetic pigments and antioxidant machineries of red pepper (Capsicum annuum L.) seedlings from gamma-irradiated seeds. *Journal of Plant Biology, 47*, 314–321.

Kim, J., Cha, S. H., Koo, C., & Tang, S.-K. (2018). The effects of indoor plants and artificial windows in an underground environment. *Building and Environment, 138*, 53–62.

Kiss, J. Z., Wolverton, C., Wyatt, S. E., Hasenstein, K. H., & van Loon, J. J. W. A. (2019). Comparison of microgravity analogs to spaceflight in studies of plant growth and development. *Frontiers in Plant Science, 10*, 1577. https://doi.org/10.3389/fpls.2019.01577

Kovalev, V. S., Manukovsky, N. S., & Tikhomirov, A. A. (2020). Bioregenerative life support space diet and nutrition requirements: Still seeking accord. *Life Sciences in Space Research, 27*, 99–104.

Kovács, E., & Keresztes, A. (2002). Effect of gamma and UV-B/C radiation on plant cell. *Micron, 33*, 199–210.

Kurimoto, T., Constable, J. V. H., & Huda, A. (2010). Effects of ionizing radiation exposure on Arabidopsis thaliana. *Health Physics, 99*, 49–57.

Lasseur, C., Brunet, J., de Weever, H., Dixon, M., Dussap, G., Godia, F., Leys, N., Mergeay, M., & Van Der Straeten, D. (2010). MELiSSA: The European project of closed life support system. *Gravitational and Space Biology, 23*(2), 3–12.

Link, B. M., Busse, J. S., & Stankovic, B. (2014). Seed-to-seed-to-seed growth and development of Arabidopsis in microgravity. *Astrobiology, 14*(10), 866–875.

Ma, N., Tian, Y., Wu, Y., & Ma, X. (2017). Contributions of the interaction between dietary protein and gut microbiota to intestinal health. *Current Protein and Peptide Science, 18*, 795–808.

Manukovsky, N. S., Kovalev, V. S., Somova, L. A., Gurevich, Y. L., & Sadovsky, M. G. (2005). Material balance and diet in bioregenerative life support systems: Connection with coefficient of closure. *Advances in Space Research, 35*, 1563–1569.

Marcu, D., Cristea, V., & Daraban, L. (2013). Dose-dependent effects of gamma radiation on lettuce (Lactuca sativa var. capitata) seedlings. *International Journal of Radiation Biology, 89*, 219–223.

Medina, F. J. (2020). Growing plants in human space exploration enterprises. *Acta Futura, 12*, 51–163.

Millar, K. D. L., Kumar, P., Correll, M. J., Mullen, J. L., Hangarter, R. P., Edelmann, R. E., & Kiss, J. Z. (2010). A novel phototropic response to red light is revealed in microgravity. *New Phytologist, 186*, 648–656.

Moffatt, S. A., Morrow, R. C., & Wetzel, J. P. (2019). Astro Garden™ aeroponic plant growth system design evolution. *ICES, 195*, 1–13.

Monje, O., Richards, J. T., Carver, J. A., Dimapilis, D. I., Levine, H. G., Dufour, N. F., & Onate, B. G. (2020). Hardware validation of the advanced plant habitat on ISS: Canopy photosynthesis in reduced gravity. *Frontiers in Plant Science, 11*, 673. https://doi.org/10.3389/fpls.2020.00673

Mousseau, T. A., & Møller, A. P. (2020). Plants in the light of ionizing radiation: What have we learned from Chernobyl, Fukushima, and other "Hot" places? *Frontiers in Plant Science, 11*, 552. https://doi.org/10.3389/fpls.2020.00552

Mousseau, T. A., Milinevsky, G., Kenney-Hunt, J., & Møller, A. P. (2014). Highly reduced mass loss rates and increased litter layer in radioactively contaminated areas. *Oecologia, 175*, 429–437.

Naito, M., Hasebe, N., Shikishima, M., Amano, Y., Haruyama, J., Matias-Lopes, J. A., Kim, K. J., & Kodaira, S. (2020). Radiation dose and its protection in the Moon from galactic cosmic rays and solar energetic particles: At the lunar surface and in a lava tube. *Journal of Radiological Protection, 40*(4), 947–961.

Parrish, C. H., Hebert, D., Jackson, A., Ramasamy, K., McDaniel, H., Giacomelli, G. A., & Bergren, M. R. (2021). Optimizing spectral quality with quantum dots to enhance crop yield in controlled environments. *Communications Biology, 4*, 124. https://doi.org/10.1038/s42003-020-01646-1

Peiro, E., Pannico, A., Colleoni, S. G., Bucchieri, L., Rouphael, Y., De Pascale, S., Paradiso, R., & Gòdia, F. (2020). Air distribution in a fully-closed higher plant growth chamber impacts crop performance of hydroponically-grown lettuce. *Frontiers in Plant Science, 11*, 537. https://doi.org/10.3389/fpls.2020.00537

Reichler, S. A., Balk, J., Brown, M. E., Woodruff, K., Clark, G. B., & Roux, S. J. (2001). Light differentially regulates cell division and the mRNA abundance of pea nucleolin during de-etiolation. *Plant Physiology, 125*, 339–350.

Romeyn, M. W., Spencer, L. E., Massa, G. D., & Wheeler, R. M. (2019). Crop readiness level (CRL): A scale to track progression of crop testing for space. In *Proceedings of the 49th International Conference on Environmental Systems*, Amsterdam.

Santos, P. P., Sillero, N., Boratyñski, Z., & Teodoro, A. C. (2019). Landscape changes at Chernobyl. In *Remote sensing for agriculture, ecosystems, and hydrology* (XXI, Vol. 11149). International Society for Optics and Photonics, 111491X

Schreurs, A. S., Shirazi-Fard, Y., Shahnazari, M., et al. (2016). Dried plum diet protects from bone loss caused by ionizing radiation. *Science and Reports, 6*, 21343. https://doi.org/10.1038/sre p21343

SharathKumar, M., Heuvelink, E., & Marcelis, L. F. M. (2020). Trends in plant science forum vertical farming: Moving from genetic to environmental modification trends in plant science. *Trends in Plant Science, 25*, 1–4. https://doi.org/10.1016/j.tplants.2020.05.012

Smith, D. J., & Sowa, M. B. (2017). Ballooning for biologists: Mission essentials for flying life science experiments to near space on NASA large scientific balloons. *Gravitational and Space Research, 5*, 52–73.

Srinivasan, C., Dardick, C., Callahan, A., & Scorza, R. (2012). Plum (Prunus domestica) trees transformed with poplar FT1 result in altered architecture, dormancy requirement, and continuous flowering. *PLoS ONE, 7*(e40715), 5.

Srinivasan, C., Scorza, R., Callahan, A., & Dardick, C. (2014). Development of very early flowering and normal fruiting plum with fertile seeds. Patent #: US8633354B2

Stankovic, B. (2018). Plants in space. In T. Russomano & L. Rehnberg (Eds.), *Into space: A journey of how humans adapt and live in microgravity* (pp. 153–170). InTech Open.

Stutte, G.W. (2016). Controlled environment production of medicinal and aromatic plants. In V. D. Jeliazkov & C. L. Cantrell (Eds.), *Medicinal and aromatic crops: Production, phytochemistry, and utilization* (pp. 49–63). ACS Publications.

Stoeva, N., & Bineva, T. Z. (2001). Physiological response of beans (Phaseolus vulgaris L.) to gamma-irradiation treatment. I. Growth, photosynthesis rate and contents of plastid pigments. *Journal of Environmental Protection and Ecology, 2*, 299–303.

Treesubsuntorn, C., Lakaew, K., Autarmat, S., & Thiravetyan, P. (2020). Enhancing benzene removal by Chlorophytum comosum under simulation microgravity system: Effect of light-dark conditions and indole-3-acetic acid. *Acta Astronautica, 175*, 396–404.

Ursino, D. J., Schefski, H., & McCabe, J. (1977). Radiation-induced changes in photosynthetic CO_2 uptake in soybean plants. *Environmental and Experimental Botany, 17*, 27–34.

Valbuena, M. A., Manzano, A., Vandenbrink, J. P., Pereda-Loth, V., Carnero-Diaz, E., Edelmann, R. E., Kiss, J. Z., Herranz, R., & Medina, F. J. (2018). The combined effects of real or simulated microgravity and red-light photoactivation on plant root meristematic cells. *Planta, 248*, 691–704.

Vandenbrink, J. P., Kiss, J. Z., Herranz, R., & Medina, F. J. (2014). Light and gravity signals synergize in modulating plant development. *Frontiers in Plant Science, 5*, 563. https://doi.org/10.3389/fpls.2014.00563

Vandenbrink, J. P., Herranz, R., Medina, F. J., Edelmann, R. E., & Kiss, J. Z. (2016). A novel blue-light phototropic response is revealed in roots of Arabidopsis thaliana in microgravity. *Planta, 244*, 1201–1215.

Vandenbrink, J. P., Herranz, R., Poehlman, W. L., Feltus, F. A., Villacampa, A., Ciska, M., Medina, F. J., & Kiss, J. Z. (2019). RNA-seq analyses of Arabidopsis thaliana seedlings after exposure to blue-light phototropic stimuli in microgravity. *American Journal of Botany, 106*, 1466–1476.

Villacampa, A., Ciska, M., Manzano, A., Vandenbrink, J. P., Kiss, J. Z., Herranz, R., & Medina, F. J. (2021). From spaceflight to Mars g-levels: Adaptive response of A. Thaliana seedlings in a reduced gravity environment is enhanced by red-light photostimulation. *International Journal of Molecular Sciences, 22*, 899. https://doi.org/10.3390/ijms22020899

Wallace, T. C. (2017). Dried plums, prunes and bone health: A comprehensive review. *Nutrients,19*, 9(4), 401. https://doi.org/10.3390/nu9040401

Wamelink, G. W. W., Frissel, J. Y., Krijnen, W. H. J., & Verwoert, M. R. (2019). Crop growth and viability of seeds on Mars and Moon soil simulants. *Open Agriculture, 4*(1), 509–516.

Wei, L. J., Yang, Q., Xia, H. M., Furusawa, Y., Guan, S. H., Xin, P., & Sun, Y. Q. (2006). Analysis of cytogenetic damage in rice seeds induced by energetic heavy ions on-ground and after spaceflight. *Journal of Radiation Research, 47*, 273–278.

Wheeler, R. M. (2017). Agriculture for space: People and places paving the way. *Open Agriculture,* *2,* 14–32.

Wyatt, S. E., & Kiss, J. Z. (2013). Plant tropisms: From Darwin to the international space station. *American Journal of Botany, 100,* 1–3.

Yue, C., Yuxin, Z., Nan, Z., Dongyou, Z., & Jiangning, Y. (2020). An inversion model for estimating the negative air ion concentration using MODIS images of the Daxing'anling region. *PLoS ONE, 15*(11). https://doi.org/10.1371/journal.pone.0242554

Zabel, P., Bamsey, M., Schubert, D., & Tajmar, M. (2016). Review and analysis of over 40 years of space plant growth systems. *Life Sciences in Space Research, 10,* 1–16.

Zabel, P., Bamsey, M., Zeidler, C., Vrakking, V., Schubert, D., & Romberg, O. (2017). Future exploration greenhouse design of the EDEN ISS project. In *Proceedings of the 47th International Conference on Environmental Systems,* Charleston.

Zabel, P., Zeidler, C., Vrakking, V., Dorn, M., & Schubert, D. (2020). Biomass production of the EDEN ISS space greenhouse in Antarctica during the 2018 experiment phase. *Frontiers in Plant Science, 11,* 656. https://doi.org/10.3389/fpls.2020.00656

Zeidler, C., Zabel, P., Vrakking, V., Dorn, M., Bamsey, M., Schubert, D., Ceriello, A., Fortezza, R., De Simone, D., Stanghellini, C., Kempkes, F., Meinen, E., Mencarelli, A., Swinkels, G.-J., Paul, A.-L., & Ferl, R. J. (2019). The plant health monitoring system of the EDEN ISS space greenhouse in Antarctica during the 2018 experiment phase. *Frontiers in Plant Science, 10,* 1457. https://doi.org/10.3389/fpls.2019.01457

Zhang, S., Wimmer-Schweingruber, R. F., Yu, J., Wang, C., Fu, Q., Zou, Y., Sun, Y., Wang, C., Hou, D., Böttcher, S. L., et al. (2020). First measurements of the radiation dose on the lunar surface. *Science Advances, 6,* eaaz1334. https://doi.org/10.1126/sciadv.aaz1334

Zhou, W. (2005). Advanced ASTROCULTURE™ plant growth unit: Capabilities and performances. SAE Technical Paper 2005–01–2840. https://doi.org/10.4271/2005-01-2840

Zhu, M., Zhang, J., You, Q. L., Banuelos, G. S., Yu, Z. L., Li, M., et al. (2016). Bio-generation of negative air ions by grass upon electrical stimulation applied to lawn. *Fresenius Environmental Bulletin, 25*(6), 2071–2078.

Zimmerman, R. (2003). Growing pains. *Air & Space Magazine.* https://www.airspacemag.com/space/growing-pains-4148507

Chapter 6
Understanding and Managing the Hazards and Opportunities of Lunar Regolith in Greenhouse Agriculture

Martin Braddock (ORCID)

Abstract Lunar settlements will include industrial agriculture, so an artificial agricultural environment will be part of the landscape and work space for early settlements. Provision of capacity and development of new capabilities will pose both opportunities and risks for engineers, scientists and project managers. Lunar regolith is the layer of particles on the surface of the Moon which has been generated by micro-meteorite impact and resembles volcanic ash. The finest particles are less than 100 μm particle size diameter and are generally referred to as dust. Particle sizes are as small as 0.01 μm and show an abundance of agglutinate glasses and metallic iron and the low electrical conductance allows grain to retain electrostatic charge. Lunar regolith appears to contain all essential minerals to support agriculture. Using lunar soil simulants, studies have shown that it is possible to grow certain plants for up to 50 days without addition of nutrient and to sow seeds to permit germination and crop harvest by the addition of organic matter.

6.1 Introduction

The concept of establishing semi-permanent or permanent colonies on the Moon and Mars is receiving much attention and is increasingly becoming a tangible possibility. Both space agencies and private enterprise have plans to establish a sustainable colony on the Moon by the end of the 2020s and on Mars in the 2030s and in the intervening years, much will be learnt from both past and future unmanned missions, and modelling from terrestrial analogue studies of confinement and self-sufficiency. There are many challenges in designing and building a safe habitat to ensure quality of life for settlers on the Moon and one of the greatest challenges is to secure a

M. Braddock (✉)
Sherwood Observatory, Mansfield and Sutton Astronomical Society, Coxmoor Road, Sutton-in-Ashfield, Nottinghamshire NG17 5LF, England, UK
e-mail: martinbraddock@rocketmail.com; Science4U.co.uk

Nottingham, UK

M. B. Rappaport and K. Szocik (eds.), *The Human Factor in the Settlement of the Moon*, Space and Society, https://doi.org/10.1007/978-3-030-81388-8_6

stable supply chain of food which provides colonists with sustenance of high quality, quantity and variety.

Provision of both food and water for space travel to sustain a colony of humans on the Moon for anything other than a short period of time is a major challenge which must be met for any duration settlement, even semi-permanent to be viable. On board the International Space Station (ISS), astronauts consume approximately 1.8 kg of food (Cooper et al., 2011) plus packaging per person per day so assuming we will establish a semi-permanent base on the Moon for six astronauts for a two-year mission, they would require approximately 8,000 kg of food in total. With the closest distance between Earth and the Moon being 363,104 kms at perigee, transporting a large weight of food this distance would be logistically very challenging using technology available today. Establishment, let alone expansion of a lunar colony totally dependent upon Earth for food would involve establishing a supply chain consisting of an armada of costly space craft, each mission having its own associated risks for success. Moreover, as it is estimated to cost approximately $4000 to launch 1 kg of material out of Earth orbit, let alone transport it to the Moon and that approximately 9 times the weight of propellant is required per kg of cargo for launch alone, it is simply not financially viable to transport a large weight such a great distance and so regularly. This situation provides scientists and engineers with both technical challenges and opportunities and drives the current set of planning assumptions that laying the foundations to rapidly develop sustainable agriculture ahead of any permanent human presence on the Moon will be critical to the success of inhabitation and the future build of a colony.

6.2   The Nature of Lunar Regolith and Terrestrial Soil. Lunar Regolith Formation and Particle Size

It is necessary in this and my subsequent chapter for the reader to have an understanding of the origin of lunar regolith and the similarities to and differences from terrestrial soil. The lunar surface is characterised by two types of terrain: the highland areas which are heavily cratered and the lunar lowlands which include the smooth maria, former craters that have flooded and filled with molten lava. Almost all the rocks at the lunar surface are igneous having formed from the cooling of lava (Heiken et al., 1991). In contrast, the most abundant rocks exposed on Earth's surface are sedimentary, which required the action of water or wind for their formation. The two most common kinds of rocks on the Moon are basalts and anorthosites. The lunar basalts are relatively rich in iron and are also rich in titanium and are found in the mare, whereas in the highlands, the rocks are largely anorthosites and are relatively rich in aluminium, calcium, and silicon. The surface of the Moon is mostly covered with regolith, a mixture of fine dust and rocky debris mostly <20 μm in size. As there is almost no atmosphere on the Moon, the surface has been bombarded by meteorites of all sizes including micro-meteorites breaking down exposed lunar bedrock and by solar and interstellar charged atomic particles over the course of billions of years (McKay et al., 1991). This has produced a loose regolith cover composed of the debris of underlying crystalline rocks, which are of different composition dependent upon location, mineral fragments, and other constituents formed under the shock-impact of the breccias, agglutinates, and glass grains. The mean thickness of the regolith covering the whole lunar surface varies from 4–5 m in the lunar maria to 10–15 m in the highlands (McKay, 1974; McKay & Ming, 1990; Taylor et al., 2005; Slyuta, 2014). Lunar soil typically refers to only the finer fraction of regolith, which is composed of grains 1 cm in diameter or less. Lunar dust generally refers to even finer materials than regolith and ranges from 1 to 100 μm and the low conductance and electrostatic charge coupled with reduced lunar gravity allowing greater airborne persistence when disturbed causes the dust to adhere to many surfaces (Kruzelecky et al., 2012; Stubbs et al., 2007; Taylor et al., 2005).

6.3   Composition

Soil on Earth suitable to support crop growth is made up of four principal components. The largest part comprises sedimentary rock particles formed by wind and water erosion. Soil also contains water and air and vitally, organic materials, including an ecosystem of organisms and micro-organisms which maintain the turnover of nutrients from humus, material found in the darker soils produced from decaying plant and animal life which acts as a natural fertilizer. As humus can hold water and air it provides a slow-release constituent improving soil quality to maximise the potential for crops to grow. Texture refers to sand, silt and clay content affecting the ability of the soil to hold air and release water and both commercial agriculture and

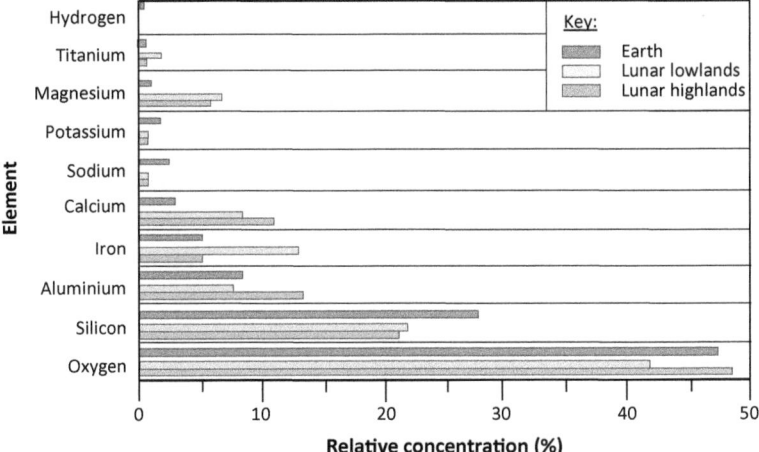

Fig. 6.1 Comparison of basic elemental composition for Earth and Lunar terrains

the global gardening leisure industry provide extensive data bases on the relationship between soil quality and different crop growth and yields and on the requirement for supplements. Fig. 6.1 shows the relative concentration of the main elements of the Earth, lunar lowlands and lunar highlands and there are clear similarities in elemental composition.

However, detailed analysis of elemental oxidation states has shown significant differences between the ratio of elemental oxides for different elements compared with those found on Earth (Korotev, 2020) together with previous findings that the acidity (pH) of simulated regolith is high at 9.6 (Wamelink et al., 2005) which may present challenges for plant growth in situ. Clearly, the absence of both extant and living organisms on the Moon precludes the existence of humus and any source of other matter which may be considered fertilizer and this will be discussed in the next section. A detailed review of the physical and mechanical properties of lunar regolith has collated data from many space missions to understand parameters such as density, porosity, cohesion and adhesion and deformation characteristics (reviewed in Slyuta, 2014) and Table 6.1 shows a comparison of densities and porosities for both surface and sub-surface samples from the Moon and from various soil types on Earth (Houston et al., 1974; Donohue et al., 1977; Carrier et al., 1991; Slyuta, 2014).

Table 6.1 Comparsion of density and porosity values for Earth soil and Lunar Regolith

Origin	Sample	Depth (cm)	Bulk density ($g\ cm^{-3}$)	Particle density ($g\ cm^{-3}$)	Porosity (%)	References
Earth	Decomposed peat	0	0.55	2.6–2.75	–	Donohue et al., 1977
	Clay	0	1.1–1.3		51–58	
	Cultivated loam	0	1.1–1.4		–	
	Tilled field	0	1.3		50	
	Sandy soil	0	1.5–1.7		36–43	
	Traffic pan	25	1.7		–	
	Surface regolith	0	1.12–1.93	2.3–3.2	–	Carrier et al., 1991, Slyuta, 2014
Moon	Sub-surface regolith	0–15	1.5		52	Carrier et al., 1991, Houston et al., 1974, Slyuta, 2014
	Sub-surface regolith	30–60	1.74		44	Carrier et al., 1991, Houston et al., 1974, Slyuta, 2014

Soil particle density tends to be lower for soils with high organic matter content (Blanco-Canqui et al., 2006). Soil bulk density is the dry mass of the soil divided by the volume it occupies which includes air space and organic materials and so bulk density is always less than soil particle density and is a good indicator of soil compaction. A lower bulk density in isolation does not necessarily indicate suitability for plant growth due to the influence of soil texture, porosity and permeability, though it is a useful indicator (Rampazzo et al., 1998). A high bulk density is indicative of either soil compaction or a mixture of soil textural classes in which small particles fill the voids among coarser particles. From the measurements shown in Table 6.1, the bulk density of both surface and sub-surface lunar regolith appear similar, albeit slightly greater than or comparable to that of Earth soils, whereas the particle density is greater for lunar regolith and values of porosity are similar for both Earth and lunar samples.

6.4   Lunar Regolith Simulants

During National and Aeronautic Space Administration's (NASA) space program between 1969 and 1972, six Apollo missions brought back 383 kg of lunar rocks, core samples, pebbles, sand and dust from the lunar surface and from 1970–1976 three automated Soviet spacecraft returned samples totalling 301 g from three other lunar sites (Rask, 2018). Many of these samples have been and continue to be used in small quantities for multiple global research projects to determine the geochemistry and many other physical properties (reviewed in Slyuta, 2014). There are insufficient amounts for larger studies to explore growth properties of multiple species of plants, and this has led to an extensive program to search for samples on Earth which may serve as lunar regolith simulants (LRSs). Some selected candidates are shown in Table 6.2 and a more detailed evaluation has been published elsewhere (Weiblen et al., 1988; McKay et al., 1994; Battler et al., 2006; Mytrokhyn et al., 2003; Hill et al., 2007; Zheng et al., 2009; Slyuta, 2014; Taylor et al., 2016; Toklu et al., 2017; Li et al., 2009; Suescun-Florez et al., 2014; Zhang et al., 2019; Engelschiøn et al., 2020).

The reader is also referred to the following website: www.simulantdb.com which provides a comprehensive data base of simulants from the Moon, Mars, a comet, asteroids and the Martian moon Phobos. All of the LRSs shown (MLS1/1, that derived from the Korosten Pluton, CAS-1, NAO-1, BP-1 and EAC-1) are credible and have physical properties similar to those of lunar regolith. In the 1990s, the LRS JSC-1 was designed to have the properties of lunar mare fit for engineering purposes and throughout the early 2000s, production of LRSs appeared more of an engineering exercise rather than one of lunar science (Taylor et al., 2016), leading to the development of JSC-1A which is regarded as a reasonable fit-for-purpose simulant widely used in laboratories around the world. Accepting that one LRS will

Table 6.2 Origin, type and replicant for selected lunar regolith simulants

Date	Regolith	Location	Source type	Lunar location Replicant	References
1988	MLS1/2	Duluth gabbro complex, Minnesota, USA	Gabbro	Highlands	Weiblen et al., 1998
1993	JSC-1	Merriam crater, San Franciso volcano field Flagstaff, USA	Basaltic lava	Mare	Mckay et al., 1994
2007	JSC-1A	Merriam crater, San Francisco volcano field Flagstaff, USA	Basaltic lava	Mare	Hill et al., 2007
2003	Korosten Pluto	Penizevitchi, Turchynka deposits, Ukraine	Anorthosite	Highlands	Mytrokhyn et al., 2003
2009	CAS-1	Sihai pyroclastics (Jinlondingzi Volcano, Jilin Province, China)	Basaltic lava	Mare	Zheng et al., 2009
2009	NAO-1	Yarlung Zangbo River in Tibet	Gabbro	Highlands	Li et al., 2009
2014	BP-1	A rock quarry San Francisco volcano field Flagstaff, USA	Basaltic lava	Mare	Suescun–Florez et al., 2014
2020	EAC-1	Siebengebirge volcanic field, Germany	Basaltic lava	Mare	Engelschiøn et al., 2020

Abbreviations JSC= Johnson Space Centre, CAS = Chinese Academy of Science, MLS = Minnesota Lunar Simulant, NAO = National Astronomical Observatories, BP = Black Point, EAC= European Astronaut Centre

not fulfil the requirement for every scientific discipline and that there is no simulant that reproduces bulk properties of the regolith or fine dust fraction (Hyatt & Feighery, 2007), it remains unclear as to which simulant is best suited to model agricultural studies and the next section will state which simulant has been used for each of the studies described.

6.5   Lunar regolith—A Viable Medium for Crop Growth?

Having become familiar with the properties of lunar regolith and understanding the constraints of LRSs as faithful replicants of lunar regolith, their suitability to sustain crop growth will be discussed.

6.5.1 Early Studies

Figure 6.2 is a step diagram of major historical milestones supporting the development of lunar agriculture from the 1970s to the present date, and Table 6.3 provides a brief summary of the key studies which will be discussed in detail.

Table 6.4, referred to in this and the following section summarises key factors including and in addition to regolith quality, where the need to consider the overall lunar environment is essential to the success of crop viability, sustainability and the dependence upon building a supply chain which could support colony expansion.

Between 1970–1974 a number of studies were conducted using material returned from Apollo 11, 12 and 14 missions (Baur et al., 1974, reviewed in Walkinshaw & Galliano, 1990; Ferl et al., 2002; Ferl & Paul, 2010). The first plant studies were conducted in 1970 as part of a quarantine study where it was shown that lunar material applied by either rubbing plants with Moon stones or supplying lunar dust as a supplement to terrestrial soil was capable of providing mineral nutrients for germinating seeds and liverworts and that no disease effects on plant growth were found (Walkinshaw et al., 1970). In 1971, a study was conducted using lunar dust from the Apollo 14 mission as a supplement to terrestrial soil and no adverse effects were observed (Walkinshaw & Johnson, 1971). It was not until 1974 that the first study of plant growth on lunar regolith was conducted investigating the uptake and translocation of nutrients in lettuce seedlings (Bauer et al., 1974) and more importantly, showing no toxic effects on plant growth. The reader is referred to the following websites which describe the early studies in more depth and details can be accessed through the NASA Technical reports server: www.hos.ifas.ufl.edu/public/lunarplan tbiology/Apollo11and12; https://ntrs.nasa.gov/search.jsp?R=20090028745

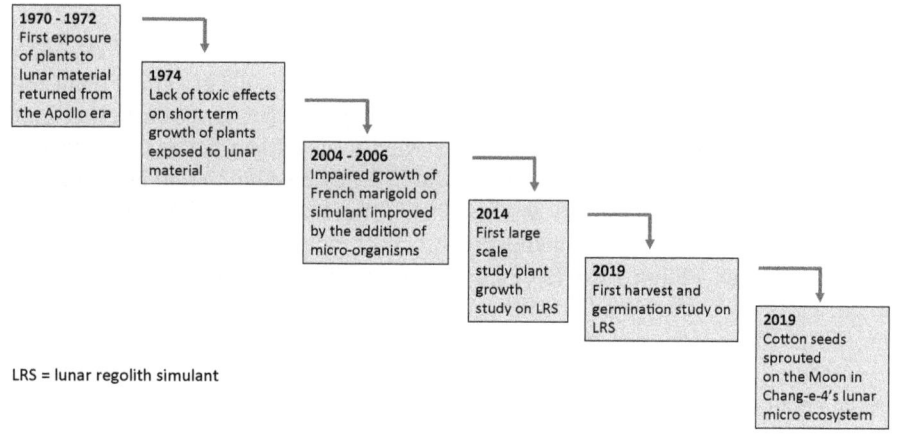

Fig. 6.2 Step diagram showing major historical milestones for lunar agriculture

Table 6.3 Summary of key studies supporting the development of lunar agriculture

Date	Regolith source	Study	References
1970	Moon (Apollo 11/12 Sample supplement)	Plant quarantine study on germinating seeds and liverwort growth	Walkinshaw et al., 1970
1971	Moon (Apollo 14 lunar Surface dust supplement)	Growth analysis of lettuce seedings	Walkinshaw & Johnson, 1971
1974	Moon (Apollo 11)	Element uptake and translocation in lettuce seedings	Bauer et al., 1974
2004	Korosten Pluton anorthosite	Growth study of *Tagetes patula* supplemented with siliceous bacterium, biocontrol agents and mycorrhizal fungi	Kozyrovska et al., 2004
2010	JSC-1A	Growth comparsion of *Arabidopsis* with terrestrial soil	Ferl & Paul, 2010
2014	JSC-1A	Growth study of 14 species belonging to 3 groups	Wamelink et al., 2014
2019	JSC-1A (with organic matter supplement)	Cultivation and seed harvesting studies for 10 crops	Wamelink et al., 2019
2019	Terrestrial soil	Growth study for potato, rape and cotton seeds	Unpublished

Abbreviation JSC= Johnson Space Centre

6.6 Plant Growth in Microgravity

The intervening years witnessed many studies conducted on the growth of plant cells in tissue culture systems as opposed to full sized plants (reviewed in Ferl & Paul, 2010) and the research emphasis further shifted to develop a better understanding of plant growth in microgravity on board the ISS (reviewed in Ferl et al., 2002; Jost et al., 2015; Kiss, 2014; Wyatt & Kiss, 2013). Between 2004, 2006 and in 2011, several studies reported investigating the effects of bioaugmentation of plants grown on the Korosten Pluton LRS (Kozyrovksa et al., 2004, 2006; Lytvynenko et al., 2006; Zaets et al., 2011) and on JSC-1A LRS (Ferl & Paul, 2010). These studies are important for several reasons. First, they are conducted on anorthosite, rock more characteristic of the lunar highlands and whose elemental composition is richer in certain elemental oxides when compared with other LRSs (Mytrokhyn et al., 2003). Secondly, studies were conducted on the French marigold *Tagetes patula,* which in addition to being tolerant to different soil conditions on Earth has medicinal and nutritional value and has a flower whose beauty evokes a positive psychological state, likely an essential feature for colonists living in self-confinement. Thirdly, recognising that terrestrial anorthosite is deficient in nutrients and that growth of

Table 6.4   Summary of key challenges, mitigation measures and opportunities for developing lunar agriculture

Factor	Challenge	Mitigation	Opportunity
Regolith quality	Replicating crop growth on Earth	Develop lunar simulants with bioaugmentation and develop better understanding of which crops grow in highest yield	Applications for desert greening on Earth
Regolith quality	Sustainability of supply chain	Utilise hydroponics, supply graded regolith or rockwool derieved from regolith as a growing medium	Application for desert greening on Earth
Water availability	Lack of liquid water	Extract water from ice trapped deep in the polar crate and develop ultra-efficient recycling capability	Technology development and application to prepare mankind for terrestrial climate change and for exploration of other new worlds
Fertiliser	Absence of natural fertiliser	Extra nutrients from recycled human waste	
Gravity	Lunar gravity is 0.16 g	Exploit selective breeding and genetic engineering	
14-day light cycle	Crops require light for photosynthesis and growth	Customise LED lighting, duration, colour, and intensity coordinated to individual species	
Exposure to ionising, cosmic radiation and solar flares	Lack of protective atmosphere and high threat to life on surface	Site greenhouse underground or growing tubes on surface buried under a protective layer of regolith. Predict sites less prone to meteorite impact	Further develop a knowledge base on tolerance of living organisms to radiation
Micro-meteorite Impact	Meteriotic material impacting at 10s km sec		Improve monitoring of candidate impacts, build and update a map of least impacted sites to inform siting of the colony
Temperature extremes	Range from 127°–173 °C during the day		Develop thermotolerant crops

Tagetes patula was abnormal (Kozyrovska, 2004), the first bioaugmentation study was conducted using a consortium of a well-defined siliceous plant bacterium and mycorrhizal fungi to seed, and inoculate the substrate which resulted in improved plant development (Kozyrovska, 2004, 2006; Lytvynenko et al., 2006). Lastly, growth of the thale cress *Arabidopsis thaliana* has been studied on the LRS JSC-1A in the presence of nutrient augmentation and compared with growth attained in regular terrestrial soil (Ferl & Paul, 2010). Taken together in all these studies, LRSs were able to support plant growth but to reduced capacity when compared with plant growth and development achieved on terrestrial soil. This finding is important as it may signify that initial plant growth functions as a primer or pioneer for future crop production establishing a closed-loop bio-regenerative life support system (BLSS) where the pioneer plants' primary role is to provide a soil of suitable fertility and to support future waves of agricultural crop production.

6.7   LRS Landmark Studies

The next two studies are landmark for several reasons. Study one (Wamelink et al., 2014) represents the first large scale study conducted on both LRS and Martian regolith simulant (MRS) for a 50-day period and without the addition of nutrients. The study was conducted for 14 plant species representing three plant groups (crop, nitrogen fixer and natural wild plant) on the JSC-1A LRS and plants which fix nitrogen were used to determine whether they may compensate for insufficient reactive nitrogen in lunar or Martian regolith. The study showed that crop growth is possible, however, LRS supported poorer plant growth when compared with growth on either terrestrial soil or the MRS JSC-1A Mars. Specifically, the nitrogen fixer common vetch did not grow on LRS and leaf formation for all plant species was poorest when grown on LRS compared with growth on MRS. In the second study, discussed in more detail, was also conducted on the LRS JSC-1A (Wamelink et al., 2019). 10 different crops were evaluated for their ability to produce edible biomass and seeds viable for germination and organic matter was added to enrich the regolith simulants, mimicking the addition from a previous harvest similarly described previously (Kozyrovska et al., 2006). The 10 different crop types studied were garden cress, rocket, tomato, radish, rye, quinoa, spinach, chives, pea, and leek. Out of the ten crops tested, only spinach was unable to grow well. Radishes and radish seeds, cress and cress seeds, rye seeds, rocket, tomatoes, and peas were harvested from plants grown on LRS, MRS and Earth soil. Generally, fruit production and seed germination were either lower or similar depending on the plant for LRS when compared with MRS and Earth soil, indicating that acceptable crop growth may be possible in regolith simulants augmented by organic matter which could be derived from a previous harvest. However, taken together the total biomass production observed for the ten different crops was significantly lower when grown on LRS as compared with MRS or Earth soil. The poorer performance of LRS to support crop growth could be attributed to a lack of nutrients, and high pH levels (Wamelink et al., 2005). The finding that nutritional supplementation may support crop growth in LRS is consistent with a recent study investigating the performance of three MRSs (Eichler et al., 2020), which challenges the assumption that MRS is capable of supporting agriculture in the absence of extraneous additives. Studies conducted with JSC-Mars-1A, Mars Mojave simulant (MMS), and Mars Global simulant (MGS-1) showed that none of these simulants were capable of supporting plant growth in the absence of nutrient supplementation; JSC-Mars-1A and MMS, but not MGS-1 (due to high pH), able to support the growth of *Arabidopsis thaliana* and *Lactuta sativa* on addition of nutrient supplementation.

6.8 Optimisation of Plant Growth on LRSs

Further studies are required to optimise plant growth on LRSs. They include development of the pioneer stage of the crop life cycle to determine how many iterations are required to produce an optimal amount of organic matter to mix with the lunar regolith in order to build a soil permissive for large scale agriculture from which sustainable harvests could be made. This becomes an in situ BLSS where the crop yield is directly related to nutrient input into the lunar soil. Studies are also required to define whether a nitrogen cycle, an essential foundation for plant growth provided by nitrogen fixing bacteria can be established. Finally, efficient water use and understanding water transport in reduced gravity is also of critical importance in building agricultural systems and also requires further research. The potential to utilise hydroponics with regolith as a growth medium is also worthy of consideration and there is an extensive literature on the topic which is beyond the scope of this chapter (reviewed in Gericke et al., 1938; Barrett et al., 2016). The use of the inert substance rock wool and other growth media (Sardare, 2013) illustrate how it may be possible to adapt hydroponics to utilise lunar regolith and again requires further investigation.

The last example caused great excitement when made public in 2019. The Chinese Space Agency's Chang'e-4 spacecraft studied the viability of different crops harboured in a lunar micro ecosystem (LME), post landing at the Von Kármán crater on the far side of the Moon (Jones, 2019). The study has not been published in the scientific literature, although it has been reported that two leaves from cotton seeds briefly sprouted before the drop in temperature to minus 190° C killed the seedlings (Jones, 2019). Seeds from potatoes and rape plants did not germinate. Although the nature of the growth medium is unclear, the LME was designed to provide Earth like conditions and this early study is a landmark as it paves the way for more detailed studies to be conducted and published when Chang'e-6 returns to the Moon in the early 2020s.

6.9 Contextualising Greenhouse Agriculture and Setting Requirements for Absolute Symbiosis

The primary aim of this chapter is to review the challenges and opportunities presented by using lunar regolith for agriculture and in this section, I will briefly review several themes all of which are interrelated and conclude that the requirement for a closed loop BLSS may be achievable with appropriate upfront understanding of risk and emergency scenario planning.

6.9.1 Desert Greening

One further opportunity relates to the potential application of terrestrial desert greening principles to the lunar landscape. Desert greening simply refers to the process of desert reclamation for irrigation to support forestry and may equally apply to landscapes which experience both hot and cold climates. In the main it relies on provision of liquid water, which may include desalination procedures for terrestrial land reclaimed from the sea. Although our attention has been focussed on lunar regolith so far, it would be remiss not to pay brief attention to other confounding issues which are presented in Table 6.4. An essential ingredient for maintenance of all life is water, however, liquid water cannot persist on the lunar surface due to the temperature extremes and any water vapour is subject to photo-dissociation and water hydrogen is rapidly lost into space. A previous report described the presence of water in anorthosite rock and proposed the existence of a wet early Moon (Hui et al., 2013). Acquisition of near infra-red reflectance spectra from the Moon Mineralogy Mapper on board the Chandrayan-1 orbiter from the central peak of Bullialdus crater and non-mare volcanic terrains at Compton-Belkovich are consistent with the existence of water in pyroclastic magma in the deep lunar interior and at vent areas suggesting there may be a natural transport pathway for water from the mantle to the surface (Milliken & Li, 2017). A recent study has suggested the existence of water on the surface of the Moon at high lunar latitudes (Honniball et al., 2020). A unique 6 μm spectral emission characteristic of the vibration of water was detected by the joint NASA/German Aerospace Centre's (DLR) Stratospheric Observatory for Infrared Astronomy. It is proposed that, for water to be present on the lunar surface, it must be protected from the lunar environment and be stored within glasses or in voids between regolith grains. Moreover, other efforts are underway to determine the chemistry behind water formation and the potential for extraction from the regolith (Zhu et al., 2019). Thus, it appears possible that a sustainable source of liquid water could be secured and this would be a major advance.

6.9.2 Other Requirements for Agriculture

A number of the basic requirements for crop agriculture such as fertilizer and carbon dioxide (CO_2), are bi-products of the colonists, and as on Earth, crops can be used to remove CO_2 and provide oxygen and food, further substantiating the concept of a BLSS and the dependency upon a fine, perhaps very fine balance of crop supply and demand. Recycling of water is well developed on the ISS and recent research has shown the potential to recycle urea, a component of urine, for regolith derived structures using 3D printing (Pilehvar et al., 2020). In addition, it is also clear that crops will show, until their adaptation, and show differential growth and transpiration in response to partial gravity as a consequence of effects on soil dynamics (Hirai & Kitaya, 2009; Maggi & Pallud, 2010a) and persistence and maintenance of a population of essential bacteria requiring aerobic or anaerobic conditions (Borer et al., 2020). What will be key, in addition to the identification, provision and harnessing of liquid water is to understand how water transport and nutrient cycles behave in reduced gravity. Studies have reported that higher water holding capacity in soil under both lunar (0.17 g) and Martian gravity (0.37 g) led to moisture content and nutrient concentrations that favoured the metabolism of various microbial functional groups which may predict Martian cropping requiring 90% less water for irrigation when compared with Earth (Maggi & Pallud, 2010a, b).

6.9.3 Lighting

Crop growth will also be determined by the duration of the light–dark cycle. As it is 14 days on the Moon, it is very unlikely crops will be able to tolerate sustained lack of natural light. A Light-emitting diode (LED) is an alternative source and there is a requirement to understand whether crops need to receive full spectrum light intensity, or lighting where the wavelength is customised to the plant's phototropic response, where for example, in some plants removing ultraviolet light produces better growth, violet light enhances colour, taste and aroma, light of blue and red wavelength increases growth rate, and light in the far red may produce crops in greater yields (Kiss, 2014; Meinen et al., 2018). As our understanding of crop growth under both model systems and in situ increases, as on Earth, there will be many opportunities to exploit selective breeding and genetic engineering whereby plants are made more tolerant to grow in extreme environments.

6.9.4   Tolerance of Environmental Extreme

Finally, the three confounding factors of exposure to both ionising and cosmic radiation, micro-meteorite impact and temperature extremes may in part provide an opportunity to develop crops better able to tolerate extreme environments. However, in such harsh conditions it will be very likely to site growing facilities for agriculture away from the surface, for example, in lava tubes (Pouwels et al., 2020), although surface based designs, based upon terrestrial analogue models have also been proposed (Maiwald et al., 2020).NASA continues to study the location pattern of lunar impacts (NASA, 2018) and data to date may be used to select a site for colony construction less likely to be affected by meteorite strike. At all times, astronauts will need to be protected from exposure to extreme environmental conditions by using appropriate astronautical hygiene control measures (Cain, 2016, 2020).

6.10   Conclusion

Taken together, a successful agricultural growing facility will require crops and colonists living in a finely balanced environment, representing a microcosm of planet Earth. It will be essential to learn from both the successes (Ehrlich, 2017) and particularly the failure (Nelson, 2018) of terrestrial isolation analogues used for the Moon and Mars. The BLSS will need to have some systems tolerance because, as on Earth, crops can cease to grow and harvests can fail. Numerous studies have explored and modelled the concept (Fu et al., 2016; Hu et al., 2010), with the Lunar Palace I conducting a 105 day experiment with 21 plant species showing recycling of water and oxygen with a 55% of food recycling capability (Fu et al., 2016). It does appear possible that under the right conditions, a BLSS may support a small number of colonists to permit emergency evacuation if systems fail in the first steps towards lunar agriculture and a risk-assessment of parameters has been reported (Cain, 2018). This latter point is especially important, and in Chap. 9, I will describe progress that has been made in understanding the principal hazards of lunar regolith and risk mitigation procedures which require the rigorous application of astronaut hygiene techniques to minimise exposure to particulates.

Summary
In this chapter, the nature of lunar regolith and the role of LRS have been discussed to set the scene for lunar agriculture. To date there have been few studies with live plants conducted on LRS and even fewer on authentic lunar regolith. This poses and immediate challenge highlighted in many publications which is do LRSs act as faithful replicants of lunar regolith and what if they do not? As with all model systems there are limitations and we should accept that LRSs can support plant growth, crop production needs to be considered on a plant-by-plant basis and the need for pioneer or primer crops to establish a lunar soil of acceptable quality. This

may imply, that the first crop harvests need to be sacrificed at least in part if not *in toto* to provide sufficient organic matter to enable acceptable crop yield. As with terrestrial agriculture, identification of water sources, irrigation and crop rotation needs to be planned and unlike the case of Earth, the behaviour of water in lunar gravity needs to be investigated in models which can be scaled. Finally, the siting of a lunar base harbouring a crop production facility warrants very serious consideration. Unlike the elegant structures of surface lunar greenhouses portrayed in scientific illustration or science fiction, the hostility of the lunar environment, driven principally by the negligible atmosphere may warrant underground location of agricultural facilities to scale from the delivery of successful testing and pilot studies to massive biomass production to feed a crew or colony. Notwithstanding these challenges, substantial progress has been made over the last 50 years and the possibilities appeal in driving forward scientific innovation to support lunar colonisation and in parallel, developing a better understanding of the utilisation and appreciation of the Earth's resources for humankind.

References

Barrett, G. E., Alexander, P. D., Robinson, J. S., & Bragg, N. C. (2016). Achieving environmentally sustainable growing media for soilless plant cultivation systems–a review. *Scientia Horticulturae, 212*, 220–234.

Battler, M., Richard, J., Boucher, D., & Spray, J. (2006). Developing an anorthositic lunar regolith simulant. In *37th Annual Lunar and Planetary Science Conference*, abstract 1622.

Baur, P. S., Clark, R. S., Walkinshaw, C. H., & Scholes, V. E. (1974). Uptake and translocation of elements from Apollo 11 lunar material by lettuce seedlings. *Phyton, 32*, 133–142.

Blanco-Canqui, H., Lal, R., Post, W. M., Izaurralde, R. C., & Shipitalo, M. J. (2006). Organic carbon influences on soil particle density and rheological properties. *Soil Science Society of America Journal, 70*, 1407–1414.

Borer, B., Jimenez-Martinez, J., Stocker, R., & Or, D. (2020). Reduced gravity promotes bacterially mediated anoxic hotspots in unsaturated porous media. *Scientific Reports, 10*, 8614.

Cain, J. R. (2016). Humans in space and chemical risks to health. *Spaceflight, 58*, 336–341.

Cain, J. R. (2018). Mars colonisation: The hazards and exposure control. *Journal of the British Interplanetary Society, 71*, 178–185.

Cain, J. R. (2020). Astronaut eye exposure to microgravity, to radiation and to lighting. *Journal of the British Interplanetary Society, 73*, 390–396.

Carrier, W. D., III., Olhoeft, G. R., & Mendell, W. (1991). Physical properties of the lunar surface. In G. Heiken, D. Vaniman, & B. M. French (Eds.), *Lunar Sourcebook* (pp. 475–594). Cambridge University Press.

Cooper, M., Douglas, D., & Perchonok, M. (2011). Developing the NASA food system for long-duration missions. *Journal of Food Science, 76*, R40–R48.

Donahue, R. L., Miller, R. W., & Shickluna, J. C. (1977). *Soils: An introduction to soils and plant growth*. Prentice-Hall. ISBN 978–0–13–821918–5.

Dubinskii, A. Y., & Popel, S. I. (2019). Water formation in the lunar regolith. *Cosmic Research, 57*, 79–94.

Ehrlich, J. W., Massa, G., Wheeler, R., Gill, T. R., Quincy, C., Roberson, L., Binsted, K., & Morrow, R. (2017). Plant growth optimization by vegetable production system in HI-SEAS analog habitat. *AIAA Space and astronautics forum and exposition* (p. 5143).

Eichler, A., Hadland, N., Pickett, D., Masaitis, D., Handy, D., Perez, A., Batcheldor, D., Wheeler, B., & Palmer, A. (2020). Challenging the agricultural viability of Martian regolith simulants. *Icarus*, 114022.

Engelschiøn, V. S., Eriksson, S. R., Cowley, A., Fateri, M., Meurisse, A., Kueppers, U., & Sperl, M. (2020). EAC-1A: A novel large-volume lunar regolith simulant. *Scientific Reports, 10*, 5473.

Ferl, R., Wheeler, R., Levine, H. G., & Paul, A. L. (2002). Plants in space. *Current Opinion in Plant Biology, 5*, 258–263.

Ferl, R. J., & Paul, A.-L. (2010). Lunar plant biology–a review of the Apollo era. *Astrobiology, 10*, 261–274.

Fu, Y., Li, L., Xie, B., Dong, C., Wang, M., Jia, B., Shao, L., Dong, Y., Liu, H., Liu, G., Liu, B., Hu, D., & Liu, H. (2016). How to establish a bioregenerative life support system for long-term crewed missions to the moon or mars. *Astrobiology, 16*, 925–936.

Gericke, W. F. (1938). Crop production without soil. *Nature, 141*, 536–540.

Heiken, G. H., Vaniman, D. T., French, B. M., et al. (Eds.). (1991). *The lunar sourcebook: A user's guide to the moon.* Cambridge University Press.

Hill, E., Mellin, M. J., Deane, B., Liu, Y., & Taylor, L. A. (2007). Apollo sample 70051 and high- and low-Ti lunar soil simulants MLS-1A and JSC-1A: Implications for future lunar exploration. *Journal of Geophysical Research, 112*, E02006.

Hirai, H., & Kitaya, Y. (2009). Effects of gravity on transpiration of plant leaves. *Annals of the New York Academy of Sciences, 1161*, 166–172.

Honniball, C. I., Lucey, P. G., Li, S., et al. (2020). Molecular water detected on the sunlit moon by SOFIA. *Nature Astronomy.* https://doi.org/10.1038/s41550-020-01222-x

Houston, W. N., Mitchell, J. K., & Carrier, W. D. III. (1974). Lunar soil density and porosity. In *Proceedings of the 5th Lunar and planetary science conference* (pp. 2361–2364).

Hu, E., Bartsev, S. I., & Liu, H. (2010). Conceptual design of a bioregenerative life support system containing crops and silkworms. *Advances in Space Research, 45*, 929–939.

Hui, H., Peslier, A. H., Zhang, Y., & Neal, C. R. (2013). Water in lunar anorthosites and evidence for a wet early moon. *Nature Geoscience, 6*, 177–180.

Hyatt, M. J., & Feighery, J. (2007). Lunar dust: Characterisation and mitigation. NASA, https://ntrs.nasa.gov/archive/nasa/casi.ntrs.nasa.gov/20080005580.pdf

Jones, A. (2019). China grew two cotton leaves on the moon. Spectrum. ieee.org/tech-talk/aerospace/robotic-exploration/china-grew-these-leaves-on-the-moon.

Jost, A.-I. K., Takayuki, H., & Iversen, T. H. (2015). The utilization of plant facilities on the international space station—the composition, growth, and development of plant cell walls under microgravity conditions. *Plants, 4*, 44–62.

Kiss, J. Z. (2014). Plant biology in reduced gravity on the moon and mars. *Plant Biology, 16*, 12–17.

Korotev, R. L. (2020). The chemical composition of lunar soil. sites.wustl.edu/meteoritesite/items/the-chemical-composition-of-lunar-soil/

Kozyrovska, N. O., Korniichuk, O. S., Voznyuk, T. M., Kovalchuk, M. V., Lytvynenko, T. L., Rogutsky, I. S., Mytrokhyn, O. V., Estrella-Liopis, V. R., Borodinova, T. L., Mashkovska, S. P., Foing, B. H., & Kordyum, V. A. (2004). Microbial community in a precursory scenario of growing Tagetes patula in a lunar greenhouse. *Space Science Technology, 10*, 221–225.

Kozyrovska, N. O., Lytvynenko, T. L., Korniichuk, O. S., Kovalchuk, M. V., Voznyuk, T. M., Kononuchenko, O., Zaetz, I., Rogutsky, I. S., Mytrokhyn, O. V., Mashkovska, S. P., Foing, B. H., & Kordyum, V. A. (2006). Growing pioneer plants for a lunar base. *Advances in Space Research, 37*, 93–99.

Kruzelecky, R. V., Brahim, A., Wong, B., Haddad, E., Jamroz, W., Cloutis, E., Therriault, D., Ellery, A., Martel, S., & Jiang, X. X. (2012). Project moondust: Characterisation and mitigation of lunar dust. In *41st International conference on environmental systems.* 2011. https://doi.org/10.2514/6.2011-5184

Li, Y., Liu, J., & Yue, Z. (2009). NAO-1: Lunar highland soil simulant developed in China. *Journal of Aerospace Engineering, 22*, 53–57.

Lytvynenko, T., Zaetz, I., Voznyuk, T., Kovalchuk, M., Rogutskyy, I., Mytrokhyn, O., Lukashov, D., Estrella-Liopis, V., Borodinova, T., Mashkovsha, S., Foing, B., Kordyum, V., & Kozyrovska, N. (2006). A rationally assembled microbial community for growing Tagetes patula L. in a lunar greenhouse. *Research in Microbiology, 157,* 87–92.

Maggi, F., & Pallud, C. (2010). Space agriculture in micro- and hypo-gravity: A comparative study of soil hydraulics and biogeochemistry in a cropping unit on earth, mars, the moon and the space station. *Planetary and Space Science, 58,* 1996–2007.

Maggi, F., & Pallud, C. (2010). Martian base agriculture: The effect of low gravity on water flow, nutrient cycles and microbial biomass dynamic. *Advances in Space Research, 46,* 1257–1265.

Maimwald, V., Vrakking, V., Zabel, P., Schubert, D., Waclavicek, R., Dorn, M., Flore, L., Imhof, B., Rousek, T., Rossetti, V., & Zeidler, C. (2020). From ice to space: A greenhouse design for moon or mars based on a protype deployed in Antarctica. *CEAS Space Journal.* https://doi.org/10.1007/s12567-020-00318-4

McKay, D. S., Fruland, R. M., & Heiken, G. H., (1974). Grain size and the evolution of lunar soils. In *Proceedings of the 5th Lunar and planetary science conference* (Vol. 1, pp. 887–906).

McKay, D. S., Heiken, G., Basu, A., Blanford, G., Simon, S., Reedy, R., French, B., & Papike, J. (1991). The lunar regolith. In G. Heiken, D. Vaniman & B. French (Eds.), *Lunar Sourcebook.* Cambridge University Press.

McKay, D. S., Carter, J. L., Boles, W. W., Allen, C. C., & Allton, J. H. (1994). JSC-1: A new lunar soil simulant. *Engineering, Construction, and Operations in Space IV,* American Society of Civil Engineers, 857–866.

McKay, D. S., & Ming, D. W. (1990). Properties of lunar regolith. *Developments in Soil Science., 19,* 449–462.

Meinen, E., Dueck, T., Kempkes, F., & Stanghellini, C. (2018). Growing fresh food on future space missions: Environmental conditions and crop management. *Scientia Horticulturae, 235,* 270–278.

Milliken, R. E., & Li, S. (2017). Remote detection of widespread indigenous water in lunar pyroclastic deposits. *Nature Geoscience, 10,* 561–565.

Mytrokhyn, O. V., Bogdanova, S. V., Shumlyanskyy, L. V. (2003). Anorthosite rocks of Fedorivskyy Suite (Korosten Pluton, Ukrainian Shield). *Current Problems in Geology. Kyiv National University, Kyiv,* 53–57.

NASA. (2018). Lunar impacts. https://www.nasa.gov/centers/marshall/news/lunar/lunar_impacts.html

Nelson, M. (2018). Pushing our limits: Insights from Biosphere 2. University of Arizona Press. https://doi.org/10.2307/j.ctt1zxsmg9

Pilehvar, S., Arnhof, M., Pamies, R., Valentini, L., Kjøniksen, A. L. (2020). Utilization of urea as an accessible superplasticizer on the moon for lunar geopolymer mixtures *Journal of Cleaner Production, 247,* 119177.

Pouwels, C. R., Wamelink, G. W. W., Musilova, M., & Foing, B. (2020). Food for extra-terrestrial missions on native soil. In *51st Lunar and Planetary Science Conference, USA, abstract* 1605.

Rask J. (2018). Lunar dust toxicity. In B. Cudnik (Eds.), Encyclopedia of lunar science. Springer, Cham

Rampazzo, N., Blum, W. E. H., & Wimmer, B. (1998). Assessment of soil structure parameters and functions in agricultural soils. *Die Bodenkultur, 49,* 69–84.

Sardare, M. D., & Admane, S. V. (2013). A review on plant without soil–hydroponics. *International Journal of Research in Engineering and Technology, 2,* 299–304.

Suescan-Florez, E., Roslyakov, S., Iskander, M., Baamer, M. (2015). Geotechnical properties of BP-1 lunar regolith simulant. *Journal of Aerospace Engineering, 28.* https://doi.org/10.1061/(ASCE)AS.1943-5525.0000462

Slyuta, E. N. (2014). Physical and mechanical properties of the lunar soil (a review). *Solar System Research, 48,* 330–353.

Stubbs, T. J., Vondrak, R. R., & Farrell, W. M. (2007). Impact of dust on lunar exploration. Goddard spaceflight center, http://helf.jsc.nasa.gov/files/ StubbsImpactOn Exploration.4075.pdf

Taylor, L. A. Schmitt, H. H., Carrier III, W. D. & Nakagawa, M. (2005). The lunar dust problem: From liability to asset. 1st space exploration conference: Continuing the voyage of discovery. Orlando, Florida, United States, pp. 71–78.

Taylor, L. A., Pieters, C. M., & Britt, D. (2016). Evaluations of lunar regolith simulants. *Planetary Space Science, 126*, 1–7.

Toklu, Y. C., Çerçevik, A. E., Kandemir, S. Y., & Yayli, M. O. (2017). Production of lunar soil simulant in Turkey. In *2017 8th International Conference on Recent Advances in Space Technologies, IEEE*, 1–5.

Walkinshaw, C. H., Sweet, H. C., Venketeswaran, S., & Horne, W. H. (1970). Results of Apollo 11 and 12 quarantine studies on plants. *BioScience, 20*, 1297–1302.

Walkinshaw, C. H., & Johnson, P. H. (1971). Analysis of vegetable seedlings grown in contact with Apollo 14 lunar surface fines. *Hortscience, 6*, 532–535.

Walkinshaw, C. H., & Galliano, S. G. (1990). New crops for space bases. In J. Janick & J. E. Simon (Eds.), Advances in new crops. Proceedings of the first national symposium on new crops: Research, development, economics, Indianapolis, Indiana (pp. 532–535).

Wamelink, G. W. W., Goedhart, P. W., van Dobben, H. H., & Berendse, F. (2005). Plant species as predictors of soil pH: Replacing expert judgement by measurements. *Journal of Vegetation Science, 16*, 461–470.

Wamelink, G. W. W., Frissel, J. Y., Verwoert, W. H. J., & Goedhart, P. W. (2014). Can plants grow on mars and the moon: A growth experiment on mars and moon soil simulants. *PLoS ONE, 9*, e103138.

Wamelink, G. W. W., Frissel, J. Y., & Verwoert, M. R. (2019). Crop growth and viability of seeds on mars and moon soil simulants. *Open Agriculture, 4*, 509–516.

Weiblen, P. W., & Gordon, K. (1988). Characteristics of a simulant for lunar surface materials. *Second conference on lunar bases and space activities of the 21st century* (p. 254).

Wyatt, S. E., & Kiss, J. Z. (2013). Plant tropisms: From darwin to the international space station. *American Journal of Botany, 100*, 1–3.

Zaets, I., Burlak, O., Rogutsky, I., Vasilenko, A., Mytrokhyn, O., Lukashov, D., Foing, B., & Kozyrovska, N. (2011). Bioaugmenation in growing plants for lunar bases. *Advances in Space Research, 47*, 1071–1078.

Zhang, X., Osinski, G. R., Newson, T., Ahmed, A., Touqan, M., Joshi, D., & Hill, H. (2019). A comparative study of lunar regolith simulants in relation to terrestrial tests of lunar exploration missions. *50th Lunar and planetary science conference LPI contribution no. 2132.*

Zheng, Y., Wang, S., Ziyuan, O., Yongliao, Z., Jianzhong, L., Xiongyao, L., & Junming, F. (2009). CAS-1 lunar soil simulant. *Advances in Space Research, 43*, 448–454.

Zhu, C., Crandall, P. B., Gillis-Davis, J. J., Ishli, H. A., Bradley, J. P., Corley, L. M., & Kaiser, R. I. (2019). Untangling the formation and liberation of water in the lunar regolith. In *Proceedings of the National Academy of Sciences of the United States of America 116* (pp. 11165–11170).

Part III
Medical and Health Perspectives on Human Lunar Settlement

Chapter 7
Consideration of the Long-Term Effects of Hypogravity

Mark Shelhamer

Abstract The effects of long-term reduced (lunar) gravity are very much unknown. The Apollo Program represents our only experience with humans in extended lunar gravity. Those exposures were brief, but few if any major medical issues were encountered that cannot be ascribed to issues other than reduced gravity. Animal experiments will be of great value in elucidating partial-g effects, but they have uncertain transfer to human responses; the best of these experiments are being performed now (2021) with rodents on an ISS centrifuge. We might expect that some of the main physiological issues that occur in extended weightlessness (e.g., ISS) will manifest also in extended lunar gravity, but this is far from certain. The reduced but non-zero gravity level on the moon may well be sufficient to halt or dramatically reduce the main aspects of physiological deconditioning seen in weightlessness. Those aspects that are not sufficiently alleviated in this way might benefit from additional exercise and intermittent exposures to higher gravity levels.

7.1 Introduction

The human body evolved and develops in a 1g environment on Earth. Many physiological functions work with or against this gravitational constant. Working against gravity provides a natural and much-needed means to maintain the integrity of muscle, bone, and cardiovascular systems. In weightlessness, astronauts have an extensive exercise program to keep these and other physiological systems healthy. In the reduced gravity of a lunar settlement, natural movements during daily activities might not provide an adequate stimulus to maintain these systems, and it is not clear to what extent exercise will be adequate to compensate for this lack, or how much will be needed. We discuss some of these issues, the potential consequences of deconditioning, and possible countermeasures.

These issues are plagued by a dearth of data. In order to properly investigate the effects of lunar gravity on humans (before actually going to the moon), the prevailing

M. Shelhamer (✉)
School of Medicine, Johns Hopkins University, Baltimore, USA
e-mail: mshelhamer@jhu.edu

117

1g on Earth must be removed. This can be done in parabolic flight (Karmali & Shelhamer, 2008) for brief durations, and in suborbital spaceflights of longer (but still relatively brief) durations. Appropriate experiments must ideally be performed in orbital spaceflight, on a human centrifuge; such a device does not exist and it is doubtful that one will exist any time soon, given the expense (cost, mass, volume, power). Thus we are left to conjecture, based on theoretical considerations and observations from extended missions in weightlessness (e.g., six months or more on ISS), what the long-term effects will be of living in lunar gravity.

As such, the spaceflight effects delineated here are drawn from a number of sources, most notably Barratt et al. (2019) and the *Human Research Roadmap* of the NASA Human Research Program (https://humanresearchroadmap.nasa.gov/). We discuss the major physiological issues due to lunar hypogravity exposure, as in a lunar settlement. The presentation is largely limited to changes and adaptive processes that might be problematic, not medical issues per se (injuries) or operational matters (teamwork, human-system interaction). We also note that lunar soil and dust (regolith) is a pernicious operational problem with medical implications, and is covered in a separate chapter in this volume. The same applies to vestibular and neural considerations in lunar gravity.

7.2   What Are the Changes, and Do They Matter?

Some of the physiological changes under consideration here reflect adaptations that are appropriate for the lunar environment. Examples include bone loss, muscle atrophy, and cardiovascular deconditioning (Lang et al., 2017; Aubert et al., 2016). Due to the lower gravity level on the moon, the normal musculoskeletal load is missing, and the body naturally conserves resources by shedding unneeded capacity in bones and muscles. Since the heart does not have to work as hard to pump against the pull of gravity to maintain blood perfusion to the head, the cardiovascular system may also degrade. As noted, the extent of these changes in long-term lunar gravity is unknown, but the changes themselves might not be detrimental, as long as the settlers experiencing them are truly *settlers*, in that they do not intend to return to Earth. If they do intend to return, then the loss of bone density, muscle mass, and cardiac capacity would certainly be detrimental.

There is a caveat to this conjecture for those remaining on the moon. It might well be the case that a certain amount of deterioration in structure and function, relative to Earth-normal, is acceptable if one is permanently settled on the moon and can avoid strenuous work outside the habitat. In the case of outside work, the elaborate suit requirements impose additional burdens of mass, impaired mobility, and mass balance (heavy backpacks for life support). This may change with advances in technology, but for the foreseeable future the suit will impair mobility and might lead to difficulty maintaining balance. Coupled with changes in vestibular function, this is a combination that might lead to falling while conducting work on the lunar surface, as was evident in the Apollo missions. If muscle strength is degraded, the ability to recover from a fall might be impaired, and if bone structure is degraded the possibility of a fracture is increased. This is not an insignificant concern in a setting where medical resources may be limited.

Related to the loss of bone density is the possibility of an increase in incidence of kidney stones (Pietrzyk et al., 2007; Kassemi et al., 2011). As calcium is lost from the bones due to unloading, it makes its way into the urinary system, where it is typically excreted. However, the high concentration of calcium in the urine can lead to the development of kidney stones, and the altered dynamics of fluid flow in a reduced-gravity environment might exacerbate this tendency. Compared to the case of an extended spacecraft journey in deep space, development of this medical condition on the moon could be treatable as on Earth, if the appropriate clinical capabilities were available.

In other cases, there are changes associated with spaceflight that are clearly detrimental, regardless of whether the person stays on the moon or returns to Earth; these changes are inherently maladaptive, for any setting. Changes in immune function have been noted in long-duration spaceflight (Crucian et al., 2020). There are a variety of immune alterations, and many appear to be benign. Two changes in particular are intriguing although they are of uncertain consequence: shedding (reactivation) of latent viruses, and an increase in skin rashes. These have, so far, been adequately addressed with no major consequences. As there does not appear to be a dramatic increase in illness in returning astronauts—upon exposure to their non-quarantined families for the first time in months—it is not clear if there are any clinically relevant concerns in this area. Another concern of uncertain consequence is changes in the gut microbiome: the multitude of microorganisms that live in the intestinal tract and contribute to a variety of physiological (and cognitive) functions (Voorhies et al., 2019). Many stressors could contribute to the changes seen in spaceflight, altered diet and medications among them.

As with the other concerns, it's not yet clear to what extent these maladaptive processes would be halted in lunar gravity (as opposed to weightlessness, where they have been most extensively observed and studied). It's also not clear to what extent they are due to hypogravity per se, or rather to the myriad other stressors associated with spaceflight: workload, isolation, confinement, altered sleep and light–dark cycles, and many others. Nevertheless, it would seem wise to monitor these systems and to be prepared for medical interventions as needed, including quarantine and isolation.

7.3 Is Lunar Gravity a Sufficient Countermeasure?

As noted, it's not clear the extent to which lunar gravity itself will serve as a countermeasure for these deconditioning effects. Astronauts on ISS exercise regularly and vigorously (by some accounts up to two hours per day). This countermeasure has been very effective in mitigating the worst effects of cardiovascular deconditioning, bone loss, and muscle atrophy. (It's unclear if maintenance of bone density as a result of exercise relates directly to maintenance of bone strength, since cortical and trabecular bone are lost at different rates and have different impacts on bone strength.) To the extent that it is desired to maintain these capabilities at or near their Earth-based norms (or at least to retard degradation in hypogravity), a regimen of aerobic and resistive exercise may be beneficial and is easily accommodated in a lunar habitat.

Another countermeasure approach is intermittent centrifugation—a form of artificial gravity (Clément et al., 2015). While it would be prohibitive to rotate an entire lunar habitat, it is perfectly feasible to provide a small centrifuge, in which settlers could experience Earth-level (or at least greater-than-lunar level) gravitoinertial forces for short periods. An intriguing possibility is to place beds on such a centrifuge, permitting some physiological processes to benefit from gravity exposure and hopefully contribute to maintenance of bone, muscle, and cardiovascular function (and perhaps others as well). This is an appealing prospect, which at the moment suffers from a problem inherent to the proposition of artificial-gravity in general: there is no data on the level or duration of such exposure that would be necessary to maintain structure and function (Lackner & DiZio, 2000).

7.4   Return to Earth?

It is perhaps worth noting that the possibility of a return to Earth will be limited, and may grow even more so over time, for lunar settlers. There are several reasons for this. First, the propulsion and spacecraft resources that would be required might better be used instead to build and support the lunar habitat. Second, development of a thriving settlement might depend on the commitment of the settlers to stay, in which case the prospect of an easy exit would be detrimental. Finally, after extended time on the lunar surface, it may well be that the adaptive processes that take place to reduce bone, muscle, and cardiovascular function under reduced gravity loading would lead to decrements of such magnitude—in the absence of intense countermeasures—that return to the normal 1g of Earth would lead to unacceptably large functional deficits. (This was discussed above.) In other words, a decision might be made at some point that the countermeasures that would be needed to enable a return to Earth are no longer worthwhile for those who intend to never return to Earth. This could be an irreversible decision. This has implications for medical evacuation (see below), but also for the psychological state of the settlers—there may be a heightened level of anxiety once it is realized that a return to Earth is unlikely if not impossible (Shelhamer, 2020).

7.5   Medical Issues

A corollary to the physiological concerns of lunar gravity is the set of medical measures that should be considered in order to alleviate or treat the adverse effects.

As noted, there may in fact be few additional concerns as a consequence of lunar hypogravity itself, and those that might occur should be "treatable" with appropriate countermeasures. Medical issues on the Apollo missions were largely related to operational or logistical issues, the relatively primitive nature of the lunar habitat (a lunar module designed primarily for transport rather than habitation), and the extremely limited resources available to the crew on the surface (what could be crammed into the lunar module). Future lunar medical issues might, in contrast, be due more to the activities undertaken by the explorers and settlers: these might be more conventional issues such as injuries and trauma, especially as missions become longer and more ambitious. In an extreme case, evacuation back to Earth might be considered, but this has its own set of problems. The accelerations and vibrations of launch and landing, the relative sparsity of spacecraft accommodations, and the difficulty of providing constant medical care during transport, all argue against evacuation. Such evacuation would take several days, and require significant resources, and the journey back to Earth would not be without risk of its own, assuming an appropriate spacecraft were even available for this use. Medical evacuation from a lunar settlement is not trivial.

How, then, to approach the possible medical issues (Antonsen et al., 2018; Canga et al., 2016; Reyes et al., 2020)? They may be mundane, but the limited capabilities for treatment and evacuation could convert even mundane issues into tragedies. One specific issue, as an example, relates to the limited lifetime of pharmaceuticals (Blue et al., 2019), which might be brought to a lunar settlement and stored for years before use; this is the type of concern that is specific to spaceflight and may require novel solutions (such as in situ synthesis) if resupply cannot be guaranteed. A comprehensive program of medical/physiological monitoring is advisable, at least in the early missions of settlement, in order to determine if there are physiological issues that are unexpected and will need to be addressed through countermeasures or enhanced medical care. At the least, a careful assessment should be made of the most likely a priori concerns, so that the most relevant medical facilities can be provided, and which can be enhanced over time based on experience. The (presumably) greater magnitudes of mass, power, and volume that will be available to lunar settlers—in comparison to the relative paucity in orbital and trans-planetary flights—should make it possible to address many more issues and even be proactive about some concerns.

7.6   Radiation

There is some doubt about the impact of radiation during extended lunar stays (Chancellor et al., 2014). By some accounts, the increased radiation from galactic cosmic rays and solar particle events, beyond the protective shielding of the Van Allen radiation belts, would render human exploration infeasible for any but the briefest excursions. Others contend that these risks are overstated, and that shielding and other precautions can mitigate what inherent risk radiation presents. Operational and engineering solutions are also available: lunar mass and especially shelters burrowed into the ground can provide more extensive shielding than on a spacecraft.

Even if biomedical risks of radiation alone can be alleviated, what's intriguing is the possibility of interactions between hypogravity and radiation. Evidence is sparse—even more than for the other systems discussed here. However, there may be detrimental effects of hypogravity (at least 0g) on mechanisms of DNA damage repair, which would be particularly troubling since these mechanisms are needed to repair damage from radiation (Moreno-Villanueva & Wu, 2019). On the other hand, there is some evidence for a beneficial synergistic effect (Kokhan et al., 2019). Again, the answers are elusive at the moment.

7.7   Animal Analogs

Hindlimb unloading is the animal model that, until inflight centrifuge experiments are more complete, provides the most relevant information on the effects of sustained reduction in body gravitational loading. This unloading is not the same as removing gravity; in these earth-bound lab experiments, gravity is fully intact at its normal level (1g). The vestibular system and various internal cellular processes still function normally; for example, convection, sedimentation, and cell–cell and organ-organ shear forces are not appreciably changed. Unloading mimics the effects of space-flight in a specific way: by removing the surface against which the limbs press as they experience gravity, thereby removing ground reaction forces from the back limbs. This removes the external stimulus that normally works to maintain bone and muscle strength. This is typically performed in rodents, hence there is uncertainty in extrapolating results to humans. (See review in Globus & Morey-Holton, 2016.) Some effects are due to reduced physical and metabolic demands. This is unlikely to be the case during lunar stays, where people will likely be quite busy, which might counter some of the possible adverse effects of gravity reduction. The most obvious and unsurprising changes are in the so-called anti-gravity bones and muscles—those predominantly in the lower body that work against gravity to maintain posture and upright stance (or the four-legged equivalent). Less appreciated are the attendant effects on metabolism, oxidative stress, and inflammation; again, these may in part be due to other stressors than just unloading.

7.8   The Apollo Experience

Apollo astronauts who walked on the moon (twelve from 1969 to 1972) reported few problems specific to the lunar experience (Scheuring et al., 2008). These were, however, short-term missions with highly select crew. Most of the concerns had to do with the severely constrained operational environment. In Apollo, crew accommodations were sparse, and amenities such as waste management and nutritional variety were secondary. These issues will presumably be corrected on longer lunar missions, which can take advantage of regular resupply, better logistics, and improvements over the last 50 years. Exposure to lunar dust in the cabin was both an operational and a medical issue, which at any rate can be addressed with proper engineering.

Apollo moonwalkers did occasionally fall while conducting surface activities. Given the possibility of altered vestibular function and tilt perception, and the possibility of bone weakening, this could be a problem, especially if suits remain limited in mobility. EVA suit injury is a concern even now. While there is no evidence for permanent or structural alterations in the vestibular system due to lunar exposure, there is evidence for alterations in tilt perception. This, and the change in footing due to surface regolith and inflexible EVA suits, might exacerbate a tendency to fall. If lunar gravity plus additional exercise is insufficient to maintain bone integrity (as well as muscle and cardiac function), then the consequences of a fall could be devastating. These interaction effects may be the most important, since they might arise from a combination of effects, each of which was thought to be adequately mitigated (Shelhamer, 2019).

It is conceivable that data from the Apollo missions might provide information on any beneficial effects of short-term exposure to lunar gravity. In each of these missions, two astronauts landed on the moon and one stayed in lunar orbit. By comparing the physiological status, after flight, of the first group (moon walkers) to the second group (lunar orbiters), it might be possible to determine if time in lunar gravity provided any countermeasure effects to the deconditioning of weightlessness in space. Much of these data are available (Johnston et al., 1975), but they are inconclusive, for several reasons. First, there is the normal large variability associated with the human response to extreme environments. Coupled with the small numbers of astronauts (six successful lunar-landing missions), this makes it difficult to discern consistent trends. Second, the time spent on the moon was brief; even for the longest mission (Apollo 17) the lunar stay was on the order of three days, out of a total mission duration of more than twelve days; any effects could be short-term and not related to adaptive processes. Finally, the different work levels involved among the crew (working on the moon was very strenuous) provides a confound that makes comparisons questionable. Nevertheless, it is the only dataset currently available on human experience in lunar gravity.

7.9 Beneficial Effects

It's worth pointing out some other detrimental biomedical effects of spaceflight, due specifically to weightlessness, that might be ameliorated by even a low ambient g level as on the moon. Several effects are a consequence of the headward shift of body fluids (blood, cerebrospinal fluid) due to the lack of a hydrostatic gradient in 0g (in orbital or deep-space flight). This has many consequences, including sinus congestion and blunted sense of taste and smell. A much more serious set of consequences is altered ocular structure and changes in visual acuity after extended spaceflight (Lee et al., 2020). According to the prevailing theory, this is a result of slightly but chronically elevated fluid pressure (intracranial pressure: ICP) in the head, as a result of the headward shift of fluids and impaired draining and outflow from the head. As an escape route, the fluid makes its way down the sheath of the optic nerve, and impinges on the back of the eyeball. This changes the shape of the eye, which changes its optical properties and hence acuity. There is also evidence of retinal damage, and the elevated pressure might lead to other neural damage as well. This is one of the most serious current problems for long-duration spaceflight, and it is an area where even a little gravity could provide a major benefit: lunar gravity may be sufficient to provide enough of a hydrostatic gradient to reduce ICP and restore normal healthy fluid distribution along the body's long axis. This serious problem of weightlessness might be a non-issue on the moon, but nevertheless monitoring should be put in place to confirm this, and g-related countermeasures implemented if needed.

Finally, we might consider a beneficial *psychological* effect of a reduced-gravity environment. The *Overview Effect* has been noted by many astronauts in orbital flight: the compelling sense of well-being that results from seeing the Earth in its majesty, without borders, from a lofty vantage point gained through hard work over many years (Yaden et al., 2016). A possible, unrecognized, contributor to this effect might be the associated experience of weightlessness, which is wondrous and unique and provides visceral confirmation that one is fortunate to be engaged in a rare and remarkable venture. On the moon, the complete lack of gravity is not present to provide that identical thrill, but reduced lunar gravity is pleasant enough in itself, as expressed by some of the Apollo moonwalkers. The view will be incredible, including the view back to Earth, which will be close enough to be significant and provide a sense of being not too far from home. These can all be critically important in providing a sense of psychological well-being (especially in comparison to a hypothetical Mars mission). This could provide a new form of Overview Effect. The reduction in stress that these factors can provide can be significant, and might help to maintain physiological function in immune, cardiovascular, and biome systems.

References

Antonsen, E. L., Mulcahy, R. A., Rubin, D., Blue, R. S., Canga, M. A., & Shah, R. (2018). Prototype development of a tradespace analysis tool for spaceflight medical resources. *Aerospace Medicine and Human Performance, 89*, 108.

Aubert, A. E., Larina, I., Momken, I., Blanc, S., White, O., Prisk, G. K., & Linnarsson, D. (2016). Towards human exploration of space: The THESEUS review series on cardiovascular, respiratory, and renal research priorities. *npj Microgravity, 2*, 1.

Barratt, M. R., Baker, E., & Pool, S. L. (Eds.). (2019). *Principles of clinical medicine for space flight* (2nd ed.). Springer.

Blue, R. S., Bayuse, T. M., Daniels, V. R., Wotring, V. E., Suresh, R., Mulcahy, R. A., & Antonsen, E. L. (2019). Supplying a pharmacy for NASA exploration spaceflight: Challenges and current understanding. *Npj Microgravity, 5*, 1.

Canga, M., Shah, R. V., Mindock, J., & Antonsen, E. L. (2016). A strategic approach to medical care for exploration missions. In *IAC-16, E3, 6, 11, x35540, 67th International Astronautical Congress*, Guadalajara, Mexico.

Chancellor, J. C., Scott, G. B., & Sutton, J. P. (2014). Space radiation: The number one risk to astronaut health beyond low earth orbit. *Life, 4*, 491.

Clément, G. R., Bukley, A. P., & Paloski, W. H. (2015). Artificial gravity as a countermeasure for mitigating physiological deconditioning during long-duration space missions. *Frontiers in Systems Neuroscience, 9*, 92.

Crucian, B. E., Makedonas, G., Sams, C. F., Pierson, D. L., Simpson, R., Stowe, R. P., Smith, S. M., Zwart, S. R., Krieger, S. S., Rooney, B., & Douglas, G. (2020). Countermeasures-based improvements in stress, immune system dysregulation and latent herpesvirus reactivation onboard the International Space Station—Relevance for deep space missions and terrestrial medicine. *Neuroscience & Biobehavioral Reviews, 115*, 68.

Globus, R. K., & Morey-Holton, E. (2016). Hindlimb unloading: Rodent analog for microgravity. *Journal of Applied Physiology, 120*, 1196.

Johnston, R. S., Dietlein, L. F., & Berry, C. A. (Eds.). (1975). *Biomedical results of Apollo*. Scientific and Technical Information Office, National Aeronautics and Space Administration.

Karmali, F., & Shelhamer, M. (2008). The dynamics of parabolic flight: Flight characteristics and passenger percepts. *Acta Astronautica, 63*, 594.

Kassemi, M., Brock, R., & Nemeth, N. (2011). A combined transport-kinetics model for the growth of renal calculi. *Journal of Crystal Growth, 332*, 48.

Kokhan, V. S., Lebedeva-Georgievskaya, K. B., Kudrin, V. S., Bazyan, A. S., Maltsev, A. V., & Shtemberg, A. S. (2019). An investigation of the single and combined effects of hypogravity and ionizing radiation on brain monoamine metabolism and rats' behavior. *Life Sciences in Space Research, 20*, 12.

Lackner, J. R., & DiZio, P. (2000). Artificial gravity as a countermeasure in long-duration space flight. *Journal of Neuroscience Research, 62*, 169.

Lang, T., Van Loon, J. J., Bloomfield, S., Vico, L., Chopard, A., Rittweger, J., Kyparos, A., Blottner, D., Vuori, I., Gerzer, R., & Cavanagh, P. R. (2017). Towards human exploration of space: The THESEUS review series on muscle and bone research priorities. *npj Microgravity, 3*, 1.

Lee, A. G., Mader, T. H., Gibson, C. R., Tarver, W., Rabiei, P., Riascos, R. F., Galdamez, L. A., & Brunstetter, T. (2020). Spaceflight associated neuro-ocular syndrome (SANS) and the neuro-ophthalmologic effects of microgravity: A review and an update. *npj Microgravity, 6*, 1.

Moreno-Villanueva, M., & Wu, H. (2019). Radiation and microgravity—Associated stress factors and carcinogensis. *REACH, 13*, 100027.

Pietrzyk, R. A., Jones, J. A., Sams, C. F., & Whitson, P. A. (2007). Renal stone formation among astronauts. *Aviation, Space, and Environmental Medicine, 78*, A9.

Reyes, D. P., Carroll, D. J., Walton, M. E., Antonsen, E. L., & Kerstman, E. L. (2020). Probabilistic risk assessment of prophylactic surgery before extended-duration spaceflight. *Surgical Innovation*. https://doi.org/10.1177/1553350620979809

Scheuring, R. A., Jones, J. A., Novak, J. D., Polk, J. D., Gillis, D. B., Schmid, J., Duncan, J. M., & Davis, J. R. (2008). The Apollo Medical Operations Project: Recommendations to improve crew health and performance for future exploration missions and lunar surface operations. *Acta Astronautica, 63*, 980.

Shelhamer, M. (2019). Maintaining crew & mission health & performance in ventures beyond near-earth space. In L. Johnson & R. Hampson (Eds.), *Homo Stellaris*. Baen.

Shelhamer, M. (2020). Human enhancements: New eyes and ears for Mars. In K. Szocik (Ed.), *Human enhancements for space missions*. Springer.

Voorhies, A. A., Ott, C. M., Mehta, S., Pierson, D. L., Crucian, B. E., Feiveson, A., Oubre, C. M., Torralba, M., Moncera, K., Zhang, Y., & Zurek, E. (2019). Study of the impact of long-duration space missions at the International Space Station on the astronaut microbiome. *Scientific Reports, 9*, 1.

Yaden, D. B., Iwry, J., Slack, K. J., Eichstaedt, J. C., Zhao, Y., Vaillant, G. E., & Newberg, A. B. (2016). The overview effect: Awe and self-transcendent experience in space flight. *Psychology of Consciousness: Theory, Research, and Practice, 3*, 1.

Chapter 8
Changes in the Central Nervous System and Their Clinical Correlates During Long-Term Habitation in the Moon's Environment

Andrew B. Newberg

Abstract The purpose of this chapter is to review the potential functional and structural effects on the brain of long-term habitation of human beings on the moon. Using current knowledge about long-term effects of space flight on the central nervous system, we can extrapolate to the potential effects of the moon environment. However, it must be realized that the moon environment is different in some key respects from the microgravity environment of space flight. Importantly, it must be determined if there will be any detrimental changes to the CNS from long term exposure to the moon environment if human beings are to plan long duration missions or establish permanent space habitats on the moon. The moon environment, like the space flight environment has many factors such as reduced gravity, electromagnetic fields, altered day-night cycles, and radiation, that may impact the structure and function of the CNS.

8.1 Introduction

With the possibility of human habitation on the moon, we must consider a number of issues regarding the impact of the moon environment on the central nervous system. At the moment, there is little specific data about living on the moon and its effects on the brain. Other than the Apollo missions, with the longest habitation on the moon being approximately three days, the only other information that could be related to long-term habitation on the moon comes from the long duration space flights on the various space station missions (including the Soviet Mir Space Station, US Skylab, and the International Space Station currently in orbit).

Because of these space station missions, there is a substantial amount of data that already exists in terms of the effect of human long-term habitation in space on the CNS, but there are differences between the microgravity environment of the space station and the various environmental factors that would be experienced on the moon.

A. B. Newberg (✉)
Marcus Institute of Integrative Health, Thomas Jefferson University, Philadelphia, USA
e-mail: Andrew.Newberg@jefferson.edu

© The Author(s), under exclusive license to Springer Nature Switzerland AG 2021   127
M. B. Rappaport and K. Szocik (eds.), *The Human Factor in the Settlement of the Moon*,
Space and Society, https://doi.org/10.1007/978-3-030-81388-8_8

Table 8.1 Environmental factors on the moon that might affect the CNS	
	Reduced weight
	Light–dark cycle changes
	Radiation and magnetic fields
	Artificial atmosphere (Oxygen Levels, Pressure, etc.)
	Crew/Personnel selection and interactions
	Personal space, work areas, recreation

Living on the moon will have a number of factors that might potentially affect the CNS and its function (Table 8.1). This review considers what is currently known about long-term effects of the space environment on the central nervous system and applies that knowledge to what might be expected with regard to long-term habitation on the moon. Such an approach is a starting point, but it will certainly be necessary to continue to acquire ongoing data once human beings are established on the moon.

Furthermore, the consideration of the effects of the moon environment might be more specific for informing future planned manned missions to Mars in which astronauts would remain in space and on Mars for as long as three years. For these plans to be realized, it must be established that human beings can safely live in these environments without adverse effects on the CNS (Clément et al., 2020; Newberg, 1994).

When it comes to general effects of the space flight environment on the CNS, the earliest studies based on the US Skylab and the Russian missions aboard Salyut and Mir, ranging from 28 days to over 300 days in space (Nicogossian, 1994) there were no gross neuropsychological changes observed in the astronauts (Buckey & Homick, 2003; Johnston & Dietlein, 1977). But there might be a number of changes that occur on a variety of levels including long-term effects on behavior, performance, and psychology, as well as functional and structural changes in the brain, cerebral metabolic changes, plasticity of neurons and neural connections, and alterations in the neurotransmitter systems.

It is the purpose of this chapter to review the effects of various space flight factors on the neurophysiological function of the brain with specific attention focused on those that would be most related to moon environment. Particular importance will be given to the neurophysiological changes that may be either permanent or expose humans to some danger over long periods of time.

8.2 General Structural and Functional Changes in the Brain Resulting from Space Flight

When it comes to the overall effect of long-duration space missions on the brain, initial data from imaging of astronauts who had been on the International Space Station reveal significant reductions in brain volume and increases in cerebrospinal fluid volumes (Hupfeld et al. 2020). Some of these changes are likely due to the microgravity environment which results in cephalic fluid shifts. Although the moon has some gravity, similar types of shifts might be expected. In addition, some changes in brain volume might be associated with other factors such as radiation exposure (see below). While some of these changes appear to begin to resolve upon return to Earth, others persist for at least one year after return (Kramer et al., 2020). An MRI study of 7 astronauts before and after time on the ISS showed that long-duration spaceflights resulted in significant crowding of brain parenchyma towards the top of the brain (Roberts et al., 2019). Moreover, changes in the left caudate correlated significantly with poor postural control. Changes in the right primary motor area/midcingulate correlated significantly with worsening performance on complex motor tasks. And change in volume of 3 white matter regions significantly correlated with poorer reaction times on a cognitive performance task. Given such findings, it is not known whether long duration space flight would be associated with general neurological dysfunction over time or possibly increase the risk for neurological disorders such as Alzheimer's or Parkinson's disease (Cherry et al., 2012). However, future research will be needed to assess the possibility of this connection.

An MRI functional connectivity study of 11 cosmonauts showed a post-flight increase in the stimulation-specific connectivity of the right posterior supramarginal gyrus with the rest of the brain (Pechenkova et al., 2019a, b). There was also a strengthening of connections between the left and right insula. There were several areas of decreased connectivity including a decrease between the vestibular nuclei, right inferior parietal cortex (BA40) and cerebellum, with areas associated with motor, visual, vestibular, and proprioception functions. There was also decreased connectivity between the cerebellum and the visual cortex and right inferior parietal cortex. Such changes implicate long duration space flight having significant effects on brain function that is associated specifically with sensory processing, but likely results in cognitive and emotional changes as well. Further studies will be needed to determine whether living in the moon's gravity, which is less than Earth's but not the microgravity of current space flight missions, would result in similar types of sensory changes.

With fluid shifts, there can also be alterations in the concentrations of various electrolytes and nutrients (Newberg, 1994). Since electrolytes such as sodium and potassium are important for proper neuronal function, it is also possible that such changes, over long periods of time, might results in altered brain function as well.

When it comes to neurotransmitter studies there have been only a few animal and human studies, and none that have measured neurotransmitter levels in specific brain regions in subjects during actual space flight. In one animal study of mice exposed to one month in space, there was decreased expression of crucial genes involved in dopamine synthesis and degradation, in addition to the D1 receptor (Popova et al., 2015). The authors indicated that such findings might support the observed spaceflight-induced locomotor impairment and dyskinesia. However, no such effects were observed within the serotonin system. In a study by Davydov et al. (1986) hypokinesia during an Earth-based head down tilt state resulted in an initial increase in serotonin activity in human subjects. This increase was followed by a decrease over a 70 day period of hypokinesia when compared to the subject's baseline levels. The authors suggested that these changes represented the brain adaptation to hypokinesis with dynamic changes occurring over the course of the adaptation period with the eventual decrease reflecting a completed adaptation to hypokinesia. Further, the decrease itself may have been in response to overall decreased afferent neuronal stimulation. In the same study, histamine levels increased throughout the hypokinesis period although the mechanism of this increase was not clearly determined. The serotonin and histamine levels were found to normalize by approximately 25 days post-hypokinesis.

An important question is how the various CNS changes to brain activity, brain volume, and neurotransmitters affect cognitive, psychological, and behavioral patterns in astronauts during long duration space flights and ultimately missions to the moon.

8.3 Sensory Changes During Long-Duration Moon Missions

Much of the early neurological research done on human exposure to space flight was on the neurovestibular system since space motion sickness (SMS) was of prime significance, especially to the US space shuttle program which had mostly short duration flights in which SMS could significantly affect productivity. The cause of SMS has been attributed primarily to the brain's attempts to adapt to the microgravity environment with resultant reinterpretation of sensory inputs regarding body position and orientation. Altered neurovestibular processes also occur both during and after long duration missions resulting in a variety of symptoms such as postural instability, dizziness, nausea, and vomiting (Gazenko, 1979).

In order to manage SMS a variety of therapeutic modalities have been employed. Mechanical devices that restrict head motion have been partially effective since they prevent head movement that might aggravated SMS symptoms (Matsnev et al., 1983). Biofeedback and other psychological measures have been used to help astronauts control their SMS symptoms (Cowings et al.,, 1982). Most commonly, pharmacological approaches consisting of anticholinergic drugs using various routes of administration have been effective (Davis et al., 1993; Homick et al., 1983). However, none of the treatment modalities has been consistently successful at reducing SMS symptoms in all astronauts and the pharmacological approaches in particular can have side effects such as fatigue, dry eyes and mouth, and even cognitive impairment (Davis et al., 1993; Nicogossian, 1994). Fortunately, symptoms from SMS usually resolve once the individual becomes adapted to the space flight environment. Adaptation of the neurovestibular system usually occurs within 5–7 days and thus SMS is less of a problem for long-duration space missions. Furthermore, SMS is less expected to be a problem during moon missions since there is a better visual up-down orientation due to seeing the moon surface versus the sky, and a better vestibular up-down orientation due to at least some gravity. However, it is possible that the moon environment might still result in some important changes in vestibular function and spatial orientation that can affect piloting and other tasks (Clark & Young, 2017).

Sensory perceptions have also been found to be altered during space flight missions. Astronauts have experienced changes in the taste, smell, and touch sensory systems (Nicogossian, 1994; Clément et al., 2020). People engaged in space flight missions have reported feelings of reduced hunger and increased fullness. In part due to fluid shifts and altered brain function, foods tend to taste bland and require more seasoning in space (Yuganov and Kopanev, 1975). Although sensory problems are somewhat common, there is a substantial amount of variability in such experiences and there have been unclear aggravating or alleviating factors, as well as variable time to onset, duration of disturbance, or the intensity of these disturbances.

Visually, one of the most prominent findings in astronauts on the ISS is a condition called spaceflight-associated neuro-ocular syndrome (SANS) which includes altered visual acuity, ophthalmologic changes such as cotton wool spots, choroidal folds, and optic disc edema (Mader et al., 2011). However, it is not clear whether the lunar gravity would result in similar changes since it is not a zero gravity environment. Another visual problem that has been reported is the intermittent perception of light flashes during flight. This phenomenon was studied on board Skylab and was shown to be caused by heavy ionized cosmic particles passing through the retina (Sannita et al., 2006). Such light flashes would still be expected in the moon environment which does not offer adequate shielding from cosmic particles.

Another effect of exposure to reduced gravitational force, which may play a role in CNS changes on the moon as well, is the adaptation process that occurs in the musculoskeletal system as well as postural changes (Paloski et al., 2006). It is known that during long duration space missions there is a significant reduction in muscle mass, especially in the weight bearing muscles, because of low gravity (Fuchs, 1980; Nicogossian, 1994; Thornton & Rummel, 1977). It might be expected that during long duration moon missions, as in long duration space flights, the brain areas responsible for directing, initiating, and coordinating the movements of affected muscle groups would be significantly altered (Roberts et al., 2019). This might be reflected by structural and functional changes in the motor and pre-motor cortices as well as in brain regions involved in motor coordination such as the basal ganglia and cerebellum. Such findings were reported in a study of 11 cosmonauts in which functional connectivity was measured before and after long duration space flight and which found significant differences in connectivity in the sensorimotor areas, visual, proprioceptive, and vestibular regions of the brain (Pechenkova et al., 2019a, b). These effects may make readaptation to Earth's 1-g environment by astronauts more difficult especially after missions lasting several years such as a moon mission. However, much work has also been done on mitigating these low gravity effects on muscle and bone through rigorous exercise countermeasures (English et al., 2019). Assuming that such measures are also performed on a long-duration moon mission, it is hoped that such effects will not be overly problematic.

8.4 Sleep Problems on the Moon

There have been both subjective reports and objective measures of sleep disturbances during short and long-duration space missions. Some individuals have required up to 12 h of sleep per day by mission end in order to ensure adequate functioning (Czeisler et al., 1991; Wu et al., 2018). The finding of sleep problems even after long periods of time in space (i.e. many months) suggests that altered sleep patterns might be a persistent issue even after acclimatization to the space environment. The impact of space flight missions on sleep appears relatively widespread in terms of personnel with up to 75% of astronauts requesting sleep medications while in space (Barger et al., 2014; Santy et al., 1988).

The most common cause of sleep disturbances is related to the altered light–dark cycle. For example, on the International Space Station, there are approximately 16 sunrises and sunsets during a 24 h period. These alterations in the light–dark cycle can substantially affect the brain's ability to maintain normal circadian rhythms. On the ISS, special methods are used to ensure that astronauts have dark areas for sleep and light areas to work in during the day regardless of the sun's position in the sky. Similar approaches will be required during long-duration habitation on the moon which has a different, and almost opposite problem with the light–dark cycle. Rather than multiple sunrises in a given 24 h period, there is one sunrise every 30 day period. For moon habitants, special areas will be needed to have dark for sleep and light to work even though the sky will be light or dark for approximately 2 weeks at a time. Changes in sleep may be caused by various other factors in the moon environment such as reduced weight, habitat noise and alarms, SMS, and unusual work shifts and workloads (Wu et al., 2018). Even astronaut anxiety and excitement have been implicated in reduced sleep (Santy et al., 1988).

Just like on Earth, impaired sleep in space or on the moon may cause secondary problems such as reduced cognition, reduced performance, increased anxiety and depression, and poorer physical health. For example, Dijk et al. (2001) reported a significant relationship between cognitive performance and loss of REM sleep. A separate study of 28 astronauts on the Mir space station found that the error rate on cognitive testing was significantly associated with deviations from the normal sleep–wake cycle (Nechaev, 2001). It has also been suggested that space flight related sleep problems might share similar mechanisms with aging related sleep problems (Czeisler et al., 1991). However, more studies are needed to determine the mechanisms underlying sleep disturbances during space flight. And such information might be crucial for helping prevent such issues during long-duration moon missions.

Sleep problems have been managed using a variety of approaches including sleeping medications, light therapy to ensure appropriate duration of dark and light phases in living/sleeping quarters, planning appropriate work schedules, and providing opportunities for good sleep hygiene (Wu et al., 2018). It is also important to carefully select crew members who are more likely to sleep well and be resilient to sleep deprivation. Similar measures and approaches would have to be considered with individuals during long-duration moon missions.

8.5 Radiation Effects on the Moon

Radiation exposure has long been considered a potentially important problem for people engaged in long duration space flight missions. Numerous Earth based studies have shown that ionizing radiation can affect most cells in the human body including those in the brain (Nicogossian, 1994; Clément et al., 2020). A number of animal studies have suggested that the long term effects of radiation exposure in space can be detrimental to the brain. However, it has been difficult to determine the effects of prolonged exposure to radiation especially in conjunction with other factors that might be experienced during long term habitation on the moon (Antipov et al., 1991; Moreno-Villanueva & Wu, 2019).

A number of animal and human studies have described the neurophysiological effects of radiation, particularly as it relates to the space flight environment (Newberg & Alavi, 1998; Clément et al., 2020). The results of such studies clearly indicate that detrimental effects occur in the brain with exposure to radiation. Furthermore, there is a need for more extensive and rigorous studies in order to assess the behavioral and performance problems that radiation exposure might cause in humans during long duration space missions. In general, studies have found that the greater the radiation dose, the poorer the performance outcomes. Several studies in mice and rats (Cucinotta & Cacao, 2019) have suggested that low dose radiation can result in cognitive deficits when the CNS is exposed to heavy ions. And there is substantial evidence that changes in spatial memory are related to damage to the hippocampus from particle irradiation (Wyrobek & Britten, 2016). Furthermore, the dopaminergic neurons might be particularly sensitive to the effects of iron-rich high charge and energy particles in galactic cosmic rays (Koike et al., 2005). This could make individuals engaged in long-duration space missions prone to dopaminergic loss and diseases such as Parkinson's. A recent analysis of a number of animal studies suggests that long duration space missions can result in a significant increased relative risk of cognitive deficits, although this was lowest for estimates of a 180 day lunar mission (Cucinotta & Cacao, 2020).

8.6 Cognitive Problems Associated with the Moon Environment

We have considered above that long duration missions to the moon might involve a number of factors that can have an impact on cognitive function. Reduced gravity, altered brain function, reduced sleep, and radiation exposure can all potentially have detrimental effects on cognition. One review of a number of published studies of how cognition might be affected by long duration space flight missions showed that there was a diverse array of potential effects but that those effects can be quite variable and affect mission crew members in very different ways (Strangman et al., 2014). Some crew members had very little cognitive effect while others had more substantial effects. Furthermore, cognitive changes occurred across a variety of domains that also affected crew members in selective ways. On the other hand, there have been a number of studies showing that cognition might remain stable through long duration missions, especially as crew become accustomed to their various mission related tasks (Strangman et al., 2014).

Part of the issue related to cognition is the brain's natural ability to adapt to new environments and demands. On space flight missions, it is likely that the brain will be able to adapt successfully and this is supported by studies of neuroplasticity showing that the brain does, in fact, adapt through alterations in the cerebellum, sensory and motor areas, and vestibular pathways (Van Ombergen et al., 2017). During such changes, cognition can worsen, particularly if specific cognitive domains are not required for adaptive purposes. However, as adaptation occurs, the efficiency and cognitive performance of astronauts is likely to improve.

In terms of longer effects of the space flight or moon base environment on cognitive decline with aging, there is insufficient data to know what risks and effects might manifest. It is possible that participants on a long-duration moon mission will be at increased risk for Alzheimer's disease or Parkinson's disease due to radiation effects, fluid and volume shifts, sleep loss, dietary changes, and other as yet undetermined factors, but at this point, there is no clear data to support such a concern.

8.7 Psychological and Behavioral Problems Associated with the Moon Environment

As a result of the various environmental factors described above, as well as crew related interaction issues, there are a number of psychological and behavioral changes that have been investigated and considered during long duration space missions that have implications for future missions to the moon. The psychological and behavioral problems include disruption of cognitive and memory functions; increased stress and anxiety induced by concern for a successful mission as well as fear of personal failure, injury or death; issues related to social isolation and decreased personal space; effects of sensory deprivation or sensory overload in a space craft design; personal problems with other crew members or familial problems; depression that could be caused by any of the above mentioned factors; and personality changes that might occur as a result of space flight factors or via intrinsic neurophysiological mechanisms (Nicogossian, 1994; Clément et al., 2020).

Interestingly, although not surprisingly, there are many interindividual differences when it comes to the psychological and behavioral responses to the space flight environment, especially regarding the isolation elements. For example, a study of six male participants in an isolation facility for 520 days to simulate a Mars mission revealed that two crew members accounted for the majority of psychological and mission related problems and approximately 85% of conflicts with other crew members

(Basner et al., 2014). These individuals also expressed the highest levels of anxiety and depression.

More detailed factors have been explored as having effects on human behavior and performance in space. Such factors include those pertaining to decision making abilities, intrinsic motivation, personal adaptability, leadership abilities, productivity and reliability, various human emotions, attitudes, mental and physical fatigue, crew composition and compatibility, psychological stability of individuals, and individual social skills. All of these factors can have a significant impact on behavior and performance. There is likewise, a long list of environmental factors that will affect human performance and behavior. These include, moon base habitability, confinement, isolation, lack of privacy, noise, artificial life support, circadian rhythm changes, hazards, and boredom.

It is also important to consider how a moon base might be designed in order to identify and mitigate potential factors that could adversely affect human psychology and behavior. To begin with, it is important to consider the mission duration and complexity, division of work, information load, and task load as these factors, if overly stressful for the crew, can lead to significant psychological or behavioral problems. It will also be important to take into consideration how decisions will be made on the moon base and the amount of crew autonomy allowable. Recreational opportunities and other approaches to ensure adequate quality of life, communication between crew members and Earth, and creative outlets can all be beneficial.

The ability and success of various therapeutic interventions is partly determined by the ability to identify psychological problems during space mission and the same will be true on the moon. A number of studies have analyzed various methods of detecting stress, diminished performance, and psychological instability. For example, speech analysis and other qualitative measures can be useful for determining the level of stress and emotional state of astronauts (Simonov & Frolov, 1973). However, the success of these approaches has not been consistent in identifying problems during space missions (Clément et al., 2020). The issue of psychological and behavioral aspects of long term moon missions will certainly need to be addressed in order to determine how and why they might occur, how to prevent them, and how to treat them if they occur. In order to accomplish these objectives, a thorough understanding of the neurophysiological mechanisms underlying these changes will be needed and this can lead to improved methods of treatment and management.

One approach to mitigating potential psychological and behavioral problems is the extensive use of testing during the selection process. However, most psychological instruments used to date have had the goal of excluding significant psychopathologies and have generally not been able to predict which individuals would manifest behavioral problems or lapses in judgment, reduced cooperation with other crew members, overt irritability, or destructive interpersonal actions (Collins, 2003). Thus, better approaches in the future might be important for selecting crew members for extended moon missions.

It is also important to continue to research various cognitive, psychological and behavioral effects of long-duration space or moon missions. This will include research studies to assess brain function as well as clinical evaluations of individuals on a moon base at regular intervals. If possible, neuroimaging might be an intriguing option for evaluating brain function during long duration moon missions. Today's neuroimaging technologies provide the opportunity to observe many of the structural and functional changes in the central nervous system in human beings as well as animals. Anatomical imaging can be performed using x-ray computerized tomography (CT) and magnetic resonance imaging (MRI) and functional imaging can be performed with fMRI as well as positron emission tomography (PET) and single photon emission computed tomography (SPECT). These techniques could be implemented to evaluated longitudinal changes in people on a moon base to determine who might require additional supportive measures or even the need to return to Earth. The use of diagnostic studies on the moon might require altered engineering of these types of scanners. MRI scanners as well as the other scanners are extremely heavy and have strong magnetic fields or require radiation. These issues would have to be managed and adapted to be able to be transported to, and used safely, on the moon. Over time, such capabilities might be possible, but unlikely in the early phases of a moon base.

8.8   Conclusions

When it comes to managing the moon's environmental effects on the central nervous system, there are a number of approaches that might be considered. Hopefully, the vestibular problems will be minimal given the ability of the brain to orient up-down more easily on the moon in comparison to the space flight environment.

Dealing with effects of radiation might be solvable in part by covering habitation modules with a sufficient layer of moon dirt to block sufficient radiation and minimize exposure. Some scholars have suggested such a possibility that might be useful for at least mitigating exposure in modules that are used for sleeping and other activities that require long-term participation. Certain areas will obviously not be able to be covered, especially those that require sunlight. However, this is one possible option to at least reduce radiation exposure. Another possibility would be to set up magnetic fields that surround the habitation modules, thereby directing radiation away from the living areas. In addition, some evidence suggests that radiation effects can be mitigated in part through diet and nutritional approaches (Turner et al., 2002). For example, natural antioxidants such as vitamins E and C and folic acid, as well as N-acetylcysteine, might be given before, during, and after exposure to radiation on the moon. Similarly, molecules such as bioflavonoids, epigallocatechin, and omega-3 polyunsaturated fatty acids might be protective as well.

Extensive personal training is also likely to be involved such as appropriate exercise, diet and nutrition, antioxidant supplements, and brain stimulation techniques. Each of these can be useful for helping the brain stay healthy during long-term habitation on the moon.

Appropriate psychological counseling will need to be available and can be administered from Earth or directly from the moon personnel. It will be important to ensure the mental and psychological health of habitants since they are likely to live there for a long period of time.

References

Antipov, B. P., Davydov, B. I., Ushakov, I. B., & Fedorov, V. P. (1991). The effects of space flight factors on the central nervous system: Structural-functional aspects of the radiomodifying effect. *Space Life Sciences Digest, 30*, 72.

Barger, L. K., Flynn-Evans, E. E., Kubey, A., Walsh, L., Ronda, J. M., Wang, W., Wright, K. P., Jr., & Czeisler, C. A. (2014). Prevalence of sleep deficiency and use of hypnotic drugs in astronauts before, during, and after spaceflight: An observational study. *Lancet Neurology, 13*(9), 904–912. https://doi.org/10.1016/S1474-4422(14)70122-X

Basner, M., Dinges, D. F., Mollicone, D. J., Savelev, I., Ecker, A. J., Di Antonio, A., Jones, C. W., Hyder, E. C., Kan, K., Morukov, B. V., & Sutton, J. P. (2014). Psychological and behavioral changes during confinement in a 520-day simulated interplanetary mission to mars. *PLoS ONE, 9*(3). https://doi.org/10.1371/journal.pone.0093298

Buckey, J. C., & Homick, J. L. (Eds.). (2003). *Neurolab Spacelab mission: Neuroscience research in space results from the STS-90, Neurolab Spacelab mission. NASA SP-2003-535.* National Aeronautics and Space Administration Lyndon B. Johnson Space Center.

Clément, G. R., Boyle, R. D., George, K. A., Nelson, G. A., Reschke, M. F., Williams, T. J., & Paloski, W. H. (2020). Challenges to the central nervous system during human spaceflight missions to Mars. *Journal of Neurophysiology, 123*(5), 2037–2063. https://doi.org/10.1152/jn.00476.2019

Cherry, J. D., Liu, B., Frost, J. L., Lemere, C. A., Williams, J. P., Olschowka, J. A., & O'Banion, M. K. (2012). Galactic cosmic radiation leads to cognitive impairment and increased aβ plaque accumulation in a mouse model of Alzheimer's disease. *PLoS ONE, 7*(12), e53275. https://doi-org.proxy.library.upenn.edu/10.1371/journal.pone.0053275.

Clark, T. K., & Young, L. R. (2017). A case study of human roll tilt perception in hypogravity. *Aerospace Medicine and Human Performance, 88*(7), 682–687. https://doi-org.proxy.library.upenn.edu/https://doi.org/10.3357/AMHP.4823.2017

Collins D. L. (2003). Psychological issues relevant to astronaut selection for long-duration space flight: a review of the literature. Human performance in extreme environments. *Journal of the Society for Human Performance in Extreme Environments, 7*(1), 43–67. https://doi.org/10.7771/2327-2937.1021.

Cowings, P. S., Toscano, W. B., Reschke, M. F., Tsehay, A. (2018). Psychophysiological assessment and correction of spatial disorientation during simulated Orion spacecraft re-entry. *International Journal of Psychophysiology, 131*, 102–112. https://doi.org/10.1016/j.ijpsycho.2018.03.001. Epub 2018 Mar 2. PMID: 29505848.

Cucinotta, F. A., & Cacao, E. (2019). Risks of cognitive detriments after low dose heavy ion and proton exposures. *International Journal of Radiation Biology, 95*(7), 985–998. https://doi.org/10.1080/09553002.2019.1623427

Cucinotta, F. A., & Cacao, E. (2020). Predictions of cognitive detriments from galactic cosmic ray exposures to astronauts on exploration missions. *Life Sciences Space Research, 25*, 129–135. https://doi-org.proxy.library.upenn.edu/10.1016/j.lssr.2019.10.004.

Czeisler, C. A., Chiasera, A. J., & Duffy, J. F. (1991). Research on sleep, circadian rhythms and aging: Applications to manned space flight. *Experimental Gerontology, 26*, 217.

Davis, J. R., Jennings, R. T., Beck, B. G., & Bagian, J. P. (1993). Treatment efficacy of intramuscular promethazine for space motion sickness. *Aviation Space Environmental Medicine, 64*(3 Pt 1), 230–233.

Davydov, N. A., Galinka, Y. Y., & Ushakov, A. S. (1986). Functional activity of the serotonin and histaminic systems in humans subjected to long-term hypokinesia. *USSR Space Life Sciences Digest, 4,* 80.

Dijk, D. J., Neri, D. F., Wyatt, J. K., Ronda, J. M., Riel, E., Ritz-De Cecco, A., Hughes, R. J., et al. (2001). Sleep, performance, circadian rhythms, and light-dark cycles during two space shuttle flights. *American Journal Physiology. Regulatory, Integrative and Comparative Physiology, 281*(5), R1647–R1664. https://doi.org/10.1152/ajpregu.2001.281.5.R1647.

English, K. L., Bloomberg, J. J., Mulavara, A. P., & Ploutz-Snyder, L. L. (2019). Exercise countermeasures to neuromuscular deconditioning in spaceflight. *Comprehensive Physiology, 10*(1), 171–196. https://doi-org.proxy.library.upenn.edu/10.1002/cphy.c190005.

Fuchs, H. S. (1980). Man in weightlessness—Physiological problems, clinical aspects, prevention, and protection. *Rivista Di Med. Aeronauticae Spaziale, 44,* 332.

Gazenko, O.G., ed. (1979). Summaries of Reports of the 6th All-Soviet Union Conference on Space Biology and Medicine. (vol. 1 and 11) Kaluga, USSR.

Homick, J. L., Kohl, R. L., Reschke, M. F., Degioanni, J., & Cintron-Trevino, N. M. (1983). Transdermal scopolamine in the prevention of motion sickness: Evaluation of the time course of efficacy. *Aviation Space Environmental Medicine, 54*(11), 994–1000.

Hupfeld, K. E., McGregor, H. R., Lee, J. K., Beltran, N. E., Kofman, I. S., De Dios, Y. E., Reuter-Lorenz, P. A., et al. (2020). The impact of 6 and 12 months in space on human brain structure and intracranial fluid shifts. *Cerebral Cortex Communications, 1*(1), tgaa023. https://doi-org.proxy.library.upenn.edu/10.1093/texcom/tgaa023.

Koike, Y., Frey, M. A., Sahiar, F., Dodge, R., & Mohler, S. (2005). Effects of HZE particle on the nigrostriatal dopaminergic system in a future Mars mission. *Acta Astronautica, 56*(3), 367–378. https://doi-org.proxy.library.upenn.edu/10.1016/j.actaastro.2004.05.068.

Kramer, L. A., Hasan, K. M., Stenger, M. B., Sargsyan, A., Laurie, S. S., Otto, C., Ploutz-Snyder, R. J., et al. (2020). Intracranial effects of microgravity: A prospective longitudinal MRI study. *Radiology, 295*(3). 640–648. https://doi-org.proxy.library.upenn.edu/10.1148/radiol.2020191413.

Mader, T. H., Gibson, C. R., Pass, A. F., et al. (2011). Optic disc edema, globe flattening, choroidal folds, and hyperopic shifts observed in astronauts after long-duration space flight. *Ophthalmology, 118,* 2058–2069. https://doi.org/10.1016/j.ophtha.2011.06.021pmid:21849212

Matsnev, E. I., Yakovleva, I. Y., Tarasov, I. K., Alekseev, V. N., Kornilova, L. N., Mateev, A. D., & Gorgiladze, G. I. (1983). Space motion sickness: Phenomenology, countermeasures, and mechanisms. *Aviation Space Environmental Medicine, 54*(4), 312–317.

Moreno-Villanueva, M., & Wu, H. (2019). Radiation and microgravity—Associated stress factors and carcinogenesis. *REACH, 13.* https://doi.org/10.1016/j.reach.2019.100027

Nechaev, A. P. (2001). Work and rest planning as a way of crew member error management. *Acta Astronautica, 49*(3–10), 271–278. https://doi.org/10.1016/s0094-5765(01)00105-9

Newberg, A. B., & Alavi, A. (1998). Changes in the central nervous system during long-duration space flight: Implications for neuro-imaging. *Advances in Space Research: The Official Journal of the Committee on Space Research (COSPAR), 22*(2), 185–196. https://doi-org.proxy.library.upenn.edu/10.1016/s0273-1177(98)80010-0.

Newberg, A. B. (1994). Changes in the central nervous system and their clinical correlates during long-term space flight. *Aviation, Space, and Environmental Medicine, 65,* 562–572.

Nicogossian, A. E. (1994). *Space physiology and medicine.* Lea & Febiger.

Paloski, W. H., Wood, S. J., Feiveson, A. H., Black, F. O., Hwang, E. Y., & Reschke, M. F. (2006). Destabilization of human balance control by static and dynamic head tilts. *Gait & Posture., 23*(3), 315–323. https://doi.org/10.1016/j.gaitpost.2005.04.009

Pechenkova, E., Nosikova, I., Rumshiskaya, A., Litvinova, L., Rukavishnikov, I., Mershina, E., Sinitsyn, V., et al. (2019). Alterations of functional brain connectivity after long-duration spaceflight as revealed by fMRI. *Frontiers in Physiology, 10*, 761. https://doi-org.proxy.library.upenn.edu/10.3389/fphys.2019.00761.

Pechenkova, E., Nosikova, I., Rumshiskaya, A., Litvinova, L., Rukavishnikov, I., Mershina, E., Sinitsyn, V., Van Ombergen, A., Jeurissen, B., Jillings, S., Laureys, S., Sijbers, J., Grishin, A., Chernikova, L., Naumov, I., Kornilova, L., Wuyts, F. L., Tomilovskaya, E., & Kozlovskaya, I. (2019b). Alterations of functional brain connectivity after long-duration spaceflight as revealed by fMRI. *Frontiers in Physiology, 10*, 761. https://doi.org/10.3389/fphys.2019.00761

Popova, N. K., Kulikov, A. V., Kondaurova, E. M., Tsybko, A. S., Kulikova, E. A., Krasnov, I. B., Shenkman, B. S., Bazhenova, E. Y., Sinyakova, N. A., & Naumenko, V. S. (2015). Risk neurogenes for long-term spaceflight: Dopamine and serotonin brain system. *Molecular Neurobiology, 51*(3), 1443–1451. https://doi.org/10.1007/s12035-014-8821-7

Johnston, R. S., & Dietlein, L. F. (1977). *NASA SP-377*. US Government Printing Office.

Roberts, D. R., Asemani, D., Nietert, P. J., Eckert, M. A., Inglesby, D. C., Bloomberg, J. J., George, M. S., & Brown, T. R. (2019). Prolonged microgravity affects human brain structure and function. *AJNR, 40*(11), 1878–1885. https://doi-org.proxy.library.upenn.edu/10.3174/ajnr.A6249.

Roberts, D. R., Ramsey, D., Johnson, K., Kola, J., Ricci, R., Hicks, C., Borckardt, J. J., Bloomberg, J. J., Epstein, C., & George, M. S. (2010). Cerebral cortex plasticity after 90 days of bed rest: Data from TMS and fMRI. *Aviation, Space, and Environmental Medicine, 81*(1), 30–40. https://doi.org/10.3357/asem.2532.2009

Sannita, W. G., Narici, L., Picozza, P. (2006). Positive visual phenomena in space: A scientific case and a safety issue in space travel. *Vision Research, 46*(14), 2159–2165. https://doi.org/10.1016/j.visres.2005.12.002. Epub 2006 Feb 28. PMID: 16510166.

Santy, P. A., Kapanka, H., Davis, J. R., & Stewart, D. F. (1988). Analysis of Sleep on Shuttle Missions. *Aviation, Space, and Environmental Medicine, 59*, 1094.

Simonov, P.V., & Frolov, M. V. (1973). Utilization of human voice for estimation of man's emotional stress and state of attention. *Aerospace Medicine, 44*, 256.

Strangman, G. E., Sipes, W., & Beven, G. (2014). Human cognitive performance in spaceflight and analogue environments. *Aviation Space Environmental Medicine, 85*(10), 1033–1048. https://doi.org/10.3357/ASEM.3961.2014

Thornton, W. E., & Rummel, J. A. (1977). Muscular deconditioning and its prevention in space flight. In R. S. Johnston & L. F. Dietlein (Eds.), *Biomedical Results from Skylab* (NASA SP-377). US Government Printing Office, Washington, DC.

Turner, N. D., Braby, L. A., Ford, J., & Lupton, J. R. (2002). Opportunities for nutritional amelioration of radiation-induced cellular damage. *Nutrition (Burbank, Los Angeles County, Calif.), 18*(10), 904–912. https://doi-org.proxy.library.upenn.edu/10.1016/s0899-9007(02)00945-0.

Van Ombergen, A., Laureys, S., Sunaert, S., Tomilovskaya, E., Parizel, P. M., & Wuyts, F. L. (2017). Spaceflight-induced neuroplasticity in humans as measured by MRI: What do we know so far? *NPJ Microgravity, 3*, 2. https://doi.org/10.1038/s41526-016-0010-8

Wu, B., Wang, Y., Wu, X., Liu, D., Xu, D., & Wang, F. (2018). On-orbit sleep problems of astronauts and countermeasures. *Military Medical Research, 5*(1), 17. https://doi.org/10.1186/s40779-018-0165-6

Wyrobek, A. J., & Britten, R. A. (2016). Individual variations in dose response for spatial memory learning among outbred Wistar rats exposed from 5 to 20 cGy of (56) Fe particles. *Environmental Molecular Mutagenesis, 57*(5), 331–340. https://doi.org/10.1002/em.22018

Yuganov, Y. M., & Kopanev, V. I. (1975). Physiology of the sensory sphere under spacesight conditions. In M. Calvin & O. G. Gazenko (Eds.), *Foundations of space biology and medicine* (pp. 571–598). NASA Scientific and Technical Information Office.

Chapter 9
Hazards of Lunar Regolith for Respiratory, Central Nervous System, Cardiovascular and Ocular Function

Martin Braddock

Abstract Lunar dust will arise as an issue for remediation in the earliest lunar settlements. The natural environment will pose risks that are becoming an important focus for research. Since Apollo 12 astronaut Alan Bean became accidentally exposed to lunar regolith in 1970, the toxic effects of Moon dust (<20 µm particle size diameter) and simulants have been extensively explored in models of animal and human cytology and pathophysiology. The absence of erosive mechanisms on the Moon such as wind and rain which are present on Earth maintain the sharp and abrasive physiochemical properties of lunar dust which can infiltrate lung tissue. Moreover, the particle size distribution and elemental composition may induce ocular and skin irritation and stimulate an allergenic cascade which may lead to cardiovascular and central nervous system inflammation. A permissible exposure limit has been established for acute exposure. Chronic exposure limits need to be established prior to life-support system mitigation of any effects of regolith on astronaut hygiene after introduction into the habitat.

9.1 Introduction

There are many scientific and logistical challenges that must be met, addressed and overcome to facilitate the next phase of human lunar exploration. The National Aeronautics and Space Administration (NASA) Artemis's programme has set the ambitious goal of returning to the Moon by 2024, although the date may require revision as assigned budgets are confirmed by the new administration in the United States (Bender, 2020). Challenges are both a combination of scientific and geo-political and will likely require concerted effort from Global Space Agencies and private enterprise to provide acceptable solutions, where outreach to the general public will play an increasingly important role, given competing demands for funding and the power of public opinion to demand action on climate change and management of

M. Braddock (✉)
Sherwood Observatory, Sutton-in-Ashfield, UK

Science4U.co.uk, Radcliffe-on-Trent, UK

© The Author(s), under exclusive license to Springer Nature Switzerland AG 2021
M. B. Rappaport and K. Szocik (eds.), *The Human Factor in the Settlement of the Moon*,
Space and Society, https://doi.org/10.1007/978-3-030-81388-8_9

the Covid-19 pandemic. Establishment of any form of semi-permanent or permanent human presence involving physical activity on the lunar surface will require management of lunar dust, which is found everywhere on the lunar surface. Lunar dust, a component of regolith has, and continues to be formed by micrometeorite impacts onto the surface of the Moon where hypervelocity collision results in the formation of sharp, abrasive agglutinates which readily adhere to most surfaces. Dust particles vary in size with the smallest being less than 10 μm (Cain, 2010). Simply using a rocket engine to lower any lander down to the lunar surface will propel rocks, particles and dust for many tens if not hundreds of kilometres posing a critical threat to humans and the technology upon which they depend for survival. This situation proffers an immediate challenge as due to electrostatic forces (Stubbs et al., 2007), dust readily adheres to space suits and will be brought into habitats including space craft after Moon walks, where it has the potential to disrupt use of instrumentation, for example, by causing visual obscuration (Gaier & Greel, 2005), respiratory, dermal and eye irritation and be toxic to astronauts. Apollo 12 astronaut Alan Bean reported (Bean et al., 1970): *'After lunar lift-off, when we were again in a 0 g environment, a great quantity of dust floated free within the cabin. This made breathing without a helmet difficult, and enough particles were present in the cabin atmosphere to affect our vision'.*

As stated by Apollo 17 commander Eugene Cernan in a technical debrief held in 1972 and who once described lunar dust as smelling like spent 'gun powder' (Gaier, 2005; Phillips, 2006; Wagner, 2006):*'I think dust is probably one of our greatest inhibitors to a nominal operation on the Moon. I think we can overcome other physiological or physical or mechanical problems except dust'.*

'Even before people step out the door of a lunar lander, the dust will be a problem'.

Jack Schmitt also an Apollo 17 astronaut who reported having a severe allergic reaction to lunar dust in 1972 said:

'Dust is the No. 1 environmental problem on the moon, we need to understand what the effects are, because there's always the possibility that engineering might fail'

Figure 9.1 shows Apollo 14 astronaut Edgar Mitchell (A) walking on the lunar surface with dust clinging to the legs and boots of his space suit (circled in red). Apollo 17 astronauts Gene Cernan and Jack Schmitt were covered in lunar dust. Gene Cernan (B) is shown resting inside the lunar module (LM) challenger after completing an extra-vehicular activity on the lunar surface and dust is clearly visible on his space suit (left panel), long johns and forehead (right panel). The normally white spacesuit of Jack Schmitt (C) is seen with grey lunar dust coating the suit (circled in red), the adherent property is associated with low electrical conductivity and maintenance of electrostatic charge. The size distribution of particles entering the lunar module has been measured and the majority of particles are within the respirable range (Linnarsson et al., 2012).

In Chap. 6, I discussed the nature of lunar regolith and the variability in depth of regolith on the Moon, summarised some of the extensive research that has been conducted into the identification and further study of lunar regolith simulants (LRSs)

Fig. 9.1 Photographs of **a** Apollo 14 astronaut Edgar Mitchell, Apollo 17 astronauts, **b** Eugene Cernan, and **c** Jack Schmitt, showing presence of lunar dust adhering to space suits (red circle). Astronaut images, credit NASA

and briefly discussed some of the limitations of the use of LRSs versus authentic lunar regolith. I will now provide an overview of some of the many studies that have been conducted to determine and quantify the toxicity associated with lunar regolith, with an emphasis on dust from non-clinical in vitro and in vivo studies and describe the astronaut reported outcomes of exposure to dust. In the next section, I will describe the size of the dust particles that are within the respirable range and which may constitute a health hazard to astronauts. It should be remembered however, that a consideration of the effect of lunar dust on human health is one component of the overall management of astronaut hygiene and that this broad field includes exposure to a wide variety of other substances within a confined environment. This has led to the determination of spacecraft maximum allowable concentrations (SMACs) of chemicals including volatiles, such as lithium hydroxide dust from cannisters, propellants and solvents and air and water filter pollutants (Khan-Mayberry et al., 2011).

9.2 The Toxicity and Health Properties of Lunar Regolith. Deposition of Lunar Regolith Into the Lungs

Characterisation of lunar dust and understanding physico-chemical properties of lunar dust and LRSs capable of penetrating the lung has been extensively investigated (e.g. Park et al., 2006; Greenberg et al. 2007; Liu & Taylor, 2008; Park et al., 2008; Loftus et al., 2010; Suescan-Florez et al. 2015; McKay et al., 1991, 2015; Reynolds, 2019). Figure 9.2 shows schematically the depth into the bronchial and alveolar structures particles are likely to penetrate with the finest particles available to transit deep into the small airways.

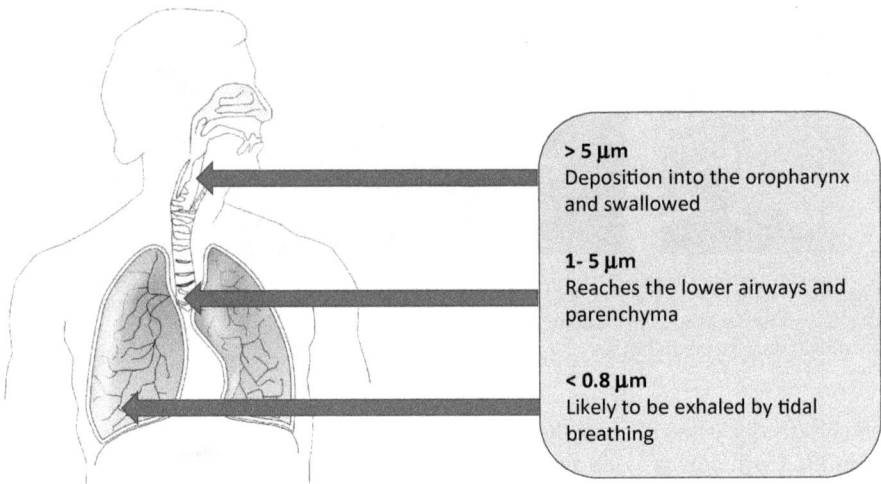

Fig. 9.2 Aerosol particle size determines the site of deposition in the lung

The mean size of regolith dust particles from the Apollo and Luna samples are between 60 and 80 μm although ultra-fine nanometre sized particles have been described. Lunar regolith contains approximately 20% dust (<20 μm diameter) and 1–2% of fine dust (Park et al., 2006, 2008) and it is estimated that approximately 10% of the very fine particulate component (<10 μm) is in the respirable range (Graf, 1993; McKay et al., 2015; Park et al., 2008; Reynolds, 2019). Figure 9.3 shows the particle diameter for lunar regolith and 3 LRSs where there appears a good correlation between the 4 samples of particle size distribution.

9.3   Setting Safe Exposures for Lunar Regolith

The Lunar Airborne Dust Toxicity Assessment Group (LADTAG) was established in 2005 (NASA, 2005, Tranfield et al. 2008). It set a starting point estimate for a range of safe exposure limits for 6 months of episodic, safe exposure from between 0.01 mg/m^3 and 3 mg/m^3 for respirable dust (<10 μm) and for short term exposure limits (1 h) that protect the crew from any acute effects from between 0.5 mg/m^3 to 10 mg/m^3 for respirable dust (<10 μm). LADTAG was tasked with designing experiments to assess the toxicity of the finest particles of lunar regolith, that is those that could or would be airbourne in the environment of a lunar module or a lunar habitat. Since 2005 many studies have been conducted in non-clinical in vitro and in vivo systems (e.g. Latch et al., 2007; Linnarsson 2012; Meyers et al., 2012; Krisanova et al., 2013; Lam et al., 2013; James et al., 2014; Horie et al., 2015; Sun et al. 2018a, b; Borisova, 2019; Caston et al., 2018; Sun et al., 2019) which has allowed the Non-Observable-Adverse-Effect Level (NOAEL) to be better constrained and

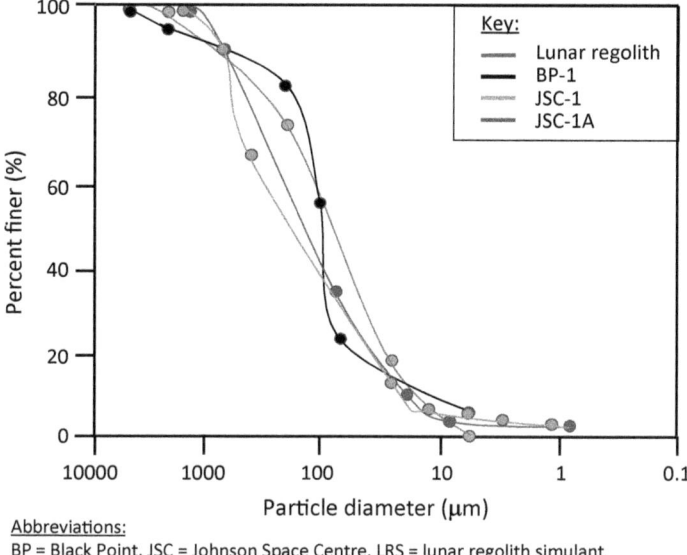

Fig. 9.3 Particle size distribution of lunar regolith and three LRSs (redrawn from data in Suescan-Florez et al., 2015)

aligned with the LADTAG proposed exposure limits. Perhaps remarkably, despite the fragile nature of the tissue, the human lung appears to perform well in partial gravity (Prisk, 2018) and although a number of studies show reduced deposition in zero or partial gravity, they suggest that fine particles (~1 μm in size) will be deposited in deeper peripheral regions of the airway. Thus it may be reasoned that the lung residence time will be longer and have potential to exert greater toxicological effects, dependent upon the nature of the particles. The next two sections will briefly review the findings of the non-clinical studies and set the in vivo studies conducted in several animal species in the context of a proposed NOAEL for humans.

9.4 Non-clinical in Vitro Studies

A number of studies have been conducted in a range of different cell types in vitro and selected studies (Caston et al., 2018; Horie et al., 2015; Krisanova et al., 2013; Latch et al., 2007; Meyers et al., 2012), are summarised in Table 9.1.

Studies predominantly used LRS, with the exception of one study (Meyers et al., 2012), which used Apollo 14 sample dust (see below) and the findings from the in vitro studies were used to set the dose for the in vivo study, thus minimising the use of lunar dust and the use of animals. Interestingly this study indicated that the dust was a mild irritant for cell growth, a finding borne out in non-clinical in vivo studies and reported as an irritant in the Apollo 16 mission which subsequently

Table 9.1 Toxicity studies: non-clinical in vitro

Regolith	Cell type	Organ	Study and principal findings	References
JSC-1	Human alveolar macrophage	Lung	Single dose LRS, MRS and reference dust treated cells for 24 h, cell viability and apoptosis measured. Involvement of dust particles with the scavenger receptor a mechanism for LRS induced toxicity	Latch et al. (2007)
Apollo 14 sample	Human stratified corneal epithelium	Eye	Single dose of dust applied to cell culture for 3, 30 and 60 min. Minimal effect on cell viability in an MTT colorimetric assay, dust classified as a mild irritant	Meyers et al. (2012)
JSC-1A	Rat brain synaptosomes	CNS	Single dose LRS treated cells for up to 10 min. Lunar dust effected nerve transmission increasing glutamate binding as a surrogate for activity	Krisanova et al. (2013)
MIN-U-SIL 5	Human A549 carcinoma	Lung	2 doses of dust, cells exposed for 24 h. Cell membrane damage, cytotoxicity, oxidative stress and inflammatory cytokines measured	Horie et al. (2015)
JSC-1A	Mouse (CAD) neuroblastoma	CNS	4 doses LRS treated cells for 1 h. Dose dependent cytotoxicity demonstrated in both proliferating and terminally differentiated cells	Caston et al. (2018)

Abbreviations

LRS lunar regolith simulant, *MRS* Martian regolith simulant, *MTT* 3-[4,5-dimethylthiazol-2-yl]2,5 diphenyltetrazolium bromide, *JSC* Johnson Space Centre, *h* hour, *d* day, *w* week, *mo* month, *SEE* safe exposure estimate, *NOAEL* no observable adverse effect level, *CNS* central nervous system, *CAD* cath a-differentiated resolved (Gaier, 2005). All other in vitro studies reported dose-dependent toxicity in cell types which represent the central nervous system (CNS) and the lung. Studies in cell types representing the CNS implied that nerve transmission may be negatively affected using glutamate binding as a surrogate biochemical marker, although no direct measurement of neuronal transmission in vivo has been reported as a consequence of this observation. Studies in cell types representative of lung tissue have informed us that measurement of lunar dust on lung function after acute exposure and lung injury after both acute and chronic exposure require careful evaluation in humans. The use of surrogate biomarkers such as exposomes (Schultz, 2019), which is the measure of all the exposures of an individual in a lifetime and how those exposures relate to health, and omics technologies which may predict early health effects is warranted (Cain, 2018). In part, this may be facilitated by the development and exploration of databases (Space Station Research Explorer, NASA, 2019) and GeneLab which is also available to the public (NASA GeneLab, 2020). GeneLab is an open repository that contains fully coordinated and curated experimental findings, raw data and meta data from omics studies of model organisms on the ISS in the Space Transportation Systems and the Russian animal capsule Bion-M1. In a later section I will discuss the potential development of early warning systems for both lung and CNS which could stimulate remedial action for the mission crew to minimise the hazard posed.

9.5 Non-clinical in Vivo Studies. Effect of Lunar Dust on Respiratory Function and Lung Injury and Determination of Safe Exposure Estimate

LADTAG scientists restricted their studies to using ~2.5 μm particle size samples from Apollo 14 termed sample 14,003,96, which was collected from the Fra Mauro Formation near Cone Crater in the Imbrian Basin near the lunar equator. This sample was used as it was present in sufficient quantity to allow a full physio-chemical evaluation (McKay et al., 2015). Table 9.2 summarises the results from animal studies.

Intratracheal instillation (Rask et al. 2013) and inhalation studies in rats (Lam et al., 2013) showed that the Apollo 14 sample was of intermediate toxicity when compare with titanium dusts (low toxicity) and quartz dusts (high toxicity) on measurements of neutrophil influx into the lung and standard histopathological measurements of lung damage such as inflammation, septal thickening, fibrosis and granulomas. Evaluation of all data from non-clinical in vivo studies and comparison with benchmarking data for reference dusts allowed a value for the safe exposure estimate (SEE) to be determined using a number of extrapolation factors. These are rat-to-human factor of 3, a time adjustment factor of 1.33 which accounts for the difference between the 6 h inhalation study exposure (Lam et al., 2013) and an 8 h working day and further factor of 6 to extrapolate from 1 month animal exposure to a potential 6 months

Table 9.2 Toxicity studies: non-clinical in vivo

Regolith	Species	Organ	Study and principal findings	References
Apollo 14 sample	Rabbit	Eye	1 dose of dust. Animals monitored from 1 to 72 h after dosing. Minor irritation of upper eyelids after 1 h which resolved over 24 h	Meyers et al. (2012)
Apollo 14 sample	Rat	Lung	3 doses of 5 different dusts (3 lunar, 2 reference). Animals dosed and assay 1 w and 1 mo after instillation. Biomarkers of toxicity and infiltrating cell count set an SEE of 0.5–1.0 mg/m^3	Rask et al. (2013), James et al. (2013a, b)
Apollo 14 sample	Rat	Lung	4 doses of lunar dust given by inhalation for 7 d. Dose dependent histopathology conducted at various times after euthanasia, NOAEL established as 6.8 mg m^3 and a minimum SEE of 0.3 mg/m^3	Lam et al. (2013)
CAS-1	Rat	Lung	2 doses of dust provided by tracheal perfusion. Animals sacrificed at +4 and +24 h post perfusion. Biochemical and histopathological markers indicated dose-dependent inflammation and acute lung injury	Sun et al. (2018a)
CAS-1	Rat	Lung	Identical study design (Sun et al. 2018a). Biomarkers of the immune response and oxidative stress were increased and acted in concert to induce lung injury	Sun et al. (2018b)
CLDS-i	Rat	Heart	3 doses of simulant given by instillation 2 × per w for 3 w. Dose dependent increase in SBP, decrease in HR and abnormal ECG types	Sun et al. (2019)

Abbreviations
h hour, *d* day, *w* week, *mo* month, *SEE* safe exposure estimate, *NOAEL* no observable adverse effect level, *CAS* Chinese Academy of Sciences, *CLDS-i* Chinese lunar dust simulant, *SBP* systolic blood pressure, *HR* heart rate, *ECG* electrocardiogram

human exposure. Taken together the NOAEL was found to lie within the range of 0.5–1.0 mg/m^3 for a mission of 6 months duration (James et al. 2013a, b; James et al., 2014; Scully et al., 2013), whereas a further study estimated a NOAEL exposure limit as 0.4 mg/m^3 for a 6-month mission (Lam et al., 2013). It may be helpful to set these values in context. The United States Occupational Safety and Health Administration has set a permissible exposure limit for respirable fused silica dust of 5 mg/m^3 during an 8 h working shift, recommending that the airborne concentration for respirable particles be kept below 3 mg/m^3 (respirable particles) and below 10 mg/m^3 (inhalable particles) for insoluble particles of low toxicity for which no threshold limit value has been established (United States Department of Labor, Occupational Safety and Health Administration, 2020).

Limitations in the current state of knowledge for lunar dust were highlighted (James et al., 2014) and include the surface reactivity of the dust which may be different in situ due to the presence of radiation from solar wind, solar flares and cosmic sources (Loftus et al., 2010). Secondly, it was acknowledged that the variability of the lunar regolith composition may produce different results and that sub-surface dust will likely be a potential confounder due to its potential different physico-chemical properties. Thirdly, acute exposure to a high level of dust over a short period of time has not been explored in detail and lastly, further work may need to be conducted on the role gravity plays in attenuating the site of deposition of the particles in the respirable range (Prisk et al. 2018).

Since the final report on lunar dust toxicity from the LADTAG group was published (James et al., 2014), there have been several further studies conducted in the rat, addressing lung injury with the LRS CAS-1 (Sun et al., 2018a, b), previously described (Zheng et al., 2009). In the first study (Sun et al., 2018a), biochemical markers observed in the bronchoalveolar fluid and pathological changes in the lung were characteristic of bronchial inflammation, neutrophil influx and septal thickening and the severity worsened with dose and time, when compared with the control, although the study was conducted with 2 doses of dust and ran for 24 h only. In the second study (Sun et al. 2018b), elevation of cytokines characteristic of the inflammatory immune response (tumour necrosis factor-α and interleukin-6) increased significantly after the 24 h but not at the 4 h timepoint and in only in the high dose dust group. Although no NOAEL or SEE dose was provided from these studies, they play an important contribution for our understanding as they use an LRS, which assists in further characterisation of tool reagents and is essential given the scarce availability of authentic lunar dust.

9.6   Effect of Lunar Dust on Cardiovascular Function

The final study referred to relates to effects of a lunar dust simulant CLDS-I (Tang et al., 2016) on cardiac function in rats who received 3 different doses (low, medium and high) of dust simulant via intratracheal administration over a 3 week period (Sun et al., 2019). In addition to the measurement of cellular, biochemical and gene expression changes suggestive of systemic inflammatory injury, cardiovascular parameters which included systolic blood pressure (SBP), heart rate (HR), heart rate variability and electrocardiogram (ECG) signals were collected and analysed. An elevated SBP and decreased HR relative to the control group was noted in rats that received the middle and high dose of CLDS-i. Rats in the middle and high dose CLDS-i groups showed abnormal ECG signals; premature ventricular contractions and ST-segment elevation which may suggest myocardial ischemia in the medium dose group and extremely abnormal discharge patterns, irregular sinus rhythm, ventricular fibrillation and ST-segment elevation in the animals which received the high dose of CLDS-i. These animals had signals which suggested supraventricular tachycardia with differential conduction, which could be very serious for humans if this was experienced in space flight or within a lunar habitat.

9.7 Astronaut Reported Events

Throughout the Apollo missions of the late 1960s and early 1970s, the presence of contaminating dust in the LM and command module (CM) imported after extravehicular activities (EVAs) has been reported as a major issue (Bean 1970, Gaier, 2005; Phillips, 2006). Although the presence of dust appears to have been less of an issue for the Apollo 14 mission (Gaier, 2005), all astronauts reported smelling dust as a consequence of inhalation and irritation to the respiratory tracts and eyes, described in Table 9.3. Jack Schmitt from the Apollo 17 missions was particularly badly affected by a severe allergic response akin to hay fever (Phillips, 2006), possibly sensitised to respond. Taken together, the effect of dust is clearly documented and no astronauts reported long term detrimental effects on their health. However, the length of the missions needs to be taken into account as the Apollo missions were of short duration and longer sustained exposure to lunar dust during semi-permanent or permanent settlement on the Moon will require setting safety exposure limits well within maximum doses predicted by short term non-clinical studies.

9.8 Management and Mitigation

On the Apollo missions, management of dust contamination was achieved by the use of a vacuum cleaner, air filtration and the use of water in wiping down clothing and instrumentation (Gaier, 2005; Wagner, 2006). In instances where EVAs had been performed, contamination of the LM was reduced by stamping boots on the entry ladder. There have been many reports on dust mitigation strategies from both National (e.g. Taylor, 2000, Taylor et al., 2005, Wagner, 2006, Hyatt & Feighery, 2007, Hyatt & Deluane, 2008, Cain, 2010, Cloutis et al., 2012, Kruzelecky et al., 2012, Creed, 2017)

Table 9.3 Astronaut reported sensory events associated with lunar dust

Date and mission	Astronaut reported event	References	Future mitigation	References
1969, Apollo 11	Contact with skin—crew dirty after removal of helmet, overshoes and gloves	Gaier (2005)	Removal of dust from space suits – potential requirement for additional airlock More efficient venting to space and wash down of lunar habitat Development of next generation dust removal systems Exploit magnetic properties of finest dust to remove from habitat Electrostatic neutralisers Self-cleaning space suit	Taylor (2000) Talyor et al. (2005) Hyatt and Feighery (2007) Curtis et al. (2009) Gaier et al. (2011) Calle et al. (2013) Manyapu (2017) Creed (2017) Lu et al. (2019) Farr et al. (2020)
	Pungent odour of lunar material			
1969/70, Apollo 12	Dust carried into lunar module leading to breathing difficulties and impaired vision	Bean et al., 1970)		
1971, Apollo 15	Smell of gun powder	Gaier (2007)		
1972, Apollo 16	Dust in the eyes. Short term irritant which resolved over time			
1973, Apollo 17	Eye and throat irritation, swollen turbinates (cartilage plates in walls of nasal chambers), hay fever symptoms for ~2 h. Astronaut may have been sensitised	Phillips (2006), Gaier (2007)		

and International working groups (e.g. NASA, 2005, International Agency Working Group, 2016, NASA, 2016, Gaier et al. 2017) and some mitigation measures are further shown in Fig. 9.4.

The management and mitigation of exposure to lunar dust is essential to prevent serious health effects; the application of the principles of astronautical hygiene can be utilised to determine measures to prevent the import of dust into the environment and by monitoring the health risks (Cain, 2011, 2014; Khan-Mayberry et al., 2011) which are summarised. Some of the many methods which may have potential for mitigation of dust adherence include design modification of surfaces (Gaier et al., 2011, 2017), exploitation of ferromagnetic properties (Taylor, 2000; Taylor et al., 2005), carbon dioxide showers at airlocks and electrostatic curtains (Hyatt & Feighery, 2007), and SPARCLE an electrostatic tool (Curtis et al., 2009), development of electrodynamic dust shields (Calle et al., 2013), a self-cleaning space suit (Manyapu, 2017), a dust

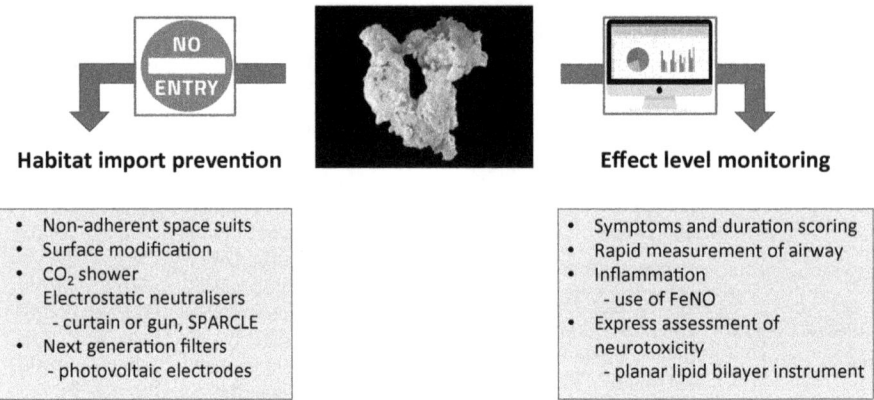

Habitat import prevention

- Non-adherent space suits
- Surface modification
- CO_2 shower
- Electrostatic neutralisers
  - curtain or gun, SPARCLE
- Next generation filters
  - photovoltaic electrodes

Effect level monitoring

- Symptoms and duration scoring
- Rapid measurement of airway
- Inflammation
  - use of FeNO
- Express assessment of neurotoxicity
  - planar lipid bilayer instrument

Fig. 9.4 Potential measures to mitigate the effects of lunar dust. Image of dust particle credit ESA

removal system using photovoltaic electrodes (Lu et al., 2019) and very recently, a proposed electron beam to neutralise electrostatic charge (Farr et al., 2020).

9.9 Early Warning Systems

Once the dust is inside the habitat, it may be possible to deploy one or more 'early warning systems' which may serve to assess the level of threat and to initiate any downstream remedial measures which may include confinement, treatment of the occupant with medicines for symptomatic relief, or in the worst case scenario, trigger the need for a rescue mission to bring the colonist back to Earth. Several examples are relevant for early detection and the first makes use of medical science and technology that is widely used on Earth for patients with asthma, which is the measurement of exhaled nitric oxide as a surrogate marker for lung inflammation (NICE, 2014). Fractional exhaled nitric oxide in exhaled breath (FeNO) was the first useful non-invasive marker of airway inflammation in asthma (Wadsworth et al., 2011) and remains the most widely used today. The non-invasive nature and the relatively easy use of the technique using internationally standardised equipment make it a useful tool to monitor airway inflammation and rationalize medical therapy in asthmatic patients, together with the traditional clinical tools of patient history, physical examination and tests of lung function. The effects of both microgravity and hypergravity on the production of FeNO have been published (Karlsson et al., 2009) and so there is the need to establish an accurate baseline under different gravity conditions prior to extrapolating the effect of dust as a proinflammatory agent. FeNO measurement studies have been conducted on the International Space Station (ISS) inside the Quest airlock (ESA, 2015). The use of this technology has recently been updated (NASA, 2019) and the full publication of scientific results, together with benchmark data under different gravity conditions is awaited. The second example of an early warning system relates to the rapid assessment of dust particles as neurotoxic agents using a planar lipid bilayer technique (Borisova, 2019). It has been known for some time that nanoparticles have been found in the human brain where the transport route is likely through the olfactory bulb (Maher et al., 2016) and along the sensory axons of the olfactory nerve directly to the CNS (Oberdörster et al., 2009). Consistent with the in vitro studies described in an earlier section (Krisanova et al., 2013), this presents both a potential threat but also an opportunity for nanomedicine and terrestrial based technology. The ability of particles to affect membrane potential, integrity of nerve terminals and consequently nerve transmission may be determined by monitoring artificial membrane conductivity using a planar lipid bilayer technique. Although extensive validation is required, the development of the technology is early stage, results from planar lipid bilayer measurements might supplement data obtained by other techniques. Given the global need to improve air quality, especially in highly urbanised areas, the need to generate a system suitable for adaptation to the ISS in the first instance and eventually lunar habitats may play a vital role in mitigating the toxicity assessments of particulate matter.

9.10 Summary

In this chapter, the nature of the dust component of lunar regolith and several LRSs have been discussed to help us seek an understanding of the potential toxicological effects on astronaut and colonist health and what mitigation measures may be taken to manage the risk. To date there have been no studies conducted with astronauts in space flight and the reports of events although anecdotal, have been consistent throughout documents assembled from the Apollo missions (Wagner, 2006). As lunar dust is more chemically reactive (Loftus et al., 2010), has jagged edges, is highly abrasive and has a larger surface area when compared with Earth dust (James et al., 2014; Linnarsson et al., 2012), deposition in the lungs may cause respiratory disease and exposure to the skin and eyes may cause ocular irritation. Long-term exposure to the dust may cause a more serious respiratory disease similar to silicosis. Indeed a number of benchmarking studies shown previously in Table 9.2 (Rask et al. 2013; James et al. 2013a, b) and elsewhere (Scully et al., 2013) have used dusts of titanium and quartz as reference dusts for low and high toxicity respectively.

Previous studies have suggested that occupational and environmental agents may contribute to the etiology of idiopathic pulmonary fibrosis and amongst others stone, sand, or metal dust (Britton & Hubbard, 2000). It would seem appropriate, given the lack of knowledge on both short and certainly long term effects of lunar dust that contact with it is minimised if not eliminated during lunar colonisation.

Non-clinical studies have served to demonstrate the toxicological effect of lunar dust in several species and a value for an SEE has been generated from all accumulated data to serve a minimum level for which standards are to be reached. Notwithstanding these challenges, substantial progress has been made in developing technology aimed at both restricting dust entry into habitats and potentially to measuring using surrogate markers the effect of lunar dust at least on lung inflammation, with the potential to explore neurotoxic effects to come. Given the importance of dust for both the viability and success of future missions including plans for lunar colonisation, NASA has made a call under its annual Breakthrough, Innovative and Game-changing (BIG) Ideas Challenge (NASA, 2020) for undergraduate and graduate students to create and implement designs for dust mitigation or dust-tolerant technologies intended for lunar applications.

Said Jack Schmitt at a Lunar Dust Symposium in 2004 (Wagner, 2006): *'A common sense, layered, engineering design defence can solve any apparent problem with dust during long-term human activity and human habitation in the lunar environment'.*

References

Bean, A., Conrad Jr., C., & Gordon, R. F. (1970). Crew Observations in Apollo 12 Preliminary Science Report, NASA SP-235; 29–38.
Bender, B. (2020). NASA gets a budget but Congress cuts moon effort. Retrieved December 23, 2020, from https://www.politico.com/newsletters/politico-space.
Borisova, T. (2019). Express assessment of neurotoxicity of particles of planetary and interstellar dust. *Npj Micorgravity, 5*, 2. https://doi.org/10.1038/s41526-019-0062-7
Britton, J., & Hubbard, R. (2000). Recent advances in the aetiology of cryptogenic fibrosing alveolitis. *Histopathology, 37*, 387–392.
Cain, J. R. (2010). Lunar dust: The hazard and exposure risks. *Earth, Moon, and Planets, 107*, 107–125.
Cain, J. R. (2011). Astronautical hygiene—A new discipline to protect the health of astronauts working in space. *JBIS, 64*, 179–185.
Cain, J. R. (2014). Astronaut health—Planetary exploration and the limitations on freedom. In: C. S. Cockell (Ed.), *The meaning of liberty beyond Earth*. New York: Springer.
Cain, J. R. (2018). The use of exposomes to assess astronaut health. *JBIS, 71*, 112–116.
Calle, C. I., Mackey, P. J., Hogue, M. D., Johansen, M. R., Yim, H., Delaune, P. B., & Clements, J. S. (2013). Electrodynamic dust shields on the International Space Station: Exposure to the space environment. *Journal of Electrostatics, 71*, 257–259.
Caston, R., Luc, K., Hendrix, D., Hurowitz, J. A., & Demple, B. (2018). Assessing toxicity and nuclear and mitochondrial DNA damage caused by exposure of mammalian cells to lunar regolith simulants. *GeoHealth, 2*, 139–148.
Cloutis, E., Rosca, J. D., Hoa, S. V., Ellery, A., Martel, S., & Jiang, X. X. (2012). Project MoonDust: Characterisation and mitigation of lunar dust. Am. Inst. Aeronaut. Astronaut. https://doi.org/10.2514/6.2011-5184.
Creed, R. (2017). Coping with dust for extraterrestrial exploration. Lunar Exploration Analysis Group (LPI Contrib. No. 2041).
Curtis, S. A., Clark, P. E., Minetto, F. A., Calle, C. J., Keller, J., & Moore, M. (2009). SPARCLE: creating an electrostatically based tool for lunar dust control. In *40th Lunar and Planetary Science Conference*, abstract 1128.
ESA. (2015). Testing astronauts' lungs in Space Station airlock. Retrieved October 11, 2020, from, http://www.esa.int/Science_Exploration/Human_and_Robotic_Exploration/Futura/Testing_astronauts_lungs_in_Space_Station_airlock.
Farr, B., Wang, X., Goree, J., Hahn, I., Israelsson, U., & Horanyi, M. (2020). Dust mitigation technology for lunar exploration utilizing an electron beam. *Acta Astronautica, 177*, 405–409.
Gaier, J. R. (2005). The effects of lunar dust on EVA systems during the Apollo missions. NASA/TM—2005-213610 Glenn Research Centre, Cleveland.

Gaier J. R., & Greel, R. A. (2005). The effects of lunar dust in advanced EVA systems: Lessons from Apollo. Lunar regolith simulant materials workshop, NASA MF 21.

Gaier, J. R., Waters, D. L., Misconin, R. M., Banks, B. A., Crowder, M. (2011). Evaluation of surface modification as a lunar dust mitigation strategy for thermal control surfaces. In *NASA/TM 2011-217230, 41st International Conference on Environmental Systems*.

Gaier, J. R., Vangen, S., Abel, P., Agui, J., Buffington, J., Calle, C., Mary, N., et al. (2017). International space exploration coordination group assessment of technology gaps for dust mitigation for the global exploration roadmap. https://ntrs.nasa.gov/archive/nasa/casi.ntrs.nasa.gov/201700 01575.pdf.

Graf, J. C. (1993). Lunar soil size catalog. NASA RP-1265.

Greenberg, P. S., Chen, D.-R., & Smith, S.A. (2007). Aerosol measurements of the fine and ultrafine particle content of lunar regolith. NASA Technical Reports Server available at https://ntrs.nasa.gov/archve/nasa/casi.ntrs.nasa.gov/20070.

Horie, M., Miki, T., Honma, Y., Aoki, S., & Morimoto, Y. (2015). Evaluation of cellular effects caused by lunar regolith simulant including fine particle. *Journal of UOEH, 37*, 139–148.

Hyatt, M. J., & Feighery, J. (2007). Lunar dust: Characterisation and mitigation. NASA. https://ntrs.nasa.gov/archive/nasa/casi.ntrs.nasa.gov/20080005580.pdf.

Hyatt, M. J., & Deluane, P. (2008). Lunar dust mitigation—Technology development. Space Technology and Applications International Forum. NASA.

International Agency Working Group. (2016). Dust mitigation gap assessment report. Retrieved October 8, 2020, from https://www.globalspaceexploration.org/wordpress/docs/Dust%20Mitigat ion%20Gap%20Assessment%20Report.pdf.

James, J. T., Lam, C.-W., Santana, P. A., & Scully, R. R. (2013a). Estimate of safe human exposure levels for lunar dust based on comparative benchmark dose modelling. *Inhalation Toxicology, 25*, 243–256.

James, J. T., Lam, C., & Scully, R. R. (2013b). Comparative benchmark dose modelling as a tool to make the first estimate of safe human exposure levels to lunar dust. NASA Technical Reports Server, 20130012803. Retrieved December 24, 2020, from https://archive.org/details/NASA_N TRS_Archive_20130012803.

James, J. T., Lam, C.-W., Scully, R. R., Meyers, V. E., & McCoy, J. T. (2014). Lunar dust toxicity: final report. NASA. humanresearchroadmap.nasa.gov/gaps/closureDocumentation/ Lunar%20Dust%20Toxicity%20FINAL%20REPORT.pdf.

Karlsson, L. L., Kerckx, Y., Gustafsson, L. E., Hemmingsson, T. E., & Linnarsson, D. (2009). Microgravity decreases and hypergravity increases exhaled nitric oxide. *Journal of Applied Physiology, 107*, 1431–1437.

Khan-Mayberry, N., James, J. T., Tyl, R., & Lam, C. W. (2011). Space toxicology: Protecting human health during space operations. *International Journal of Toxicology, 30*, 3–18.

Krisanova, N., Kasatkina, L., Sivko, R., Borysov, A., Nazarova, A., Slenzka, K., & Borisova, T. (2013). Neurotoxic potential of Lunar and Martian dust: Influence on Em, proton gradient, active transport and binding of glutamate in rat brain nerve terminals. *Astrobiology, 13*, 679–692.

Kruzelecky, R. V., Brahim, A., Wong, B., Haddad, E., Jamroz, W., Cloutis, E., Therriault, D., et al. (2012). Project MoonDust: characterisation and mitigation of lunar dust. In *41st International Conference on Environmental Systems*, 2011, Portland, Oregon. https://doi.org/10.2514/6.2011-5184.

Lam, C.-W., Scully, R. R., Zhang, Y., Renne, R. R., Hunter, R. L., McCluskey, R. A., Chen, B. T., Castranova, V., Driscoli, K. E., Gardner, D. E., McClennan, R. O., Cooper, B. L., McKay, D. S., Marshall, L., & James, J. T. (2013). Toxicity of lunar dust assessed in inhalation-exposed rats. *Inhalation Toxicology, 25*, 661–678.

Latch, J. N., Hamilton, R. F., Jr., Holian, A., James, J. T., & Lam, C.-W. (2007). Toxicity of lunar and Martian dust simulants to alveolar macrophages isolated from human volunteers. *Inhalation Toxicology, 20*, 157–165.

Linnarsson, D., Carpenter, J., Fubini, B., Gerde, P., Karlsson, L. L., Loftus, D. J., Prisk, G. K., Tranfield, E. M., & van Westrened, W. (2012). Toxicity of lunar dust. *Planetary & Space Science, 74,* 57–71.

Liu, Y., & Taylor, L. A. (2008). Lunar dust: Chemistry and physical properties and implications for toxicity. In *NLSI Lunar Science Conference* 2072.

Loftus, D. J., Tranfield, E. M., Rask, J. C., & McCrossin, C. (2010). The chemical reactivity of lunar dust relevant to human exploration of the Moon. *Earth, Moon, and Planets, 107,* 95–105.

Lu, Y., Jiang, J., Yan, X., & Wang, L. (2019). A new photovoltaic lunar dust removal technique based on the coplanar bipolar electrodes. *Smart Materials and Structures, 28,.* https://doi.org/10.1088/1361-665X/ab28da

Maher, B. A., Ahmed, I. A. M., Karloukovski, V., MacLaren, D. A., Foulds, P. G., Allsop, D., Mann, D. M. A., Torres-Jardon, R., & Calderon-Garciduenas, L. (2016). Magnetite pollution nanoparticles in the human brain. *Proceedings of the National Academy of Sciences of the United States of America, 113,* 10797–10801.

Manyapu, K. K. (2017). Spacesuit integrated carbon nanotube dust mitigation system for Lunar exploration. Theses and Dissertations 2278. https://commons.und.edu/theses/2278.

McKay, D. S., Heiken, G., Basu, A., Blanford, G., Simon, S., Reedy, R. French, B., & Papike, J. (1991). The lunar regolith. In G. Heiken, D. Vaniman, & B. French (Eds.), *Lunar sourcebook.* Cambridge University Press.

McKay, D. S., Cooper, B. L., Taylor, L. A., James, J. T., Thomas-Keprta, K., Pieters, C. M., Wentworth, S. J., Wallace, W. T., & Lee, T. S. (2015). Physicochemical properties of respirable-size lunar dust. *Acta Astronautica, 107,* 163–176.

Meyers, V. E., Garcia, H. D., Monds, K., Cooper, B. L., & James, J. T. (2012). Ocular toxicity of authentic lunar dust. *BMC Opthalmology, 12,* 26.

NASA GeneLab. (2020). Retrieved December 24, 2020, from https://genelab.nasa.gov/.

NASA. (2005). Lunar Airborne Dust Toxicity Advisory Group (LADTAG). Consensus opinions and recommendations. Retrieved October 8, 2020, from https://www.nasa.gov/centers/johnson/pdf/486003main_LADTAG15Sep05MtgMinutes.pdf.

NASA. (2016). International Space Exploration Coordination Group Assessment of Technology Gaps for Dust Mitigation for the Global Exploration Roadmap. Retrieved October 8, 2020, from https://ntrs.nasa.gov/citations/20170003926.

NASA. (2019). Understanding asthma from space. Retrieved October 11, 2020, from https://www.nasa.gov/mission_pages/station/research/news/b4h-3rd/hh-understanding-asthma-from-space.

NASA. (2020). The 2021 Big idea challenge: Dust mitigation technologies for lunar applications. Retrieved October 11, 2020, from http://bigidea.nianet.org/.

NICE. (2014). Measuring fractional exhaled nitric oxide concentration in asthma: NIOX MINO, NIOX VERO and NObreath. Retrieved October 11, 2020, from https://www.nice.org.uk/guidance/dg12/resources/measuring-fractional-exhaled-nitric-oxide-concentration-in-asthma-niox-mino-niox-vero-and-nobreath-pdf-1053626430661.

Oberdörster, G., Elder, A., & Rinderknecht, A. (2009). Nanoparticles and the brain: Cause for concern? *Journal of Nanoscience and Nanotechnology, 9,* 4996–5007.

Park, Y., Liu, K., Kihm, D., & Taylor, L. A. (2006). Micro-morphology and toxicological effects of lunar dust. In *37th Annual Lunar and Planetary Science Conference,* abstract 2193.

Park, J., Liu, Y., Kihm, K. D., & Taylor, L. A. (2008). Characterisation of lunar dust for toxicological studies. I: Particle size distribution. *Journal of Aerospace Engineering, 21,* 266–271.

Phillips, T. (2006). Apollo chronicles: the mysterious smell of moondust. NASA. Retrieved October 5, 2020, from https://www.nasa.gov/exploration/home/30jan_smellofmoondust.html.

Prisk, G. K. (2018). Effects of partial gravity on the function and particle handling of the human lung. *Current Pathobiology Reports, 6,* 159–166.

Rask, J., (2013). The chemical reactivity of lunar dust influences its biological effect in the lungs. In *Lunar and Planetary Science Conference,* p. 3062.

Reynolds, R.J (2019). Human health in the lunar environment. In Y. H. Chemin (Ed.), *Lunar science.* https://doi.org/10.5772/intechopen.84352.

Schultz, I. R., Cade, S., & Kuo, L. J. (2019). The dust exposome. In S. Dagnino & A. Macherone (Eds.), *Unraveling the exposome—A practical view* (pp. 247–254). Springer Publishers.

Scully, R. R., Lam, C. W., & James, J. T. (2013). Estimating safe exposure levels for lunar dust using benchmark dose modelling of data from inhalation studies in rats. *Inhalation Toxicology, 25*, 785–793.

Space Station Research Explorer. (2020). Retrieved December 24, 2020, from https://www.nasa.gov/mission_pages/station/research/experiments/explorer/.

Stubbs, T. J., Vondrak, R. R., & Farrell, W. M. (2007). Impact of dust on lunar exploration. Goddard Spaceflight Center. http://helf.jsc.nasa.gov/files/StubbsImpactOn Exploration.4075.pdf.

Suescan-Florez, E., Roslyakov, S., Iskander, M., & Baamer, M. (2015). Geotechnical properties of BP-1 lunar regolith simulant. *Journal of Aerospace Engineering, 28*,. https://doi.org/10.1061/(ASCE)AS.1943-5525.0000462

Sun, Y., Liu, J.-G., Zheng, Y.-C., Xiao, C.-L., Wan, B., Guo, L., Wang, X.-G., & Bo, W. (2018a). Research on rat's pulmonary acute injury induced by lunar soil simulant. *Journal of Chinese Medical Association, 81*, 133–140.

Sun, Y., Liu, J. G., Kong, Y. D., Sen, H. J., & Ping, Z. X. (2018b). Effects of lunar soil simulant on systemic oxidative stress and immune response in acute rat lung injury. *International Journal of Pharmacology, 14*, 766–772.

Sun, Y., Zhang, L., Liu, J., Zhang, X., Su, Y., Yin, Q., & He, S. (2019). Effects of lunar dust simulant on cardiac function and fibrosis in rat. *Toxicology Research, 8*, 499–508.

Tang, H., Li, X., Zhang, S., Wang, S., Liu, J., Li, S., Li, Y., & Lv, Z. J. (2016). A lunar dust simulant: CLDS-I. *Advances in Space Research, 59*, 1156–1160.

Taylor, L. A. (2000). The lunar dust problem: A possible remedy. In *Proceedings of Space Resources Roundtable II* (p. 71).

Taylor, L. A. Schmitt, H. H., Carrier III, W. D., & Nakagawa, M. (2005). The lunar dust problem: from liability to asset. In *1st Space Exploration Conference: Continuing the Voyage of Discovery*, Orlando, Florida, United States (pp. 71–78).

Tranfield, E., Rask, J. C., Wallace, W. T., Taylor, L., Kerschmann, R., James, J.T., Khan-Mayberry, N., & Loftus, D. J. (2008). Lunar airborne toxicity advisory group (LADTAG) research working group (RWG). In *NLSI Lunar Science Conference*, abstract 2125.

United States Department of Labor, Occupational Safety and Health Administration. (2020). OSHA Occupational Chemical Database. Silica, fused, respirable dust. Retrieved December 24, 2020, from https://www.ohsa.gov/chemicaldata/chemResult.html?recNo=442.

Wadsworth, S. J., Sin, D. D., Dorscheid, D., & R, . (2011). Clinical update on the use of biomarkers of airway inflammation in the management of asthma. *Journal of Asthma and Allergy, 4*, 77–86.

Wagner, S. A. (2006). The Apollo experience lessons learned for constellation dust management. NASA/TP-2006-213726.

Zheng, Y., Wang, S., Ziyuan, O., Yongliao, Z., Jianzhong, L., Xiongyao, L., & Junming, F. (2009). CAS-1 lunar soil simulant. *Advances in Space Research, 43*, 448–454.

Part IV
Social, Political, and Legal Frameworks for Lunar Settlement

Chapter 10
International Treaties and the Future of the Nation-State Concept on the Moon

José Antonio Jurado Ripoll

Abstract Public International Law has included outer space since the 1960s as an "object" of regulation. The Outer Space Treaty (1966) states that the use of outer space, including the Moon "shall be carried out for the benefit and in the interests of all countries", underlining the commitment of all parties to make use of the Moon "exclusively" for peaceful purposes. This and other legal instruments address issues such as the possibility of establishing manned and unmanned lunar-stations, whether this placement would create a right of ownership over the surface and resources or not, adoption of measures to safeguard the life of persons on the Moon, *inter alia*. But not all (space-faring) States are parties to all agreements. This is one of the main obstacles of this regulatory regime, concerning the Moon and other celestial bodies. The existence of a stable lunar community may reinforce the old concept of the Nation-State on Earth.

10.1 Introduction

When I knew the title of this volume, it was clear to me that it was a book about law: if the "human factor" includes at least two people or one person relating to one thing (assuming the existence of other people somewhere), the need for regulation arises.

Rules exist to regulate the relationship between people; it is true that rules also govern "relations" between people and things, but only as far as other people's interests might be affected.

If we consider a lunar settlement, it is easy to understand that a proper legal system is required for it to be established, to operate, to solve possible conflicts and, if necessary, to be extinguished or modified. But this legal system will depend on the type of settlement we are talking about.

J. A. Jurado Ripoll (✉)
Málaga, Spain
e-mail: jurado-ripoll@gmx.es

© The Author(s), under exclusive license to Springer Nature Switzerland AG 2021
M. B. Rappaport and K. Szocik (eds.), *The Human Factor in the Settlement of the Moon*, Space and Society, https://doi.org/10.1007/978-3-030-81388-8_10

10.2   International Treaties

Outer space (including the Moon) has been an "object" of regulation by Public International Law ever since human activity in outer space became the scene of opposing political, geostrategic or economic interests.

The basic rules in space law are contained in the *international treaties on outer space* issued within the United Nations (UN):

- The "Outer Space Treaty" (OST): Treaty on Principles Governing the Activities of States in the Exploration and Use of Outer Space, including the Moon and Other Celestial Bodies, UN General Assembly Resolution 2222 (XXI) of 19 December 1966.
- The "Rescue Agreement": Agreement on the Rescue of Astronauts, the Return of Astronauts and the Return of Objects Launched into Outer Space, UN General Assembly Resolution 2345 (XXII) of 19 December 1967.
- The "Liability Convention": Convention on International Liability for Damage Caused by Space Objects, UN General Assembly Resolution 2777 (XXVI) of 29 November 1971.
- The "Registration Convention": Convention on Registration of Objects Launched into Outer Space, UN General Assembly Resolution 3235 (XXIX) of 12 November 1974.
- The "Moon Agreement" (or Treaty): Agreement Governing the Activities of States on the Moon and Other Celestial Bodies, UN General Assembly Resolution 34/68 of 5 December 1979.

The status and signatories of each treaty can be checked in the annual report of United Nations Office for Outer Space Affairs (UNOOSA, 2020).

Additional UN General Assembly resolutions are also noteworthy, including:

- The "Declaration of Legal Principles Governing the Activities of States in the Exploration and Uses of Outer Space", Resolution 1962 (XVIII) of 13 December 1963, as a basis of the OST.
- The "Broadcasting Principles", Principles Governing the Use by States of Artificial Earth Satellites for International Direct Television Broadcasting, Resolution 37/92, of 10 December 1982.
- The "Remote Sensing Principles", Principles Relating to Remote Sensing of the Earth from Outer Space, Resolution 41/65, of 3 December 1986.
- The "Nuclear Power Sources Principles", Principles Relevant to the Use of Nuclear Power Sources in Outer Space, Resolution 47/68, of 14 December 1992.
- The "Declaration on International Cooperation in the Exploration and Use of Outer Space for the Benefit and in the Interest of All States, Taking into Particular Account the Needs of Developing Countries", Resolution 51/122, of 13 December 1996.

In addition, the sources of space law are supplemented by bilateral or multilateral treaties that may directly or indirectly affect the subject (including some issues of

Public and Private International Law), by the rules of each international organisation where applicable (e.g. European Union law) and by the domestic law of each State. Within the scope of its competences, the reports and documents of UN Committee on the Peaceful Uses of Outer Space (COPUOS) are of interest.

As regards a lunar settlement, I shall refer especially to the OST and to the Moon Treaty, notwithstanding any provisions in other treaties that may be applicable according to circumstances, and notwithstanding any other rules or documents that may be applicable (for example, in the context of COPUOS: the "UN Guidelines for the Long-term Sustainability of Outer Space Activities" adopted by the COPUOS, 2019).

The OST has been accepted by more than a hundred States; it provides the basic framework on international space law and general legal basis for the peaceful uses of outer space (Gutiérrez Espada, 1999).

The first Article of the treaty states: "The exploration and use of outer space, including the Moon and other celestial bodies, shall be carried out for the benefit and in the interests of all countries, irrespective of their degree of economic or scientific development, and shall be the province of all mankind."

As a logical consequence of this initial statement:

- Astronauts shall be regarded as envoys of mankind (Article V).
- The exploration and use of outer space may be carried out "by all States without discrimination of any kind, on a basis of equality" (Article I[2]), although of course in practice the technological capacities of different States result in a natural discrimination among them in this field.
- To implement the above-mentioned exploration and use, including scientific research, the international cooperation is required (Article I[3] and especially Article IX, which establishes the "principle of cooperation and mutual assistance", also incidentally reflected in Articles III, V, X, XI, XII).
- If exploration and use of outer space is to be made through international cooperation and for the benefit of all countries, it is reasonable that it should always be done in accordance with international law (Articles I[2] and III). This declaration implies the affirmation that space law is part of international law and the former must be applied within the general framework of the latter: "Space law is not a self-contained regime" (Marchisio, 2018).
- As the exploration and use of the Moon and celestial bodies is for the benefit of all mankind, it must be made without harmful contamination (Article IX), and exclusively for peaceful purposes (Articles III and IV). This implies that States are not allowed to place nuclear weapons or other weapons of mass destruction, and that the establishment of "military bases, installations and fortifications" is forbidden. However, this does not prevent "The use of military personnel for scientific research or for any other peaceful purposes" (see in this volume Stewart & Rappaport, Chap. 11) or the use of "any equipment or facility necessary for peaceful exploration" (Article IV): Although it could at some point cover the use of even explosives for such purposes, unless this contravenes other provisions of the treaty, it is not so simple, especially in view of Article IX.

- The OST also sets out the liability of States for damage caused by their space objects (Article VII), and for national space activities "whether such activities are carried on by governmental agencies or by non-governmental entities" (Article VI).
- Finally, one more consequence of the initial statement: "Outer space, including the Moon and other celestial bodies, is not subject to national appropriation by claim of sovereignty, by means of use or occupation, or by any other means" (Article II). It is remarkable that this Article underlines the prohibition of national appropriation or claim of sovereignty not only by use or occupation, but also "by any other means": it seems to be an attempt to avoid legal artifices to circumvent the prohibition, from which we can deduce the importance that the treaty gives to this issue. However, according to Article VIII, "Ownership of objects launched into outer space, including objects landed or constructed on a celestial body, and of their component parts, is not affected by their presence in outer space or on a celestial body or by their return to the Earth."

The *"Moon Agreement"* (or Treaty) is more specific and could therefore be thought to be the main document to refer to. However, it presents a major problem in terms of implementation: it has been signed by 11 States and there are less than 20 States parties (18 States, on 14th January 2021, following the information given by United Nations Treaty Series: UNTS, 2020). The United States (US), the Russian Federation, the People's Republic of China and some of the members of the European Space Agency (ESA), are not among the signatories; India and France have signed, but not ratified this treaty yet. In April 2020, the executive order "on Encouraging International Support for the Recovery and Use of Space Resources", signed by the US President, states that "the United States does not consider the Moon Agreement to be an effective or necessary instrument to guide nation states regarding the promotion of commercial participation in the long-term exploration, scientific discovery, and use of the Moon, Mars, or other celestial bodies."

The limited international support for this Treaty, however, does not prevent us from finding in it important issues to tackle if we are thinking about a possible lunar settlement, either to take them on or to discard them.

Regarding its provisions, the Moon Treaty emphasises the principles of the OST, but adds interesting issues and particular aspects of its implementation to the Moon:

The Agreement reaffirms the premise of the use and exploration of the Moon for the benefit and in interest of all countries on the basis on equality (Articles 4 and 6) and always in accordance with international law (Articles 2 and 6).

With regard to the principle of cooperation and mutual assistance (Article 4[2]), it is specified by mentioning that "States Parties agree on the desirability of exchanging scientific and other personnel on expeditions to or installations on the Moon to the greatest extent feasible and practicable" (Article 6[3]); the reference to "other personnel" allows us to think of both civilian and military personnel, in accordance with the applicable treaties (see again Chap. 11). In addition, the Moon Treaty provides for measures to ensure compatibility among different missions of various States (Articles 5[2] and 9[2]).

The use of the Moon for peaceful purposes and the prohibition of the establishment of nuclear weapons or weapons of mass destruction (Articles 2 and 3) are supplemented by the prohibition of "Any threat or use of force or any other hostile act or threat of hostile act on the Moon" (Article 3[2]). To guarantee this, it is further stipulated that all equipment, vehicles and facilities on the Moon "shall be open" to other States parties (Article 15).

The exploration and use of the Moon must be made without harmful contamination or imbalance in the environment (Article 7); it is therefore logical that Article 14 sets out the liability of States for national space activities "whether such activities are carried on by governmental agencies or by non-governmental entities" (as stated in OST), which imposes an obligation on States to supervise the activities of non-governmental entities on the Moon (Article 14[1]).

Concerning the prohibition of national appropriation of the Moon by claims of sovereignty (Article 11[2]), the Agreement specifies that this prohibition applies to the subsurface and natural resources (Article 11[3]), and Article 11(1) includes an important statement absent from the OST: "The Moon and its natural resources are the common heritage of mankind". This wording is also used in other international treaties that govern common spaces (UN Convention on the Law of the Sea, 1982, Article 136: "The Area and its resources are the common heritage of mankind"); perhaps the fact that it was not employed in the OST (1967) is due to the subsequent evolution of space law and international law on this area (Rodríguez Carrión, 1998).

This is interesting in relation to natural resources and the possibility of exploiting them; on this point, the Treaty distinguishes between resources and the profits (or benefits) obtained from their exploitation, which seems to open slightly the door to commercial interests in this field. On the one hand, the Agreement states that lunar natural resources are the common heritage of mankind and should not be appropriated (Articles 11[1] and 11[2]). On the other hand, it establishes the right to use the Moon without discrimination and to exploit its resources (Articles 11[4] and 11[5]), but to regulate this exploitation of natural resources, the Treaty obliges States to "establish an international regime", i.e. they are committed to reach a new agreement regulating such exploitation (Article 11[5]), which must comply with the purposes of Article 11(7) (including rationality in exploitation and "equitable sharing by all States Parties in the benefits derived from those resources, whereby the interests and needs of the developing countries, as well as the efforts of those countries which have contributed either directly or indirectly to the exploration of the Moon"). Sector agreements are therefore possible to regulate specific resources, the benefits that can be obtained and their distribution. (A brief reference to the Artemis Accords [2020] will be made later).

The Treaty includes the right of each State to establish manned or unmanned stations (Article 9) with its own personnel, facilities and equipment under its own jurisdiction even if they are on the Moon (Article 12), which is an example of *extraterritoriality* (a fiction that implies that [part of] the law of a State applies to places not located on its territory, excluding the application of possible local law, what entails extending the territorial application of its sovereignty to certain events occurring in places outside its territory). However, in cases of emergency, the equipment, vehicles, etc. may be used by personnel of other States, since the safety and health of people on the Moon is a crucial concern, to which all efforts should be directed (Articles 10 and 12[3]).

This raises an interesting distinction among ownership of the ground, the subsurface and the facility or structure located on the ground (or underground): if the station is placed on lunar surface, the facility or structure would belong to the State of origin, while the ground and subsurface would remain a common heritage of mankind; this would not change in the case of establishing underground stations.

10.3   What Can We Do with All of This?

What kind of lunar settlement can we expect to see under this regulation (apart from recognising the aforementioned weakness of the Moon Agreement)? Perhaps something similar to what we have with the International Space Station (ISS), but on firm ground and with a little more natural gravity.

The ISS is an example of a permanently inhabited station in outer space; it is also an example of international cooperation: people of various nationalities live and work there, sharing equipment and infrastructure for scientific and commercial purposes, implementing programmes that require joint action and all being located in outer space, that is, in a space over which no State is sovereign. For all these reasons, the legal regime of the ISS faces problems that could be similar to those of a lunar station under the current legal framework.

The legal regime established for the ISS was founded on compliance with the four international treaties on outer space mentioned above (our "*big four*": OST, Rescue, Liability, Registration), except the Moon Treaty, so the legal regime of the ISS could also constitute a further example of how to overcome the latter's weakness when addressing the task of regulating an eventual lunar settlement.

Furthermore, as the legal regime of the ISS had to meet the particular needs arising from the establishment, operation and further development of the station, it was necessary to add to the legal framework of the previous treaties an Intergovernmental Agreement (IGA), Memoranda of Understanding (MOUs, signed by NASA and the cooperation agencies of each State), contemplating the possibility of additional Implementation Agreement (IA), if necessary.

As for the IGA, it is an agreement that links the signatories internationally, regardless of its nature in domestic law of each State (Faramiñán Gilbert, 2018). The initial IGA was signed in 1988 and was replaced by a new one signed on 29 January 1998 in order to incorporate Russia: a new agreement was necessary, as the initial IGA was a closed agreement between parties, with no accession clause (Faramiñán Gilbert, 2018). The IGA establishes the basic framework of cooperation required for the implementation and operation of the ISS, dealing with some matters of public law (Articles 18 and 22) and others of private law (Articles 9[3] and 21):

Article 1 (object and scope) affirms the civilian nature of the ISS and its purpose for peaceful scientific, technological and commercial use.

Article 2 refers to our *"big four"* (OST, Rescue, Liability and Registration), according to which the IGA must be implemented, and makes it clear that nothing in the IGA could constitute "a basis for asserting a claim to national appropriation over outer space or over any portion of outer space".

Articles 7, 21 and 22, especially, establish the following legal regime in terms of general operation: for management, although each party is responsible for its own programmes and plans, the US (through NASA) has a leading role (for example, it is responsible for "overall programme management and coordination", "overall system engineering and integration", "overall safety requirements and plans", etc.) Article 21 implements the fiction of extraterritoriality for intellectual property, while Article 22 establishes for criminal jurisdiction the rule of personal application, which also implies extraterritorial implementation of the law: States parties exercise their criminal jurisdiction for acts committed by their own nationals in ISS (however, a certain "guarantee clause" is provided for the State affected by a criminal act committed by the national of another State party, enabling the affected State to exercise its criminal jurisdiction over it, in certain cases). This kind of regulation (referring to national jurisdictions depending on connecting elements) may be considered an example of cooperation-based regulation, while integration-based regulation would require the establishment of a particular ad hoc regime by agreement among States (however, Article 9[2], which contains a special rule, and Article 16, which we will see later, could be more in line with an initiative towards integration-based regulation); it is not always easy to reach such agreements in international law, so it may be a challenge for the legal regime of an eventual lunar settlement.

Article 16, in conjunction with Articles 2 and 17, provides for a special rule ("Cross-Waiver of Liability") in cases of "protected space operations". This may be considered a specific exception to the (general) international liability regime established in the treaties, being applied (this special rule) among the signatory parties as *lex specialis* (in preference to *lex generalis*) and in favour of the exploration and exploitation objectives set for the ISS.

What interests us about this Article is not its content, but what it represents: it is an example of adapting a general legal framework (international liability in outer space) to a particular situation that has special features (the ISS in its case or the lunar station, in ours), although it does not go much further than the first step: deciding on the applicable law (the second step would be to create that particular law applicable in such cases).

Replicating the ISS model to the case of a lunar settlement would also require a specific (similar) IGA to establish the operational bases for cooperation and the particularities to be applied on the Moon, and MOUs among the space agencies involved, regardless of any complementary or sector-based agreements that might be necessary; all in accordance of course with our *"big four"*.

I believe such a need for specific agreements for a lunar settlement exists, regardless of whether we consider the application or not of the Moon Treaty: this Treaty itself, as we have seen, expressly provides for additional agreements (for example on international liability and on exploitation of natural resources, Articles 14[2] and 11[5], even if it is an inhabited station of a single State, Article 9[1]), apart from others resulting from the implementation of the Treaty (Article 6[3]: exchange of personnel between the parties). The fact of applying or not applying the Moon Treaty will not prevent additional international agreements, but rather will imply that these additional agreements focus more or less on purely operational and implementation issues or also include provisions of more material and generic scope, respectively; notwithstanding that in either case it is necessary to address specific issues of a lunar settlement which may differ from those of the ISS, and which justify one or several specific or sector agreements: ground-surface and subsurface use regime, presence of military personnel, exploitation of resources, energy, etc.

And now it is appropriate to briefly mention the Artemis Accords, signed on 13 October 2020 by the representatives of the US (NASA) , Australia, Canada, Italy, Japan, Luxembourg, the United Arab Emirates, and the United Kingdom. These Accords establish among the parties "Principles for cooperation in the civil exploration and use of the Moon, Mars, Comets, and Asteroids for peaceful purposes". The signatories to these Accords "reaffirm their commitment to the Outer Space Treaty" (section XI) and address in section X natural space resources, whose extraction and use "should be executed in a manner that complies with the Outer Space Treaty", while creating in section XI controversial "safety zones". Regardless of whether or not they fulfil the provision made by Article 11[7] of the Moon Treaty for the exploitation of lunar resources (among the signatories of the Artemis agreements, Australia also signed and ratified the Moon Treaty), the fact is that the Artemis Accords only contain a general framework of principles and provide for necessary implementation through MOUs and additional agreements (section II).

10.4   What Do We Really Want to Do?

If we take the approach resulting from all the above, we would manage to have a "(international or not) space station" on the Moon: the only thing that would change conceptually with respect to the ISS would be that our lunar station would be placed on solid ground, but as no State exercises exclusive ownership or sovereignty over this ground (the ownership regime is the same as that of outer space, according to the OST), the scenario would be similar. In fact, we could provide the lunar station with purposes more focused on the exploitation of resources (as allowed or agreed) or we could make the lunar station not only with civilian (like the ISS), but also with military personnel for scientific or other peaceful purposes. But this would merely involve adding some specific agreement to cover these issues; the legal framework of reference for the inhabitants, however, would be the same as for the ISS or very similar: they would continue to be people whose lives are genuinely on Earth, even if they spend periods of time on the lunar station.

To go further in a lunar settlement, the key is just there: in asking where do these people (the inhabitants) really have their lives? Where is their centre of vital interests? And before that question: what do we want to do on the Moon (or what can we do, according to the status of technology)? What do we want the lunar settlement for?

The answer to these questions will shape the model of settlement we want, including the design of its specific legal regime, as this will determine whether we can configure the lunar settlement we need from the existing legal framework in the treaties (adding an IGA or additional agreements) or, on the contrary, if it is not sufficient and we should aspire to attempt more substantial changes, for example in the legal regime for real property.

Indeed, the question of ownership is significantly important (cf Velázquez Elizarrarás, 2013). It is addressed by the OST, the Moon Treaty, but also by the IGA for ISS and by Artemis Accords (the latter to state in section X that the States parties consider that "the extraction of space resources does not inherently constitute national appropriation" under OST). Let us suppose that the prohibition of appropriation on the lunar surface and subsurface is maintained; then the lunar settlement would be condemned to a permanent division between ownership of the surface and subsurface, and ownership of whatever is on the surface (even if it is attached to it), as described above (Sect. 10.2, last paragraph).

Of course, this might affect a hypothetical lunar real property market: This division of ownership would mean that the value of the facility, building or what was built on the surface or underground would (a priori) be lower, both to sell it and to obtain financing through it, since the land would have to be discounted from the total value of the property. But the relevant fact is that the owners of structures or facilities placed on the lunar surface (or underground) would be enjoying in an exclusive way a portion of lunar land that would not legally own them (but common heritage of mankind): Would they have to pay for it? If so, to whom would they have to pay? Could they acquire in the course of time any sort of permanent right of use to such land by placing their buildings or facilities on it? And who would be able to claim the land from them? It poses another problem: Who manages the lunar land? Who decides what may and may not be done on it (or with it)? In short, an authority should be established to decide, to manage the land and, in the end, to dispose of it (without owning it either).

But this is only a very small example. The challenges are quite similar in many other areas: depending on the type of settlement we want, we will have to think about whether or not to establish public services such as security, water, electricity and other supplies, waste collection, transport and road infrastructure, education, health, justice, administration, etc. Some things can be dealt with "along the way" (for example, which products for the lunar population can be produced on site [ISRU: In-Situ Resource Utilization] and which cannot), but others must be well organised beforehand (providing oxygen or an adequate atmosphere for the whole population to live in). This may also imply the emergence of new or more specific "Human Rights" (right of access to certain technology, required for daily oxygen supply, for example), since circumstances determine the needs to be meet and the level of importance of the latter (on the Moon, air supply will condition freedom, work and life in these cases: Stevens, 2015), but it may also imply the need to establish a specific legal regime for other issues such as the location of companies there or a separate tax regime, etc. In short, *everything will depend on the type of settlement we want to make* or we can make (or that finally results from the evolution of the originally founded).

10.5   A Few Words on the Nation-State Concept on the Moon

The concept of the Nation-State (which today could be considered to be overcome by the phenomenon of globalisation: Arriola Echániz, 2019) requires a political, legal or institutional structure (the State) that guarantees (founded on the social contract theory or whatever is preferred) the necessary conditions for the exercise of subjective rights to a stable population, whose members are united by certain common cultural ties (Nation) and which is settled on a delimited territory, over which the State's jurisdiction and sovereignty is exercised. This "home-made" definition (despite its shortcomings) allows us to distinguish the characteristic elements of the Nation-State: political, legal or institutional authority, a population with common cultural elements (language, tradition, religion, history…) and a territory (Ribó & Pastor, 1996).

The approach to a hypothetical lunar settlement from the perspective of the Nation-State concept depends on the answer given to the question raised in the previous section: *what kind of settlement are we talking about?* (What do we really want to do?).

On the one hand, we have the most reasonable model of lunar settlement in the short to medium term: a model based a priori on current international law (although more or less important things may be added or modified subsequently), for scientific or commercial purposes and whose members would maintain their centre of vital interests on Earth. For this model of settlement (multinational or otherwise), a system of organisation imported from the Earth would be used, which might either be an extension of the founder State's own system, or might be an organisational model agreed among several founder States from the outset (since even this last scenario would contain some reference to the legal regime of each State with respect to each other's nationals, who would remain essentially citizens of their own Nation-State, where they would maintain their centre of vital interests).

This model reinforces the concept of the Nation-State of the founder States (Earth countries), since through the fiction of extraterritoriality, the sovereignty and jurisdiction of a given Earth State is extended in its implementation beyond its borders: on the Moon. A typical feature of the Nation-State is the capability to ensure the implementation of its own law (its legal system) throughout its territory; in addition, the State that succeeds in getting its law implemented beyond its physical borders extends its influence in the international sphere and thus carries out a sort of "colonisation" (not always based on the expansion of territories, but of spheres of influence, for which the application of a State's law in a territory or area it intends to dominate is crucial, as has been historically proven: Merry, 1991); it is an effect that can perhaps be generated (albeit only partially), for example, in favour of the US thanks to Article 7 of the IGA for the ISS as far as its additional or pre-eminent tasks are concerned: if the US is responsible for "overall planning", it has the opportunity to transfer to the ISS issues pertaining to its own legal system by means of such planning.

On the other hand, a serious approach to the concept of the Nation-State applied to an independent lunar settlement is difficult outside the realm of science fiction (some examples in Baxter, 2015). It should be remembered that the Nation-State implies political, legal or institutional authority, a population with common cultural ties and a well-defined territory in which to apply jurisdiction and sovereignty: political authority is not common to emerge from nowhere without territory, history or population; the population must share common cultural ties, and a mere common feeling of discontent towards the political authority in force is not enough (see, for example, the Roman servile wars: Gerrish, 2016); as for territory, it must be a territory where jurisdiction and sovereignty can be exercised effectively and independently, which also entails the capacity to defend it against aggressions.

Moreover, all this would require a *complete legal system,* which would also regulate more than just relations of people who share work and coexist in a common place because of that work (in the way of the ISS); we would now need a legal regime also for people who may not be involved in that work and who develop legal relationships among themselves in various spheres, public and private.

In any case, going further, thinking of a lunar settlement with real autonomy and independence (including possible procedures more or less similar to decolonisation, declarations of independence, etc.) would ultimately result in the application of the doctrine of recognition of States or governments and, in the end, it would provoke that our settlement ceases to be seen as an "object" of international law and becomes a "subject" of international law.

But, returning to the sense of the previous question, would anyone really want to live on the Moon and feel it as its own land enough to encourage dreams of independence and fight for it? Well, as always, that will depend on human needs and interests in the future…: this is something that has always characterized the "human factor" on Earth, and possibly also the "human factor" on the Moon.

References

Arriola Echániz, N. (2019). ¿Más allá del Estado nacional? Una revisión de la doctrina cosmopolita. *Revista de Estudios Políticos, 183,* 243–259. https://doi.org/10.18042/cepc/rep.183.09.

Baxter, S. (2015). The birth of a new republic: Depictions of the governance of a free Moon in science fiction. In C. S. Cockell (Ed.), *Human governance beyond earth* (pp. 63–79). Springer. https://doi.org/10.1007/978-3-319-18063-2_6.

COPUOS. (2019). Guidelines for the long-term sustainability of outer space activities of the committee on the peaceful uses of outer space. *Report of the Committee on the Peaceful Uses of Outer Space.* Retrieved January 2, 2021, from https://www.unoosa.org/oosa/en/oosadoc/data/documents/2019/a/a7420_0.html.

Faramiñán Gilbert, J. M. (2018). The International Space Station: Legal reflections. *Ordine internazionale e diritti umani, 5*(supplement), 49–54.

Gerrish, J. (2016). Monstruosa species: Scylla, Spartacus, Sextus Pompeius and Civil War in Sallust's histories. *The Classical Journal, 111*(2), 193–217. https://doi.org/10.5184/classicalj.111.2.0193.

Gutiérrez Espada, C. (1999). La crisis del derecho del espacio, un desafío para el derecho internacional del nuevo siglo. *Anuario Español de Derecho Internacional, 15*, 235–273. https://revistas.unav.edu/index.php/anuario-esp-dcho-internacional/article/view/28509.

Marchisio, S. (2018). Setting the scene: Space law and governance. *Ordine internazionale e diritti umani, 5*(supplement), 55–65. http://www.rivistaoidu.net/sites/default/files/Volume%20speciale%202018%20completo.pdf.

Merry, S. E. (1991). Law and colonialism. *Law & Society Review, 25*(4), 889–922. https://doi.org/10.2307/3053874

Ribó, R., & Pastor, J. (1996). La estructura territorial del Estado. In M. Caminal Badía (Ed.), *Manual de Ciencia Política* (pp. 451–469). Tecnos.

Rodríguez Carrión, A. J. (1998). *Lecciones de Derecho Internacional Público*. Tecnos.

Stevens, A. H. (2015). The price of air. In C. S. Cockell (Ed.), *Human governance beyond earth* (pp. 51–61). Springer. https://doi.org/10.1007/978-3-319-18063-2_5.

UNOOSA. (2020). *Status of international agreements relating to activities in outer space as at 1 January 2020*. Retrieved January 2, 2021, from https://www.unoosa.org/documents/pdf/spacelaw/treatystatus/TreatiesStatus-2020E.pdf.

UNTS. (2020). *Status of the Moon agreement*. Retrieved February 22, 2021, from https://treaties.un.org/Pages/ViewDetails.aspx?src=TREATY&mtdsg_no=XXIV-2&chapter=24&clang=_en.

US President. (2020). *Executive order on encouraging international support for the recovery and use of space resources*. Retrieved January 2, 2021, from https://www.whitehouse.gov/presidential-actions/executive-order-encouraging-international-support-recovery-use-space-resources/.

Velázquez Elizarrarás, J. C. (2013). El derecho del espacio ultraterrestre en tiempos decisivos: ¿estatalidad, monopolización o universalidad? *Anuario Mexicano de Derecho Internacional, 3*, 583–638. https://revistas.juridicas.unam.mx/index.php/derecho-internacional/article/view/439.

Chapter 11
The Military on the Moon: Geopolitical, Diplomatic, and Environmental Implications

Richard A. Stewart⊙ and Margaret Boone Rappaport⊙

Abstract The authors review the geopolitical situation now existing that could lead to a presence of military bases on the Moon and more generally, in cis-lunar space. They suggest some of the effects that a military presence on the Moon might have on lunar settlement. They discuss the roles of Western militaries in defense, diplomacy, and a potential role in environmental protection of the lunar environment. In the second part, the history of lunar military base planning is reviewed briefly, and then the authors take the reader farther into the future, describing the nature and configuration, as well as some of the equipment needs, for a lunar base in the next stage of lunar settlement.

11.1 Concept, Roles, and Rationale for a Military Function in Cis-Lunar Space

11.1.1 Introduction

According to our colleagues in the field of astronomy (including authors Impey and Corbally in this volume), Earth's Moon likely formed over four billion years ago. The latest theories include a violent collision of Earth with another planetary body, which ripped away part of the early Earth's mantle, from which the Moon was then formed (cf Haviland, Chap. 3, on the physical nature of the lunar surface). Since that time, much has happened, including the organic evolution of all species now living. Human social organization has grown increasingly complex, and the nation-states now existing did not always exist, nor did their alliances. Groups of nation-states form and give way, and others form in a long-term evolution of human social life on Earth, and now, in space.

R. A. Stewart
7612 Bear Forest Road, Hanover, MD 21076, USA

M. B. Rappaport (✉)
400 E. Deer's Rest Place, The Human Sentience Project, LLC, Tuscon, AZ 85704, USA
e-mail: msbrappaport@aol.com

This chapter is presented in two parts. In the first part, we attempt to capture a broad view of the existing geopolitical roles that are developing in cis-lunar space at the beginning of the early centuries of human space exploration. Leaders, alliances, and client states will likely change over the hundreds and thousands of years in the future. However, the existing nations of the United States (US) and China, as well as their allies, appear at the present in broad competition on the world stage. Their relationship will have many formative effects on the arrangements of people and installations on the Moon's surface and in surrounding space. Therefore, our chapter assumes that nation-state alliances may well change, but the competition between Earth's existing nation-states will substantially affect settlement of the Moon, in one way or another. For this reason, we devote our attention to it in the following pages.

In the second part of this chapter, we describe an anticipated, Western lunar military base, its physical components, staff, and the rationales for their arrangement.

11.1.2   A Military Role in Lunar and Martian Exploration

It would be naïve to suggest that earthbound political relationships will not be transferred, to some extent, to the settlement of the Moon. There are chapters in this book and elsewhere that argue for a Moon that is military-free if that is feasible. Positions are described in other volumes (e.g., Arnould, 2019; Haqq-Misra, 2019) that suggest the boundaries of nation-states of Earth should not be transposed to the lunar surface and to cis-lunar space. These sentiments should, to an extent, find their way into the legal language of treaties between Earth-based powers for use of the lunar surface and surrounding space. Negotiations promise to be lengthy and complicated, and previous approaches to date are summarized elsewhere in this volume (The reader is referred to Chap. 10 in this volume, for a more in-depth treatment of Treaties impacting settlement, in general, and the military on the Moon, by Jurado-Ripoll). At the present, there are no existing treaties that address the use of militaries on the Moon. Here, we explore various types of involvement of Western militaries on the Moon and in surrounding cis-lunar space: geopolitical, strategic, diplomatic, rescue and recovery, and environmental responsibilities.

Space and other planetary objects like the Moon and Mars have always posed both an exploratory challenge as well as potential threat to the nations on Earth from space-capable military powers. With the rapid advancement of technology adaptable to space, this threat grows ever closer, temporally and organizationally, and requires serious consideration by world leaders and military establishments. Yet, in the current political and economic climate, the idea of establishing a military presence on the Moon would likely be met with skepticism, criticism, fear of warmongering, satirical comment, and hand wringing. However, the situation may soon change.

The reality is that there are geopolitical and national security risks, as well as space-based national and global security threats that will probably lead sooner, rather than later, to a Western military presence on the Moon. That is our focus here. Establishment of, for example, a US base would involve a set of national security missions, which would join the existing goal of supporting the US National Aeronautics and Space Administration (NASA) in its exploration of the Moon and its ability to extend space travel to Mars. The recent creation of the US Space Force (USSF) as a separate military service appears to presage this, but publicly, so far, USSF and the US Department of Defense (DoD) have not publicly defined a military mission for the Moon. In an article that asks, "Is Earth-moon space the US military's new high ground?" Peter Garretson of the American Foreign Policy Council, notes the absence of an overall deep-space strategy (in David, 2020), and to the date of this writing.

Despite a lack of clarity about the future role of the Moon in Western defense strategy, NASA and US Space Force are working together in ways that could lead toward lunar military missions. In 2020, NASA announced a new memorandum of understanding, in which NASA and DoD commit themselves to "a broad collaboration… in areas including human spaceflight, US space policy, space transportation, standards and best practices for safe operations in space, scientific research, and planetary defense" (USSF & NASA, 2020). This joint statement clearly hints at future collaboration in lunar operations and exploration.

NASA's Artemis program is described further in the second part of this chapter. It has the goal of landing the first woman and next man on the Moon in 2024, as the beginning of a sustained presence to explore the lunar surface using innovative technologies (NASA 2020a).NASA will collaborate with its commercial and international partners to establish sustainable exploration of the Moon by 2030, and use the lessons learned to prepare for a mission to send astronauts to Mars (Roberts, 2020). The Artemis mission will be conducted in three stages, which begin by sending robotic systems to the Moon for stays as long as seven days, during Phase I. Robots and humans will search for and extract resources such as water, which can be converted into other usable products, especially oxygen for human consumption and fuel. With precision landing technologies and innovative means to move around, moon crew will explore new regions of the lunar surface.

The longer-term plan is to establish a Moon base as a waypoint for sending astronauts to Mars. In order to better explore the Moon and Mars, NASA's Jet Propulsion Laboratory (JPL) has developed an improved four-wheeled rover named, DuAxel. Its rear can anchor itself to the ground so that the front goes free on two wheels, and a tether holds the pieces together so the front section can rappel down a steep slope. JPL robotics technologist Nesnas states that, "DuAxel opens up access to more extreme terrain on planetary bodies such as the Moon, Mars, Mercury, and possibly some icy worlds, like Jupiter's moon Europa" (in Kooser, 2020). These types of developments are exciting in themselves, but they also suggest capabilities that could be acquired in the future by Western militaries.

11.1.3 Geopolitical Implications for Military Roles on the Moon

As leaders and administrations come and go in all nation-states on Earth, their points of view, plans for settlement of the Moon, and desire to establish military bases there will vary. At the time of this writing, a specific Moon mission remains undefined by US Space Force, although general goals of establishing an American base have been voiced openly by some officials. Furthermore, some activities indicate that American and other Western governments are, if not taking positive steps, considering the idea of establishing a military presence on the Moon. For example, in 2020, a report entitled, "State of the Space Industrial Base 2020: A Time for Action to Sustain US Economic and Military Leadership in Space", was released by a virtual workshop of 150 "thought leaders" from industry, government, and academia. It was hosted partly by the US Air Force (SSIB, 2020). The report emphasizes military challenges in space and the need to control critical, so-called "choke points" in cis-lunar space and on the Moon, itself. It describes "a race to the great wealth of lunar resources which will fuel the greater space economy and enable future exploration and settlement in the solar system" (SSIB, 2020: 53). Once again, settlement of the Moon is perceived as a stepping-stone, or a gateway, to exploration of Mars and the remainder of the solar system. It is also conceived as a storehouse of great natural resources. When these understandings become apparent, it becomes ever more obvious why there would be a competition to settle Earth's Moon.

Readers should note that the US-Western stance generally does not link space exploration and military functions. These nations see exploration and military programs as focusing on different goals—political, economic, exploratory. However, it is important to acknowledge that the pursuit of exploratory, military, and economic goals will occur at the same time. For Western crew and eventually settlers, the Moon will likely be a science research center, a military depot, and probably a testing field for mining methods and equipment to be used farther afield in the asteroids. There will be broad multinational efforts to maintain a peaceful cis-lunar environment, and this is both desirable and appropriate. There is much at stake in maintaining peace in outer space and on the Moon.

The SSIB (2020) report realistically assumes that an Earth-Moon-Mars environment is a competitive one, and that an important function of militaries is to assist in maintaining peaceful, unarmed competition. The report foresees an expanding cis-lunar economy that will rely upon the USSF to provide security with a stabilizing military presence, to protect and enable US, Western, and world commerce across cis-lunar space. It is important that this expanded role be multifaceted, that it project confidence and that it lowers "the perception of risk by providing stabilizing presence, surveillance, aids to navigation, and help when required" (SSIB, 2020: 11). The United States, as represented in both NASA and DoD, are aware of the value of lunar resources that could "enable cheaper mobility for civil, commercial and national security applications…" and that help to provide "access to asteroid resources and Mars…" (SSIB, 2020: 11). The report does not hesitate in forewarning that, "the first nation to establish transportation infrastructure and logistics capabilities serving GEO [geospace] and cis-lunar space will have superior ability to exercise control of cis-lunar space and in particular the Lagrange points and the resources of the Moon" (SSIB, 2020: 17). The Lagrange points are strategically important gravitationally stable locations in space where probes and vehicles can park and remain without using much fuel. The Lagrange points are important for exploratory, economic, communication, transportation, and military purposes.

Geopolitical security considerations focus on the present competition between the US, China, and an increasing number of other nation-states, producing a need to plan for a military presence in cis-lunar space. Yet, "The US is not alone in planning to return humans to the Moon or expanding the use of space. China has announced its intention to do so by 2035…" and China is "committed and credible in its pledge to become the leading, global super-power, to include space, by 2049 marking the 100th anniversary of the People's Republic. A key component of China's strategy is to displace the US as the leading power in space and lure US allies and partners away from US-led space initiatives… Nor is China alone. An expanding number of nations are emerging as space competitors and potential adversaries." (SSIB, 2020:17).

11.1.4 Effects of Cis-Lunar Competition and Cooperation Among Earth Nations

There are varied opinions about both the advisability and the dangers of transferring the nation-state concept to the Moon. From a certain perspective, it is difficult to imagine how this could *not* happen, especially given the competition now ongoing among Earth's nations. One can fashion treaties that voice hopes for peace and the absence of militaries on the Moon, but it is not clear how enforceable they are. (For a more indepth description of treaties involving outer space and the Moon, See Chap. 10 in this volume). Therefore, given the present competitive environment of space programming on Earth, let us take a look at some of the ongoing space program efforts by the main competitors, China and the US, and then discuss how they could affect the settlement of the Moon. As noted, the US and China have both announced their plans to settle on the Moon and there are also unclear statements from various leaders about establishing a presence for their militaries. It is likely that these statements will become more clear in the coming months and years. The following discussion assumes a variety of partner nations will also participate, but the primary nations are China and the US.

It is important to acknowledge that efforts by both China and the US, their partners as well as other nations, have already begun. Principal among these other nations are Japan, the European countries alone or through the European Space Agency, and Israel, which are all making noteworthy advances in space. China is active in advancing its lunar position. For example, China's Chang'e 4 Mission landed a robotic rover, Yutu, on the far side of the Moon and returned it to Earth (Wall, 2019).

Chinese space officials have also discussed building a crewed "lunar palace" near the Moon's south pole as part of a program to send astronauts to Mars.

China's activities give the appearance of the pursuit of peaceful space exploration, habitation, and resource mining. However, the longer-term, geopolitical reality of China's and some other nations' efforts may be quite different. As China's economy has grown and its middle class has expanded, it has become an important member of the global community, both economically and politically. With this new influence, China's program efforts outside its own borders have grown, partly to obtain resources for its own population and for client states. China's political influence all over the world has increased, and the US State Department and other Western governments are actively monitoring Chinese programs both on Earth (as in Africa), and in space.

Many analysts doubt that the ambitions of China, Russia, and their client states are entirely benevolent, and this extends to their programs on Earth, in space, and on the Moon. The literature on the emergence of the competition between China and Western powers, and both its economic and political ramifications, outlines the values and beliefs beneath this global, Earthside competition (e.g., Reuters, 2017; Africa Times, 2018; Dabus et al., 2019; Campbell & Rapp-Hooper, 2020; Keeley, 2020; Meltzer, 2020; Taylor 2020a, b, c; USCESRC, 2020). With the real possibility of extending China's influence to the Moon, it is understandable that Western nations begin to plan a military presence on the Moon, whether it eventuates, or not. The functions of their militaries are not simply defensive or aggressive as in a classic model, but also diplomatic, with an important rescue capability, and a potential for leadership in environmental protection in cis-lunar space.

It is likely that the US military, supported by other Western partners, will have a substantial presence on the Moon. The environment is challenging, with no atmosphere, one-sixth gravity, and no rescue services unless devised and provided by the first inhabitants. The military may have a rescue and recovery function on the Moon, as it has in most quarters on Earth. When it comes to the safety of crew, scientists, and other settlers, it is important to remember how well funded and well supplied the Western militaries are. The Earthside competition among China, Russia, Western nations, and their allies may be transferred, no matter how much some people hope this is not true. However, before that plays out, humans on the Moon will be challenged as never before to simply survive. In the earliest settlements, we have no doubt that militaries, including those who compete amongst themselves, will exercise a rescue and recovery function from the very start, and they will save many lives of different nationalities.

11.1.5   Defensive and Aggressive Military Roles on the Moon and in Cis-Lunar Space

The history of armed conflict on Earth suggests some basic functions that militaries might assume on the Moon. In the second part of this chapter, we review some of the existing features of relationships among the likely crew settling in the first military Moon bases. They may be separate from or contiguous to non-military habitations. Defensive military functions would almost certainly extend to these non-military facilities, and to scientific facilities and staff. The location of the latter would likely depend on specific scientific requirements. Observatories, for example, may be located on the far side of the Moon. Militaries may have the only vehicles, staff, and crew who can travel to some locations on the Moon.

Both NATO and the Western powers in alignment with the US, Australia, Japan, and India (referred to as the "Quad") have begun taking Chinese threats quite seriously, including their expansion, exploitation, ability to corrupt and coerce, and their threat to cybersecurity. NATO Secretary-General Jens Stoltenberg has warned that China's emergence as a superpower shifts the global balance of power, so NATO's focus could become more global. India, which in the past has eschewed military pacts and remained a leader of non-aligned nations, has conferred with the US on the topic of strengthening security (Gertz, 2020).

In view of China's geopolitical ambitions, it would be naïve to propose that Chinese settlement of the Moon will not be accompanied by military capabilities. While the US has formal collaboration between NASA and US Space Force, there is a clear separation in their non-military and military missions. The Chinese have no hesitation in combining military and non-military functions, and this integration will likely be reflected in their programs on the Moon. An Earthside example of combined purposes is found in the Chinese launch of military missiles from Chinese merchant ships (Davis & Jones, 2020). It is important to ask: *How* will their planned cis-lunar presence involve and integration of military and non-military?

Western nations, in taking Chinese military threats in the lunar sphere quite seriously, have begun reorienting their military strategies toward encountering Chinese expansion in space and on the Moon. Western planning will include at least a defensive capability on the Moon. To do otherwise would be sacrificing the safety of Western scientists, settlers, and military staff and crew. We note that a potential threat from China in space does not necessarily exclude future threats from other powers, like Russia or a rogue militarized force that has space capabilities. However, other nations such as Russia appear much farther behind than China. Still, China, at the time of this writing, appears to offer the most serious foreseeable threat to Western lunar exploration, settlement, and scientific research on the Moon. The impediment would be very unfortunate, since Western and Chinese astronomers, for example, are routinely involved in cooperative research now on Earth.

Both Western nations and China are now actively preparing to send astronauts to the Moon for an extended human presence. The question remains as to whether military capabilities will accompany these missions. It is a question that cannot be ignored, given its impact on global and cis-lunar security. The Chinese have clearly expressed a policy of expansion, and the Chinese military's desire to venture into space and to the Moon. If this threat is borne out, Western nations, particularly the US, will likely implement a military capability of its own as a deterrent, and to position itself defensively. Before events reach that point, the question of whether a

Western military presence on the Moon can be averted through diplomacy should be fully explored.

11.1.6 Strategic Planning Considerations

US Space Force would develop contingency plans to balance a Chinese military intention by establishing an inhabited base by the 2030s. The Chinese would, as on Earth, have a military capability associated with their outposts. They would be most threatening if they intended to control the strategic Lagrange points and deny them to Western nations. Controlling Lagrange points could support military capabilities such as launching weapons to control cis-lunar space and establish additional military posts at critical lunar locations. In response to the Chinese, Western nations (especially US Space Force) would evaluate contingency plans and eventually make preparations to establish a military presence of its own. This would be similar to Western nations' military deterrence in East Asia and Africa.

Granted, envisioning missions to establish a military lunar presence is speculative, and our purpose here is primarily to point out the strategic factors that may change the settlement of the Moon. Any discussion of potential missions for a U.S. military presence on the Moon are, at this point, purely speculative and based on logical and reasonable assumptions. It is not unreasonable to speculate how this would affect existing NASA and other Western space agencies' capabilities that support human habitation on the Moon, exploration of it and Mars, and especially, scientific research. Military support is not usually the first consideration of scientists who plan and budget for studies of the universe, the Moon as a planetary body, or the natural resources found there, but scientists and their staff could well be dependent upon them in emergencies. There will initially be no law enforcement on the Moon, except as provided by individual nations or projects. There will be no Fire Department to call, and the environment of the Moon, itself, is dangerous for human habitation. When one adds to the dangers of no air, no food, and invasive dust, the dangers of other humans who can do harm, military back-up could prove quite valuable. Other help may be very far away.

11.1.7 Military Roles in Diplomacy and Environmental Protection on the Moon

The establishment of a permanent human base on the Moon presents diplomatic challenges under international law. Efforts at diplomacy make the best sense in the context of treaties and other agreements between nations, and military staff on the Moon could well work toward the settlement of disputes within these guidelines. Today, many issues involve the control of weapons in space. In 2008, China and

Russia submitted a draft treaty to the United Nations known as the "Treaty on Prevention of the Placement of Weapons in Outer Space and of the Threat or Use of Force against Outer Space Objects" (PPWT). The draft was opposed by the US because of security concerns, although the treaty affirmed a nation's right to self-defense. In 2014, the United Nations General Assembly passed two resolutions on preventing an arms race in outer space which was addressed to all nations.

The Moon has not been overlooked in international agreements that call for diplomatic involvement. For example, "The Agreement Governing the Activities of States on the Moon and Other Celestial Bodies"—better known as the "Moon Treaty"—was ratified by 18 nations as of January 2019. It called for turning the jurisdiction of all celestial bodies (including orbits around them) over to the participant countries. All activities would conform to international law, including the United Nations Charter, which governs military actions of US members through the Security Council. However, the treaty has not yet been ratified and it is not likely to be approved by countries including the US, China, Russia, Japan, and the member nations of the European Space Agency, which all have plans to self-launch human spaceflights. The Moon Treaty does not now appear to have relevance under international law. (see Chap. 10 in this volume).

The United Nations' Outer Space Treaty prevents a nation from claiming territory on another planetary body, although this could be modified in the future. There remains a lack of clarity over whether anyone can own lunar resources, and the issue of approval for mining projects remains murky (Anderson et al., 2019). While there have been repeated international efforts to govern military activities in outer space and, to some extent, on the Moon, our point here is that governance of outer space and the Moon extend the reach of what is considered "diplomacy" among nations on Earth. Military officers in Western and other military forces who serve around the globe are routinely trained to act also as diplomats and operate, to an extent, in concert with their local embassies. One can imagine that those functions might well develop on the Moon, especially if competitor-nations interact. Military staff could act as arbiters between different nations on the Moon, and the administrators of the moon bases they inhabit.

The ominous aspect of the effectively defunct Moon Treaty is the increasing probability that China and possibly the US will move actively to establish a permanent military presence on the Moon—either with their allies, partners, and client nations, or not. To forestall these actions, there may be bilateral treaties or agreements to restrict these plans. If one nation moves aggressively to establish a military presence, other nation-states would then be motivated to follow suit. There is also the possibility that a competitor nation might move to stop the development of an opponent's actions through negotiation. Were the US to take the lead in openly planning a meaningful military presence on the Moon, China would be motivated to support a treaty to prevent it. At the present, it seems unlikely the US would take action of this type, without severe provocation. On Earth, there are diplomatic avenues to forestall an escalating confrontation, through the United Nations Security Council, NATO, and other partners, including the Quad (US, Australia, Japan, and India).

One wonders whether these types of diplomatic protections will extend to the Moon, and if they do, what form they will take. Communications would not be so easy, and on-site crew and staff might well have to attempt diplomacy with little support. Among these staff could well be members of the armed forces of one or more nations. These are the uncertainties that will flavor international relations while multiple nations seek to gain access to space, and eventually to the Moon, itself.

Once moon bases are established, Western nations will have an opportunity to serve as leaders in protecting the lunar surface, perhaps through their militaries. A moon base would necessarily involve heavy equipment and materials, and military bases would involve ordnance, protective structures for habitation, and transportation facilities, lines, and depots. They would also need to be engineered to minimize environmental impact to the Moon's surface. This could be achieved in several ways, including keeping the size and footprint of the base small or placing work habitats underground, as envisioned long ago in Project Horizon, in 1959. Environmental protection could involve using solar power to the extent possible, avoiding surface construction such as missile silos and perimeter fencing, and not directly participating in mineral exploration or mining.

In the following sections on the design, logistical, and staffing issues of establishing a US military moon base, environmental values and goals can be seen to emerge. The base we envision in the following pages can serve as either a model for environmental protection, or not. Our view is that planning the base with the environment in mind makes good sense for both present and future populations who inhabit Earth's Moon.

11.2 A Lunar Military Base: Mission, Design, and Equipment

11.2.1 A Brief History of Moon Base Plans

The US Department of the Army's 1959 lunar base planning study, Project Horizon, was implemented under President Dwight D. Eisenhower. It is surprisingly visionary, noting, "The lunar outpost is required to develop and protect potential United States interests on the moon; to develop techniques in moon-based surveillance of the earth and space, in communications relay, and in operations on the surface of the moon; to serve as a base for exploration of the moon, for further exploration into space and for military operations on the moon if required; and to support scientific investigations on the moon" (USDA Vol. I, 1959: 4).

Project Horizon envisioned a lunar outpost to support ten to twenty military personnel on a sustained, self-sufficient basis with equipment allowing survival and some construction activity. The design anticipated an expansion of the facilities, and a resupply and rotation of personnel. The location of the base would favor communications with Earth and space travel between the two. The mission of the base was to provide defense for itself against attack. The Project Horizon study considered moon-based weapons systems that could be used against Earth and space targets, and the base was to serve as a "strong deterrent to war." The design provided defensive positions that provided protection against temperature extremes and incoming meteorites. The use of solar and nuclear power was considered, in order to extract oxygen and water from the natural environment (USDA Vol. I, 1959: 6). The plan included buried, cylindrical, double-walled lightweight Titanium metal tanks, ten feet in diameter and twenty feet in length, for living quarters. The reuse of empty propellant containers as living quarters was part of the design (USDA Vol. II, 1959: 31–32). Quarters with two air locks would provide "a suitable atmosphere" from insulated tanks of liquid oxygen or nitrogen. To power the outpost, the study envisioned small nuclear power plants, 300–400 feet from personnel quarters with radiation shielding (USDA Horizon Vol. II, 1959: 27–28). This was before environmentally suitable solar panels existed.

Project Horizon was rejected by President Eisenhower when responsibility for America's space program was given to NASA, a civilian agency. Its missions have remained civilian since that time.

The Lunex Project was initiated in 1958, by the US Air Force, with the intention of establishing an underground moon base by 1968 (USAF, 1961). The main difference between this project and the Apollo missions that followed was that Lunex involved a vehicle that would land all crew on the Moon. Apollo missions used a separate module for ascent, with a command module and service module left in lunar orbit with one astronaut, while others descended to the surface. It is interesting that the design of the Apollo missions initially involved the direct ascent concept of Lunex.

11.2.2   Artemis Moon Base Planning

The Artemis Program is NASA's existing plan for an American moon base at the time of this writing (cf NASA, 2020a; 2021). It is important to remember that the mission of NASA is non-military, and that its programs and planning focus on civilian space exploration, colonization, and support for science and scientists off-world. While staff and crew who live there would be largely from the US, it is inconceivable that other Western nation-states would not have visiting or resident crew, scientists, administrators, and diplomats, as well. The space program to date has been especially open to participation by all nationalities with a variety of specializations.

As presently conceived, Artemis would have little if any defensive capability of its own. However, if any moon base–Artemis or a later design 2020 was threatened by another nation-state, such as China, an obvious mission of US Space Force would be deterrence, with the goal of protecting NASA's and perhaps other lunar bases and their operations.

For practical logistics, a US military presence on the Moon would likely be co-located with NASA, but physically separate with different and wider missions. This arrangement would provide perimeter protection, internal security, threat detection and intelligence, and additional non-military technical capabilities, such as medical services, alternative communications, and rescue-and-recovery crew and equipment. US Space Force would, to an extent, leverage NASA's lunar capabilities rather than reinvent new ones, except as needed for military missions, for example: customized military weapons suitable to the lunar environment, protected and hardened facilities for crew and weapons storage, specialized lunar suits, intelligence gathering techniques and equipment, specialized vehicles for transport and patrolling, different and more sensitive communication equipment, and a command structure that would have plans and training for threat response of a variety of types.

11.2.3 Military Mission Types

Human operation in cis-lunar space refers to a presence on the surface of the Moon, in facilities located beneath the surface, and in the surrounding space, which includes the Lagrange points. In fact, cis-lunar space reaches out to include the entire off-world space around the Earth and its Moon. US Space Force missions in cis-lunar space can be anticipated to a certain extent from past military mission profiles and knowledge of geopolitical considerations discussed earlier in this chapter. A set of missions occurring in the entirety of cis-lunar space would naturally imply coordination with many agencies, organizations, and other military forces, in the pursuit of a variety of missions.

Now that Space Force is a separate military force, it appears that one consideration is to use the Moon to train crew for missions to Mars. For both military and non-military purposes, this is quite logical because of the Moon's lower gravity, lack of atmosphere, and exposure to high radiation. Likely missions would be to protect NASA's and other Western states' moon bases and to prevent denial of the Lagrange points in space–a move that could be used by adversaries to prevent full access to lunar landing sites and eventually, to large space-based construction projects of mammoth ships for missions to the asteroids and outer planets. Protecting Lagrange points would require a lunar force projection capability, i.e., a military show of strength. This type of demonstration promises to be controversial and vigorously debated, although it is difficult to see, at the present, how it can be entirely avoided.

Other military missions might include intelligence collection and surveillance of others' lunar and space activities through stationary collection sites, or patrolling in lunar rovers, or from aerial drones and space-based orbital satellites. These missions would provide visual, digital, photographic, electronic, and telemetry surveillance data. Lunar intelligence analysis may be limited at first and data may be transmitted via encrypted telemetry to Earth-based facilities for analysis and reporting.

A further important mission type could involve both Space Force and NASA, in the tracking of asteroids and comets that could threaten the Earth or Moon. It has become clear over the past decade, as these types of threats whiz by at rapid velocities, and that the detection of possibly dangerous asteroids and comets has not yet approached a sufficient level. Coordination of Moon-based and Earth-based tracking capabilities, as well as Moon-based observatories and space surveillance systems could enhance early warning systems enormously.

It is worth considering the use of the Lunar Reconnaissance Orbiter and the Lunar Crater Observation and Sensing Satellite (both, robotic spacecraft operated by NASA) for intelligence surveillance, although the mixture of civilian and military purposes may be questioned. Still, for the foreseeable future, the number of sensitive technologies capable of producing surveillance-quality data in cis-lunar space will be limited, and how the combination of military and non-military missions will be achieved is not yet clear. Another approach is to develop military surveillance technologies that mimic those of NASA, for example, deploying a Space Force version of the planned NASA Lunar Gateway with specialized sensors and electronic, imaging, infrared, and radar surveillance systems. This might keep military missions separate from NASA's Lunar Gateway.

There are technologies that do not now have military purposes, but that could, in the future. For example, NASA's Double Asteroid Redirection Test, known as DART, is a spacecraft that will launch in 2021, on course to deliberately smash into an asteroid in the fall of 2022 (NASA, 2020b). This is a set of technologies to prevent a hazardous asteroid from impacting Earth. Redirection of the asteroid is achieved by deliberate kinetic impact deflection (Skibba, 2019). Such a system could have military applications, although it does not now.

Other military missions might include reporting any unusual and unidentified space phenomena to the Department of Defense Unidentified Aerial Phenomena Task Force, or using NASA's exploration of lunar minerals to determine if such minerals could have a military application.

11.2.4   Military Base Design

Beginning in this section, we take the reader beyond the initial stages of establishing a military presence on the Moon for a small number of crew, as the civilian program, Artemis, would initially do. NASA plans for a fixed habitat at the Artemis Base Camp that can house up to four astronauts for a month-long stay. The concept includes a modern lunar cabin, a rover, and even a mobile home. They are designing lunar habitation and associated environmental control and life support systems, including outer-structure options, i.e., rigid shells, expandable designs, and hybrid concepts, but they would lack ballistic hardening needed by USSF (NASA, 2020c). Nevertheless, much of this NASA technology could be shared with Space Force, and either supplemented or augmented.

The requirements we describe here take us into the next stage of development to describe a somewhat larger and later military presence, its specialized transportation, telecommunications, and spacesuit design needs; specialized weaponry for the Moon; and staffing a military installation on the Moon.

Military missions on the Moon and in cis-lunar space will form a foundation for the design of a military lunar base and they will determine its basic functions.

These include, minimally, housing, feeding, medical care, and supervision of military personnel assigned there. Beyond that, a military base will require greater protection and defensive hardening in comparison to, for example, a co-located Artemis-like civilian moon base. Portions of a military base would likely be underground to provide ballistic protection, as well as better temperature and solar insulation.

Components of a military base would necessarily include storage of weapons; emergency power; emergency water recycling and filtration systems; armed and armored rover garages and maintenance facilities for them; weapons launching platforms; emergency medical triage areas; housing for command, control, and communications systems; and intelligence capabilities. Some activities could involve military crew sharing NASA's habitats, for example, co-located cafeterias, kitchens, food storage, base power, wire voice and telephone communications, wireless communications, and routine medical and dental facilities.

11.2.5   Lunar Military Transportation, Telecommunication, and Space Suit Design

For transportation on the lunar surface, Space Force might use NASA's proposed lunar terrain vehicle (LTV), an unpressurized, or open-top vehicle that astronauts can drive in their spacesuits for about 12 miles from a moon base site. The LTV could also operate autonomously and drive on pre-programmed paths. NASA plans to use a pressurized rover to enable lunar surface exploration and enable astronauts to travel in their regular clothing as opposed to wearing spacesuits all the time. This would be more comfortable when exploring larger areas of lunar terrain. Spacesuits would be worn when collecting samples. This type of vehicle–if hardened with mounts for specialized weapons, and possibly controlled remotely–would suit military purposes during ground surveillance patrols, or when responding to a threat situation. Alternative lunar rovers are the Mars Perseverance Rover and the DuAxel rover being developed for Mars missions to explore deep craters and able to traverse rugged terrain (Kooser, 2020).

For Artemis, NASA is designing an improved spacesuit with better mobility, functionality, modern communications, and a more robust life support system than its Apollo predecessors (NASA, 2020a). These spacesuits could be modified for military purposes, with encrypted inter-team communications, body video, illumination and target identification using visual infrared, and computerized modeling. To provide ground communications across a large area, NASA contracted with Nokia to build and deploy a 4G-LTE (Long-Term Evolution) wireless network to provide high speed data transmission to control lunar rovers, navigate lunar geography in real time, and stream videos. The network could serve military needs for high-speed encrypted data communications.

The 4G network would eventually be upgraded to the more powerful 5G and adapted to withstand extreme temperatures, radiation, and even rocket landings and launches, which vibrate the moon's surface. The 5G network would use much smaller cells than those on Earth, with a shorter range, but also require less power and be easier to transport (Dean, 2020). A Space Force moon base would likely require a perimeter surveillance system with video feeds, area illumination, motion and infrared detection to signal intruders, but not be enclosed chain link fence with concertina or barbed wire.

A lunar military base would require advanced communications and automated system applications with a minimal IT footprint to perform its operational missions. The limitations of Moon-to-Earth data transmission require storing and processing as much data locally as possible. Telemetry and radio frequency communications with Earth–which are useful for voice and limited video–would have limited data throughput and so, could not efficiently operate interactively at high data rates with Earth-based applications. The questions are the extent to which new technologies can overcome these limitations, and how much they will inevitably render a military base alone as never before on Earth.

Military data systems could operate on VM (virtualized machine) servers in a private Cloud to minimize power and footprint. The databases would likely be loaded before launch so that only incremental data uploads would be required when operational. The base would also likely use NASA software for managing launches and space travel. A Space Force lunar base would require electrical power and would likely leverage NASA's solar panel technology and a nuclear fission surface power unit that can continuously provide 10 kW of power although several of these might be required to meet base needs (NASA, 2020a; Wall, 2020).

11.2.6 Specialized Weaponry for the Moon

To perform its basic missions and provide security for American (and indeed, all Western personnel and eventually, visitors), a lunar military base would need specialized weapons. Conventional hand-held weapons can function on the Moon but firing a 5.56 rifle or even a 9 mm handgun within a lunar enclosure could rupture a rigid hull and cause dangerous decompression and loss of life. A more practical alternative, aside from nightsticks or Tasers, might be a small caliber handgun firing shotgun-type pellets that are effective at short range. Conventional hand-held weapons could be operated outside of the enclosure, but because of the limitations of spacesuit gloves, would likely require customized triggers, trigger guards, and weapon stocks. Lower gravity and no wind resistance might make a smaller hand weapon such as a Heckler & Koch MP5-A3 9 mm submachine gun (as used by the US Secret Service) with a laser designator more practical and easier to carry.

For air defense, a mounted Stinger system would be practical, as well as anti-tank missiles like AMG-114R, Spike or Javelin missiles for anti-vehicle defense. A US Air Force program named, Self-Protect High Energy Laser Demonstrator (SHiELD), is designed to protect older Air Force fighters from incoming air-to-air or surface-to-air missiles (Mizokami, 2020). However, it could be adapted and rendered stationary, or vehicle-mounted to target ground or hovering threats. Unlike guns, lasers have an almost infinite ammunition supply, although this may differ on the Earth and Moon, where power may be uncertain or in short supply until moon bases are well established.

11.2.7 Lunar Military Base Staffing

Base staffing would necessarily align with military mission types. This logically implies round-the clock, seven-days-per-week staffing of important intelligence, operations, communications, security, and logistical functions. Hypothetically, a base crew might consist of:

- A Base Commander with overall responsibility for base operation, sustainment, communications, security, personnel welfare, and interaction with NASA.
- A Base Executive Officer as second-in-command and responsible for administrative and human resources management.
- An Operations Officer and an Assistant Operations Officer, who oversee all operational functions, evaluate intelligence information, and plan for security and other operational missions.
- An Orbit/Lander/Rocket Engineering Team of three engineers tied closely to their NASA counterparts to coordinate orbital operations, docking, and orbit-lunar transport.
- An Intelligence and Surveillance team of at least three specialists, who monitor security surveillance feed, receive and process intelligence and operational information, and report to the Operations Officer.
- A Special Weapons Team of three-four weapons specialists who remotely operate from inside a hardened base bunker, and who have defensive and special-purpose weapons, perhaps like the asteroid-deflecting DART system, the SHiELD Laser anti-missile system, and defensive anti-air missile systems such as a rotatable Stinger missile.
- An IT/Communications Team of at least three to four specialists to monitor and maintain communications systems, perform system administration, troubleshoot IT applications, workstations and servers, and process and report messages.
- A Security Officer responsible for overall security, weapons manning, and leading a small security team of five to six armed security personnel trained to respond using SWAT-like tactics, weapons employment, and bomb disposal.
- A Logistics Officer responsible for managing and coordinating all supplies, vehicle maintenance, engineering, power, armory (weapons and ammunition functions), and medical functions. Under the Logistics Officer, there might be a supply specialist, mess (food preparation) specialist (unless this function was shared with NASA), a medical officer and assistant (unless shared with NASA), an

ordnance/armory/weapons specialist, and one or two military engineers to maintain the power systems, install/repair defensive/perimeter surveillance systems, repair equipment and facilities and perform de-mining, if required.

This staffing configuration would total around 30 staff, which is not much larger than envisioned for the USAF Lunex Project.

11.2.8 Role of Artificial Intelligence (AI) in Reducing Staff Needs and Saving Lives

Staffing requirements could be reduced by developing Artificial Intelligence (AI) software that could reduce the need for continual human oversight. This could be accomplished by automating the processing of threat detection sensor data, alerting staff, and even automating a weaponized response to an immediate detected threat.

AI could play an even broader role in lunar operations. The harsh, airless, dusty, barren, and cratered lunar environment poses challenges to many lunar military operations, especially those that are constrained by force size and the need for specialized capabilities. To effectively surveil and control any portion lunar surface, Space Force and Western allies, in collaboration with NASA, will be required to innovate with new technologies and new weaponry that can support the ability to surveil, protect, patrol, and respond to security threats in the large lunar environment. Technologies based on AI robotics, drones that are modified for lunar operations, advanced weapons, and small efficient power sources designed for lunar operations must be developed and tested. These advanced capabilities should enhance and expand on emerging terrestrial technologies, as well as technologies developed by NASA for civilian lunar, Martian, and farther exploration of space.

It is noteworthy that the US Army Telemedicine and Advanced Technology Research Center (TATRC) is developing a robotic manipulation system to assist combat medics in the field (RBR, 2017). Medical evacuation of military personnel or, for example, civilian scientists who fall grievously ill, will be limited by availability of transport and the sheer distance of the Moon from the Earth. AI robotic capabilities have the potential for saving many lives in the new and dangerous lunar environment that humans are about to enter.

11.3 Conclusion

Assuming that US political will stays steady and necessary funding is forthcoming, NASA is committed through its Artemis Program to establish an extended human presence on the Moon, and to explore the Moon and use its moonbase to support later travel to Mars. It is important to note that US Space Force has not yet publicly stated any mission or plans to establish a US military presence on the Moon. Potential lunar or cislunar actions, or strong indications by China, in particular, or other space-capable nation-states, to create a military presence on the Moon that can threaten US lunar operations or cis-lunar security could very rapidly change that. This also may be affected by current and future international treaties and resolutions governing military activity on the Moon. Any discussion of a possible US lunar military base remains very speculative at this stage and assumes a rationale will emerge (very likely if China pursues their lunar intentions) and diplomatic actions and agreements do not preclude American action. The ability to move quickly, should a relevant impetus occur, is helped by the technological developments of the Artemis Program which could be leveraged by US Space Force. To determine such a national security need, the U.S. must continue to monitor very closely China's ongoing initiatives on the Moon and in cis-lunar space, in view of China's aggressive international economic and military expansion, and threatening actions.

Disclaimer This chapter does not represent the views or opinions of KBR, Inc.

References

Africa Times. (2018). China-Africa security forum concludes in Beijing. Online, July 11. https://africatimes.com/2018/07/11/china-africa-security-forum-concludes-in-beijing/

Anderson, D., Hunt, B., & Mosher, D. (2019, September 24). NASA's $30 billion Artemis missions will attempt to set up a moon base. *Business Insider*. https://www.businessinsider.com/nasa-art emis-moon-base-apollo-space-rocket-sls-2019-9

Arnould, J. (2019). Colonising Mars. A time frame for ethical questioning. In K. Szocik (Eds.), *The human factor in a mission to mars. space and society.* Cham: Springer. https://doi.org/10.1007/978-3-030-02059-0_7

Campbell, K. M., & Rapp-Hooper, M. (2020, July 15). China is done biding its time; The end of Beijing's foreign policy restraint? *Foreign Affairs.* https://www.foreignaffairs.com/articles/china/2020-07-15/china-done-biding-its-time

Dabus, A., Badu, M., & Yao, L. (2019, May 27). China's Belt and Road reaches Latin America. *Brink News.* https://www.brinknews.com/chinas-belt-and-road-reaches-latin-america/

David, L. (2020, September 18). Is Earth-moon space the US military's new high ground? *Microsoft News.* https://www.msn.com/en-us/news/technology/is-earth-moon-space-the-us-militarys-new-high-ground/ar-BB19aLTg?ocid=se

Davis, M., & Jones, C. L. (2020, October 14). Chinese test shows potential for ballistic missiles on merchant ships. *The Maritime Executive.* https://www.maritime-executive.com/editorials/chinese-test-shows-potential-for-ballistic-missiles-on-merchant-ships

Dean, G. (2020, October 19). NASA gave Nokia $14.1 million to build a 4G network on the moon. *Business Insider.* https://www.businessinsider.com.au/nasa-nokia-4g-network-moon-2020-10

Gertz, B. (2020, October 28). US, India step up defense ties with a wary eye on China. *The Washington Times.*

Haqq-Misra, J. (2019). Can deep altruism sustain space settlement? In K. Szocik (Eds.), *The human factor in a mission to mars.* Cham: Springer. https://doi.org/10.1007/978-3-030-02059-0_8

https://www.nasa.gov/sites/default/files/atoms/files/nasa_ussf_mou_21_sep_20.pdf

Keeley, M. (2020, August 8). China and Taiwan both send military to South China Sea as tensions grow. *Newsweek.* https://www.newsweek.com/china-taiwan-both-send-military-south-china-sea-tensions-grow-1523840

Kooser, A. (2020, October 13). NASA prototype rover can split in two, could climb down deep Mars craters. *CNET*. https://www.cnet.com/news/nasa-prototype-rover-can-split-in-two-could-climb-down-deep-mars-craters/

Meltzer, J. P. (2020). China's digital services trade and data governance: How should the United States respond? *Brookings*. https://www.brookings.edu/articles/chinas-digital-services-trade-and-data-governance-how-should-the-united-states-respond/

Mizokami, K. (2020, November 11). The air force is putting death rays on fighter jets. Yes, Death Rays. *Popular Mechanics*. https://www.popularmechanics.com/military/aviation/a34632516/air-force-fighter-jet-death-rays/

USSF & NASA (jointly published). (2020, September 21). *Memorandum of understanding between the National Aeronautics and Space Administration ad the United States Space Force.*

NASA. (2020a). Lunar living: NASA's Artemis base camp concept, NASA Blog, October 28, 2020. https://blogs.nasa.gov/artemis/2020/10/28/lunar-living-nasas-artemis-base-camp-concept/

NASA. (2020b). Planetary defense: DART. https://www.nasa.gov/planetarydefense/dart

NASA. (2020c, September 22). NASA Press Release 20–091: "NASA, US space force establish foundation for broad collaboration." https://www.nasa.gov/press-release/nasa-us-space-force-establish-foundation-for-broad-collaboration

NASA. (2021). Artemis Program. https://www.nasa.gov/artemisprogram

RBR—Robotics Business Review. (2017, January 9). Combat medics to get robotic help from RE2 Grant. https://www.roboticsbusinessreview.com/ health-medical/combat-medics-get-robotic-help-re2-grant/

Reuters. (2017, August 1). China formally opens first overseas military base in Djibouti. https://www.reuters.com/article/us-china-djibouti/china-formally-opens-first-overseas-military-base-in-djibouti-idUSKBN1AH3E3

Roberts, J. (2020, September 23). Preparations of next moonwalk simulations underway and underwater. NASA. https://www.nasa.gov/feature/preparations-for-next-moonwalk-simulations-underway-and-underwater

Skibba, R. (2019, May 7). NASA's DART mission will try to deflect a near-Earth asteroid. *Astronomy*. https://astronomy.com/news/2019/05/nasas-dart-mission-will-try-to-deflect-a-near-earth-asteroid

SSIB—State of the Space Industrial Base. (2020). A time for action to sustain US economic and military leadership in space. Report from a virtual solutions workshop hosted by NewSpace New Mexico and the defense innovation unit, US air force, and the Air Force Research Laboratory (AFRL).

Taylor, G. (2020a, October 5). Hawks Push US to confront rising China. *The Washington Times*.

Taylor, G. (2020b, October 7). Pompeo pitches Asian allies on plan to counter China. *The Washington Times*.

Taylor, G. (2020c, December 22). U.S., China charge toward 'full-blown cold war' as Beijing becomes more aggressive. https://www.washingtontimes.com/staff/guy-taylor/

USAF—US Air Force. (1961). Lunar expedition plan: Lunex. Air Force Systems Command, Space Systems Division.

USCESRC—US-China Economic and Security Review Commission. (2020, February 20). Agenda: Hearing on 'China's military power projection and U.S. national interests'. https://www.uscc.gov/hearings/chinas-military-power-projection-and-us-national-interests

USDA—US Department of the Army. (1959). Project Horizon; A U. S. army study for the establishment of a lunar outpost, Vol. I, Summary and supporting considerations; Vol. II, Technical considerations and plans. Army Chief of Ordnance and Research and Development.

Wall, M. (2019, January 5). China just landed on the moon's far side—and will probably send astronauts on lunar trips. *Space.com*. https://www.space.com/42914-china-far-side-moon-landing-crewed-lunar-plans.html

Wall, M. (2020, December 17). Trump signs space policy directive-6 on space nuclear power and propulsion. *Space.com*. https://www.space.com/trump-space-policy-nuclear-power-propulsion

Chapter 12
A Right to Return to Earth? Emigration Policy for the Lunar State

James S. J. Schwartz

Abstract I argue that citizens of future lunar states should enjoy the right to emigrate. The lethality of the space environment may result in lunar settlements pursuing oppressive and illiberal norms, policies, and laws in order to resolve societal problems. A right to emigrate, which for the foreseeable future is tantamount to a right to return to Earth, has individual and societal value. For individuals, the right establishes a legal path for lunar citizens wishing to flee from averse social or political circumstances. For the lunar state, protecting the right to emigrate disincentivizes the pursuit of policies which give rise to desires to flee. After presenting the case for the right to emigrate, I respond to several objections, including an objection derived from the "brain drain" debate over terrestrial migration, as well as the objection that protecting the right to emigrate would be financially ruinous for the lunar state.

12.1   Introduction

In all of the excitement surrounding the prospect of human settlement of the Moon, it is easy to forget that it will be incredibly difficult for humans to eke out an existence there. The Moon is devoid of immediately available breathing air and drinking water, and lunar settlers will have to devote a considerable portion of their energies to the provision and distribution of the basic necessities of life. Whoever is tasked with overseeing life support production and distribution systems might succumb to the temptation to use their position to extort others by controlling the flow of breathable air or consumable water. Privacy may be virtually nonexistent because pervasive surveillance and security enforcement may be necessary to ensure that airlocks are not misused in ways that lead to depressurization events, and that life support production and distribution machinery is not vandalized or tampered with. And the need to maintain a population that is both large enough to sustain the settlement but not so large it strains life support systems will place possibly unwelcome bounds on the reproductive autonomy of settlers. So, we may find that lunar societies adapt in

J. S. J. Schwartz (✉)
Department of Philosophy, Wichita State University, 1845 N Fairmount, Wichita, KS 67260, USA
e-mail: james.schwartz@wichita.edu

© The Author(s), under exclusive license to Springer Nature Switzerland AG 2021   193
M. B. Rappaport and K. Szocik (eds.), *The Human Factor in the Settlement of the Moon*,
Space and Society, https://doi.org/10.1007/978-3-030-81388-8_12

ways that run contrary to liberal conceptions of the ideal life.[1] Perhaps the rights and privileges that humans have become accustomed to, especially humans living in contemporary Western democracies, are maladaptive for lunar or other space societies. However, there are strategies open to us that could reduce the odds that the lunar state opts to employ illiberal or authoritarian practices, and in this chapter I shall argue for one such course of action, viz., that the right to emigrate be included among the basic liberties afforded to lunar citizens.

While I have discussed the extent of rights or entitlements for space-dwellers elsewhere,[2] this chapter is not a general discussion of what is owed to space settlers, but focuses narrowly on the right to emigrate. Further, while this chapter draws on the philosophical literature on freedom of immigration and open borders, I will not defend freedom of immigration or open borders per se. Instead, the reasons I will offer in support of lunar citizens' right to emigrate reference the particular circumstances that are likely to arise in lunar societies and what it would mean for a human life to go well or poorly living in such circumstances. Thus, the conclusions I reach about emigration concern lunar settlement in particular as opposed to space settlement generally. While it is possible that what I have to say here applies *mutatis mutandis* to Martian or other space settlement efforts, determining whether this is so is beyond the scope of this chapter.

To circumscribe matters further, my primary interest in this chapter concerns the liberties and entitlements of those living in *permanent* lunar societies. It does not matter for my purposes whether these are independent states or settlements supported by terrestrial states. What matters for my analysis is that the citizens of these societies face the prospect of spending the remainder of their lives on the Moon, have raised children or plan to raise children there, or are themselves natives of the Moon. In other words, my concerns rest primarily with the people who will call the Moon a *home* as opposed to a mere *construction site, workplace* or scientific or military *outpost.*

The essence of my argument is that a right to emigrate is a necessary form of insurance against the violation of other rights that lunar settlers–indeed all humans–should be entitled to, most importantly: freedom of expression, religious liberty, vocational autonomy, and bodily and reproductive autonomy. Not only does protecting the right to emigrate provide an escape route for those whose basic rights are being violated, it also pressures the lunar state to ensure that other rights violations are minimized, if not eliminated entirely. It also encourages the state to build and maintain institutions that make the prospect of living on the Moon attractive and rewarding, as opposed to a constant struggle for survival.

In these respects, my arguments rest squarely in the liberal tradition, and throughout I will assume the correctness of a broadly liberal conception of the good life. Moreover, this chapter is more an exploration of possible norms and ideals for

---

[1] See the contributions to Cockell (2015, 2016a, b) for discussion.

[2] See Greenall-Sharp et al., (2021), Schwartz (2016, 2018, 2020).

space settlement than it is roadmap for establishing those norms or ideals,[3] although pragmatically-oriented objections are considered in Sects. 12.4 and 12.5. Fortunately or unfortunately, space ethics is a nascent field, and is in as much need of ideation of moral principles and institutional arrangements as it is of implementation strategies. I take seriously Joseph Carens's exhortation that "[w]e want institutional arrangements that will enact and reflect our principles of justice, not principles of justice that simply reflect our institutions arrangements" and that "[i]f we begin with the moral obligations that we have within existing institutions and arrangements and allows those to set limits to our moral horizons, we will simply reproduce and legitimate whatever moral defects they contain" (2013, 259). In the present context, I take seriously the oft-repeated refrain that space settlement presents an opportunity for societal progress (Yorke, 2016), and I would add that this involves, *inter alia*, reexamining assumptions about what humans are owed as a matter of course. Simply going into space will not automatically make anyone's life better, now or in the future. It matters not only *what* we choose to do in space, but *how* we choose to do it. We are only just beginning to consider what it would mean to pursue a truly *humanitarian* approach to space settlement.

12.2 Freedom of Emigration: From the Earth to the Moon

In order to see why lunar states should protect the right to emigrate, it will be helpful to rehearse some of the terrestrially-oriented rationales for acknowledging this right. To begin with the matter of precedent, the right to emigrate is asserted in Article 13 of the United Nations' *Universal Declaration of Human Rights* and in Article 12 of the UN *International Covenant on Civil and Political Rights*. Thus, the right to emigrate is already recognized by the international political community, and therefore it is not simply consonant with, but moreover it is partially constitutive of existing international political norms. But of course, we should not confuse precedent for justification.

One rationale for the right to emigrate is derivative of more general reasons for ensuring freedom of movement and self-determination, which is the protection of other, more basic rights. As Alan Dowty argues,

> Freedom of movement is also vital to fulfillment of other basic human rights and needs, including marriage and family life, an adequate standard of living, education and employment, the pursuit of creative opportunities, and the practice of religion. There are few human rights, important or trivial, whose enjoyment is not easier when international travel is possible. (1989, pp. 15–16)

This argument may present a relatively weak case for rights to emigrate from states closer to being ideal states, i.e., ones that protect the basic rights of their citizens,

---

[3] To use language more familiar to political theorists, this chapter is more an example of ideal theory than it is of realpolitik or non-ideal theory.

offer them an adequate range of opportunities, etc. But, as Dowty notes, the force of the argument is clear in non-ideal circumstances:

> The right to leave is most important when other rights are denied. In that case, it becomes the ultimate defense of human dignity... Historically, the final recourse against intolerable conditions has been flight... Denial of the right to flee is therefore a more basic insult to human dignity than anything save loss of life itself, for it eliminates the means of escaping all other forms of persecution and injustice... It is, in other words, the *ultimum refugium libertatis* – the last refuge of liberty–when other rights are threatened. (ibid., p. 16)[4]

The right to emigrate, then, is of the essence for persons living in states that are rife with persecution, which do not protect the basic rights of their citizens, do not provide citizens with opportunities for accessing the basic necessities of life, etc., because it provides a legal escape route, at least for those for whom flight is the only reasonable strategy for avoiding persecution, starvation, or worse.

Nevertheless, as Dowty notes, there is an ancillary, consequentialist reason for supporting the right to emigrate, viz., as a counter to repressive and coercive forms of governance. "So long as the option of leaving remains open," Dowty states, "the regime must confront sources of dissatisfaction" (ibid., p. 19). A state that prohibits emigration effectively holds its citizens hostage to whatever forms of persecution or injustice it visits upon them. A state permitting emigration, meanwhile, would have an incentive to reform if it is acting in ways that generate desires to leave among the citizenry. The importance of the right to emigrate, then, is especially great within states that are experiencing pressures–either internal, external, or environmental–for curtailing the rights of their citizens.

A further justification for the right to emigrate is that it touches deeply on the liberal ideal of governance by the consent of the governed. As Michael Blake argues, "[l]egitimate governments have no right...to insist on coercively maintaining *permanence* in their relationships with their subjects" (2014, p. 525). Many factors affect a person's ability to exit a state and take up residence elsewhere, including their financial resources and legal obligations, as well as the existence of willing receiving states. Moreover, there are many circumstantial reasons that might compel someone to remain somewhere they are unhappy. But it is one thing to say "I cannot move because I want to stay close to my family" or "I'm staying here because I need my job". It is another thing to say "I cannot move because my government won't let me leave". Emigration restrictions represent *legal* impediments to renouncing state membership–but one cannot speak meaningfully of *consenting* to the laws of a state from which one has no *legal path* to exit.

---

[4] C.f. Blake (2014), who provides a similar argument in opposition to *immigration* restrictions for refugees.

12.3   Provisional Lessons for Lunar States

I have described three basic reasons for acknowledging the right to emigrate: (1) it is necessary for preserving basic rights and escaping injustice; (2) it incentivizes responsible governance; and (3) it is necessary for governance by the consent of the governed. *Prima facie*, each reason applies to lunar states.

Citizens of a lunar settlement might experience many forms of injustice or curtailment of basic rights: They might be coerced unacceptably into specific vocations (to provide essential services for the lunar state), particular places of residence (to maximize efficiency of life support systems), or even forced to participate against their will in reproductive acts (in order to secure a minimum viable, genetically diverse population). They might find themselves persecuted against on the basis of ethnicity, gender, sexual orientation, religion, or disability. They might find that they or their children have inadequate access to opportunities for pursuing lives worth living, especially with respect to opportunities for rest, recreation, creativity, romantic companionship, or spending time with family. They may find that access to life's basic resources becomes too expensive or too unreliable for forming and carrying out long-term plans. They may find that existence in a lunar state amounts to a constant struggle for survival. On the Moon, where there will be continual pressures to prioritize efficiency and security over other concerns, the right to emigrate will serve as a necessary outlet for those who have decided the only reasonable recourse is exiting the lunar state.

Furthermore, citizens of lunar states may encounter problems like those just mentioned, not simply as the result of the ebb and flow social currents, but also as the result of deliberate policies set in place by the ruling authorities, perhaps from a perception of the fragility of the lunar state given the wide range of threats to its existence. Thus, the lunar state could easily qualify as the kind of state that is *most* in need of incentives to ensure that the lives of its citizens go well rather than poorly. And of course, if citizens of a lunar state have no legal path to emigrate, they effectively become hostages of the lunar state, undercutting the idea that they are being governed under principles which they have consented to (in one form or another).

The pressures which recommend a right to emigrate will exist in lunar states. Thus, I would like to conclude, provisionally, that lunar states ought to protect the right to emigrate. And, as I will argue over the next two sections of this chapter, this position holds against several sources of criticism.

12.4   The "Brain Drain" Objection to the Right to Emigrate

One of the most widely-discussed objections to the right to emigrate is the "brain drain" objection, which calls attention to the sometimes harmful consequences of emigration. Of particular concern are the effects of the emigration of skilled workers from states suffering critical shortages in essential services, for instance the emigration of healthcare workers from developing states with critical shortages of nurses and physicians. Emigration of skilled workers exacerbates problems in the short-term by rendering sending states less able to fulfill their obligations to provide essential services to their citizens. Emigration also deprives sending states of taxable income and specialized knowledge which, over the long-term, renders these states less able to establish or maintain institutions capable of fulfilling their obligations to provide essential services for their citizens. Unrestricted emigration threatens to draw human capital, intellectual resources, as well as financial resources *away* from where these resources are needed the most. Thus, emigration restrictions, provided they are implemented in ways that help struggling states satisfy their duties to meet the basic needs of their citizens, may be ethically permissible.

It should be easy to see why the lunar state might come to worry about a problem akin to brain drain. The hostile lunar environment will ensure that the survival of a lunar state will always be a tenuous affair, with disaster just one failed airlock, one failed air pump, one underperforming hydroponic crop yield, etc., away. An obvious solution to this problem would be to ensure that there is always a sufficiently large and diverse pool of human capital–in the form of oxygen production staff, water production staff, food production workers, maintenance workers, healthcare workers, etc. Shortages of individuals in any of these critical areas might leave the lunar state unable to respond to a crisis, or even to maintain basic life-support systems during non-crisis periods, depending on how narrow its normal operating personnel margins are. Thus, if emigration is permitted, especially of essential services workers, then the lunar state will face a constant risk of destabilization through the loss of human capital.

Nevertheless, if we examine this objection in more detail, an important disanalogy emerges between terrestrial skilled emigration and lunar emigration, which unveils lunar emigration as less likely to admit of restrictions. For the sake of brevity, my foil here will be Gillian Brock's defense of emigration restrictions (Brock & Blake, 2015), which in any case provides a good preview of the reasons why a number of other scholars have also argued that emigration restrictions can sometimes be justified.[5] Brock's focus is emigration of skilled workers from regions in which the emigrees' skills are in vital demand, e.g., the emigration of healthcare workers from states suffering massive healthcare worker shortages. She maintains that, so long as certain conditions obtain, it would be permissible to enact reasonable emigration restrictions, possibly including periods of compulsory service.

Brock identifies a series of background conditions that circumscribe the range of acceptable compulsory service regimes. These conditions include: that there is evidence that skilled migration would exacerbate deprivation and undermine the state's attempts to meet its citizens' needs; that the state has invested "in training skilled workers to provide for their citizens' needs and to promote beneficial development"; that those seeking skilled training are made aware of this information and made aware that compulsory service will be expected of them; and that the loss to the state resulting from departure would not be adequately compensated by other means (e.g., remittances, taxes) (ibid., 101–2). Beyond these background conditions, the

---

[5] *Cf.* Song (2019) and Oberman (2013).

two most salient framing assumptions that Brock makes are that (a) a substantive case can be made that potential emigres are to some degree *responsible* for remedying the situation facing their home state, and that (b) the state in question is exercising its power *legitimately* when imposing its emigration restrictions (ibid., 102–4). The legitimacy of a state requires that it "sufficiently respect[s] human rights, especially core civil and political ones" (ibid., 85), that its power "must be exercised in ways that show concern for the needs of citizens, such as by providing core public goods essential for a decent life (and doing what is necessary to provide secure access to these, such as planning for citizens' well-being)" (ibid., 86), and that it refrains from persecuting its citizens (ibid.).

In summary, if the state is in dire need of skilled workers, and if citizens willingly seek out state-funded training fully aware of the fact that a period of compulsory service will be expected of them, then by completing their training these persons have agreed to a reciprocal obligation to provide their services for the benefit of the state. Provided that the state is using its power legitimately, it is within its rights to prevent these skilled workers from emigrating until they have completed their compulsory service.

For the sake of argument, I would like to grant that Brock's position is correct. What I would like to challenge is that anything resembling a reciprocal duty to remain on the Moon to perform skilled labor could arise without undermining the *legitimacy* of the lunar state.

It is far from clear that, in lunar states, the receipt of training in essential services will be optional for citizens in the way that it is optional for citizens of terrestrial states. In general, members of terrestrial states are not *required* to become nurses, refuse collectors, physicians, technicians, etc., and thus they are under no compulsion to seek out any training that might engender compulsory service obligations. This is in part because most terrestrial states enjoy a surplus of human capital. On Earth there are usually more than enough individuals available to perform essential societal functions, even if, unfortunately, it is rare that enough individuals wish or are able to seek training and employment in all important labor sectors.

For the lunar state, this proves *at most* that emigration restrictions on essential service workers might be permissible when citizens' decisions to acquire specialized training are voluntary in the vast majority of cases. This would only be possible in lunar states with populations large enough so that all essential services can be provided by the uncoerced vocational choices of the citizenry. At the same time, such a state could more easily survive (or even fail to notice) the departure of some of its essential service workers, weakening any claim it might have to imposing emigration restrictions, especially restrictions mandating the indefinite stay of potential emigres.

In a smaller lunar state, few citizens if any will enjoy the opportunity to perform *non-essential* work. The state's survival might even depend on the laboring of every citizen (which could include the requirement to participate in reproductive activities in order to maintain a minimum viable population base). Here there is no longer any effective option to *not* seek the kind of training that the state takes to engender a reciprocal obligation of compulsory service. But if every citizen is essential, then none may be permitted to leave. And at this point, I would offer that we are no longer describing a *legitimate* state, that is, one which has done the necessary work for ensuring that, e.g., its citizens enjoy core civil and political rights. This is because there are good reasons for holding lunar states to *higher* standards of legitimacy compared to terrestrial states.

Terrestrial states are not generally *substitutable* goods. The benefits conferred by state membership are not limited to protection from violence or the provision of basic necessities, but also include access to valued connections to human communities, a shared history, and ties to the land. These relationships are often *geographically situated*, that is, they make essential references to particular geographic locations or regions. Out of respect for the importance of these relationships, we adopt presumptions against forced relocations, since this alienates peoples from deeply valued components of living a human life, which in almost all cases involves belonging to a particular nation or community. Thus, where there are people living in need, we should prioritize bringing help to them, as opposed to requiring them to travel to however far away help is already available. People in need will have established, valued relationships with their homes, and these are relationships that we should not ask them to uproot unless there is no reasonable alternative available. In this respect, part of a state's claim to *legitimacy* is the fact that it exists in a particular place at a particular time, with the connections that it has with its people and has had with its people throughout its history.

The central disanalogy is that no lunar state already exists. No people of the Moon yet exist who have developed deep community and geographical ties to their homeland. Moreover, while a sense of community identity can be forged relatively quickly in extreme situations (Ntontis et al., 2017), nevertheless a variety of factors influence whether the resulting community becomes therapeutic or corrosive, i.e., becomes enabling or disabling of social cohesion and mutual assistance (Lidskog, 2018). In other words, we cannot simply assume that lunar settlers will be inclined to cooperate with one another to solve major societal problems. This is not the kind of buck we can afford to pass.

So, our question here is not what to do about struggling people who have already formed morally compelling ties to a particular place, region, or home. Our question is not about *reforming* the lunar state. *Instead, our question concerns what should be expected of the founders and builders of the lunar state. Our question concerns how much must be done to prove that we are genuinely ready to create a new, off-world society.*[6] After all, the level of foresight and support given to lunar settlements will impact the evolution of the lunar state and its culture. If injustices, persecution, and curtailment of liberties are to become norms of life in the lunar state, then this will be because this path was left (more) open as a result of the decisions of the founders.

A productive path forward would be to devise innovative proposals for guaranteeing that the lunar state has access to the resources it needs to deal with the problems it is likely to face, and to deal with these problems in ways that do not require unreasonable curtailments of individual liberties. While we should continue searching for

---

[6] See Schwartz et al. (2021) for a discussion of what else we might ask as part of a "humanitarian review" of space settlement proposals.

and may find engineering solutions to some problems (Cockell, 2019), policy and institutional strategies will also be needed.

If we rightly anticipate that labor shortages will be one of the major problems that the lunar state will face, then we should recognize that the lunar state must protect the right to emigrate to ensure that it does not opt for policy solutions that unjustly coerce labor from its citizens. A commitment to avoid coercive employment policies simply becomes constitutive of the legitimacy of the lunar state. However, as Brock would remind us, essential services cannot be provided without skilled workers working within reliable institutions.[7] But instead of viewing the maintenance of reliable institutions as a justification for curtailing emigration, we should instead see the construction of reliable institutions as a *prerequisite* for founding a legitimate lunar state.

Rather than asking whether lunar states may permissibly restrict emigration, we should instead ask if *we may permissibly create* a society (especially one we expect to be impoverished in many ways) that restricts emigration. Similarly, in place of asking whether lunar states may permissibly enforce restrictions on religious, vocational, or reproductive autonomy, we should instead ask if *we may permissibly create* a society with both the incentive and the power to enforce such restrictions. To deny that such things are permissible is not to demand the impossible or even the improbable. Strategic investments in research related to human space settlement–not simply technological, biomedical, and psychological research–but also anthropological, sociological, philosophical, and political science research–all promise invaluable insights as we contemplate creating a lunar society. And, given time, these disciplines will produce reliable answers to question such as: What emigration rates could a lunar state expect to face in various contexts? How does population size correlate with perceptions that one has an adequate range of vocational opportunities? Attempting to create a space society without knowing the answers to these questions would be like trying to build a rocket and fly a mission before knowing anything about the shape and mass of its payload. These investigations, while very much in their infancies, are not luxuries, but simply critical parts of doing our due diligence as we make decisions about lunar settlements–decisions which could affect generations of lives.

12.5   Additional Objections and Replies

12.5.1   What if Earth Closes Its Doors to Immigrants?

One issue with a right to emigrate is that its exercise ultimately depends on the existence of a willing receiving state, as the freedom to leave means little when there is nowhere else to go.[8] If lunar citizens have the right to emigrate and return to Earth,

---

[7] See also Shacknove (1985) and Sager (2014) on the importance of good institutions as a prophylactic against emigration.

[8] See Ypi (2008), Stilz (2016), and Wellman (2016) for discussion.

what good would this be if terrestrial states refuse admission to lunar emigrants? If there are multiple lunar states, or multiple space-based states, then lunar emigrants will have options other than Earth-return. But there may be no adequate alternatives to Earth-return. For this reason, terrestrial states must be prohibited from denying admission to lunar immigrants. No state should be permitted to found (or help found, or approve a launch license for persons seeking to found) a lunar state, without also accepting responsibility for accepting immigrants and refugees from this lunar state. Thus, retaining the capacity to absorb lunar immigrants should factor into the preparations a state makes[9] as it prepares to settle the Moon.

12.5.2 Will It Be Too Expensive? Who Shoulders the Burden?

Freedom of emigration may be too expensive or resource-intensive to maintain. While Moon-to-Earth transit is much less energy intensive, and presumably, much less expensive than Earth-to-Moon transit, the price of the trip could still fall beyond the means of most lunar citizens. Thus, we might expect the lunar state to bear the costs associated with lunar emigration. But supporting an indefinite capacity for sending lunar citizens to Earth could fall beyond the means of the entire lunar state, especially if it is struggling to maintain basic life support for its citizens. Even in stabler lunar states, if the state must shoulder the burden of providing transportation to Earth, this represents a burdensome opportunity cost. What good is a freedom that few citizens or states could afford to uphold?

First, we should not assume that all emigration costs will fall on either the settlers or the lunar state itself. We might insist that Earth-return capabilities be built into the lunar settlement as an expected cost shouldered by any desiring to found a lunar settlement, in much the same way that we require sailing vessels to come with lifeboats. (If lifeboat-style spacecraft are used for this purpose, they could also function as temporary refuges during emergency situations.) Settlement growth, either by immigration or reproduction, could be tied to increases in the emigration (or "lifeboat") capacity of the settlement, rather than being tied simply to increases in habitat size and life support capacity.

The issue of cost also demonstrates the wisdom of settling the Moon only after the development of a reliable Earth-Moon transportation network. If there is regular, frequent transportation between Earth and the Moon, then much of the necessary transportation infrastructure for lunar emigration will already exist, and emigrants could simply hitch rides on spacecraft already planned for Moon-to-Earth trips. Humanitarian-minded spaceship designers might even view spare crew capacity as

---

[9] Or requires private-sector settlement initiatives to make!

a design requirement for cislunar transports to ensure there are always berths for emigres.[10]

12.5.3 Will It Be Physiologically Possible?

Lunar citizens, especially those reared on the Moon, may not be physiologically capable of leaving the Moon to live on Earth due to Earth's much stronger gravity. While advances in medical and biotechnology may provide therapeutic solutions to this problem,[11] it may be that physical adaptation to lunar gravity is irreversible. While this renders ineffectual immigrating to Earth, it would still leave immigrating to other space states a potential remedy, if any other space states exist. But it simply may not be possible for citizens of the lunar state to emigrate, no matter how pleasantly or unpleasantly things are going for them. For this reason–that is, because lunar citizens might be held hostage *physiologically* by the Moon–it is even more important for us to design reliable societal institutions prior to founding a lunar state. We must view the creation of a lunar settlement as a serious undertaking, one that we are only ready to engage in when we have shown that we understand and are prepared to shoulder the many responsibilities and burdens that come with the creation of a new society. *We must ensure that the Moon is a pleasant place to live precisely because for some, especially those reared on the Moon, exit may not be an option.*

12.6 Conclusion

In endeavors like space settlement, we must keep in mind that our goal is not simply to perpetuate the human species, but to advance and grow human lives and human cultures. It is not space settlement *per* se that is desirable, but instead the growth and flowering of human societies and cultures in space. We will not make progress on these goals if we are stuck in a mindset that focuses first and foremost on cost reductions and on forcing space travelers to survive with the bare minimum. As Carens reminds us in a different context, "[e]ven if we must take deeply rooted social arrangements as givens for purposes of immediate action in a particular context, we should never forget about our assessment of their fundamental character. Otherwise we wind up legitimizing what should only be endured" (2013, 229). An attitude which regards protections of basic liberties for space-dwellers as profligate luxuries

---

[10] A similar reply is possible in defense of a right to emigrate from any settlement in cislunar space, since settlement-to-Earth transit times and costs should be comparable to Moon-to-Earth transit costs.

[11] For discussion regarding the ethics of using biotechnology to enable space settlement, see the contributions to Szocik (2020).

is a clear example of an attempt to legitimize what should only be endured–a confined existence in a hostile world from which one cannot exit.

While many (especially in the United States) mythologize humble beginnings on open frontiers, we should not forget that, in addition to causing countless atrocities committed against Indigenous Peoples, America's founding and expansion also produced many failed settlements rife with suffering (Limerick, 1992). While the first lunar settlement will not encounter any indigenous cultures, it will still be accountable to its own people, and moreover, it will be accountable to all of humanity for the role it plays in creating a human future in space. How many times can we afford to *fail* at space settlement?

We should set the bar high instead of low when it comes to offering lunar settlers lives that are worth living. Because the higher the bar is that we overcome when planning and constructing the first lunar state, the better able lunar society will be to provide for its citizens and to ensure that their lives go well rather than poorly. For, as Tony Milligan implores,

> …whoever we bring into being should have at least the *opportunity* for some sort of good life (on a complex understanding of the latter) even if suffering figures as a component part of such a life just as it figures as a component part of our lives and most human lives. It should be a life, in short, which the agent themselves could readily accept as meaningful in spite of suffering and meaningful and worthwhile in its own right rather than being simply a part of someone else's grand plan. (2016, p. 16)

We should take every reasonable step available to us to ensure that anyone living in a lunar state has an opportunity for some sort of a good life. Supporting a right to emigrate is part of ensuring that such an opportunity persists, no matter whether the lunar state succeeds or fails in satisfying its other obligations to its citizens.

Acknowledgements I thank Tony Milligan as well as two reviewers for commenting on previous versions of this chapter.

References

Blake, M. (2014). The right to exclude. *Critical Review of International Social and Political Philosophy, 17*, 521–537.
Brock, G., & Blake, M. (2015). *Debating brain drain: May governments restrict emigration?* Oxford University Press.
Carens, J. (2013). *The ethics of immigration.* Oxford University Press.
Cockell, C. (Ed.). (2015). *The meaning of liberty beyond Earth.* Springer.
Cockell, C. (Eds.). (2016a). Human governance beyond Earth: Implications for freedom. Springer
Cockell, C. (Eds.). (2016b). Dissent, revolution and liberty beyond Earth. Springer
Cockell, C. (2019). Freedom engineering–using engineering to mitigate tyranny in space. *Space Policy, 49*, 101328.
Dowty, A. (1989). The right of personal self-determination. *Public Affairs Quarterly, 3*, 11–24.
Greenall-Sharp, R., et al. (2021). A space settler's bill of rights. In O. Chon-Torres et al., (Eds.), *Astrobiology: Science, ethics, and public policy,* forthcoming. Wiley-Scrivener

Lidskog, R. (2018). Invented communities and social vulnerability: The local post-disaster dynamics of extreme environment events. *Sustainability, 10*, 4457.

Limerick, P. (1992). Imagined frontiers: Westward expansion and the future of the space program. In R. Byerly (Ed.), *Space policy alternatives* (pp. 249–261). Westview Press.

Milligan, T. (2016). Constrained dissent and the rights of future generations. In C. Cockell (Ed.), *Dissent, revolution and liberty beyond Earth* (pp. 7–20). Springer.

Ntontis, E., Drury, J., & Amlôt, R., et al. (2017). Emergent social identities in a flood: Implications for community psychosocial resilience. *Journal of Community and Applied Social Psychology, 28*, 3–14.

Oberman, K. (2013). Can brain drain justify immigration restrictions? *Ethics, 123*, 427–455.

Sager, A. (2014). Reframing the brain drain. *Critical Review of International Social and Political Philosophy, 17*, 560–579.

Schwartz, J. S. J. (2016). Lunar labor relations. In C. Cockell (Ed.), *Dissent, revolution and liberty beyond Earth* (pp. 41–58). Springer.

Schwartz, J. S. J. (2018). Worldship ethics: Obligations to the crew. *Journal of the British Interplanetary Society, 71*, 53–64.

Schwartz, J. S. J. (2020). *The value of science in space exploration.* Oxford University Press.

Schwartz, J. S. J., Wells-Jensen, D., & Traphagan, J., et al. (2021). What do we need to ask before settling space? *Journal of the British Interplanetary Society, 74*, 140–149.

Shacknove, A. (1985). Who is a refugee? *Ethics, 95*, 274–284.

Song, S. (2019). *Immigration and democracy.* Oxford University Press.

Stilz, A. (2016). Is there an unqualified right to leave? In S. Fine & L. Ypi (Eds.), *Migration in political theory: The ethics of movement and membership* (pp. 57–79). Oxford University Press.

Szocik, K. (Ed.). (2020). *Human enhancements for space missions: Lunar, martian, and future missions to the outer planets.* Springer.

Wellman, C. (2016). Freedom of movement and the rights to enter and exit. In S. Fine & L. Ypi (Eds.), *Migration in political theory: The ethics of movement and membership* (pp. 80–101). Oxford University Press.

Yorke, C. (2016). Prospects for Utopia in space. In J. Schwartz & T. Milligan (Eds.), *The ethics of space exploration* (pp. 61–71). Springer.

Ypi, L. (2008). Justice in migration: A closed borders Utopia? *The Journal of Political Philosophy, 16*, 391–418.

Part V
Philosophical, Educational, and Religious Perspectives

Chapter 13
Lunar Settlement, Space Refuge, and Quality of Life: A Prevention Policy for the Future of Humans on Luna

Konrad Szocik⊙

Abstract While there are many arguments in favor of a human mission to the Moon, there are also many arguments in favor of the concept of a lunar space refuge. This chapter introduces the ethics of a space refuge. The potential benefits of a refuge on the Moon compared to a colony on Mars are discussed, with particular attention to differences in ethics and bioethics. Lunar settlement seems to be less challenging ethically and bioethically because human enhancements, which are often seen as obligatory for Mars missions, may not be required for lunar settlement. In this chapter, these issues are not considered in a traditional framework of space environmental ethics (i.e., in terms of intrinsic and instrumental values). Instead, issues are examined within a context that states humans should not colonize space because of their environmentally deleterious moral attitudes and behaviors on Earth. We also discuss an environmental ethical framework for future colonization of the Moon as a type of "prevention policy" to avoid environmental destruction and a similar existential risk for humans on the Moon.

13.1 Introduction: The Ethics of Space Refuge

In this chapter, the idea of a lunar settlement is discussed in terms of a possible space refuge. There are many good reasons to seriously consider a space refuge in general—as a great project of humanity—and a lunar space refuge in particular. In this sense, this chapter argues in favor of any space refuge or, optimally, as many space refuges as possible, taking into account current human technological progress and financial feasibility. Current technological possibilities are a crucial factor. It makes a significant difference whether humanity has the technology to create multiple bases on the Moon, Mars, or large ship bases (this may be a scenario that is possible over

K. Szocik (✉)
Yale University, Interdisciplinary Center for Bioethics, New Haven, US
e-mail: konrad.szocik@yale.edu

University of Information Technology and Management in Rzeszow, Sucharskiego 2 street, Rzeszow, Poland the next 50 years), or whether humanity is going to send "only" a few astronauts for reconnaissance (which is the current NASA scenario for the Moon in 3 years and maybe Mars in the 2030s). For the philosophy of space refuge and the studies in global catastrophic risks, the more shelters for humanity, the better, of course, to provide evacuation opportunities for as many inhabitants as possible. Experts in the risk studies give strong arguments in favor of the fact that at least a portion of humanity will survive basically any catastrophe such as a super volcanic eruption or a nuclear war in appropriate earthly shelters (for review, see Szocik, 2019).

However, every discourse on the mass evacuation of humanity to a space base encounters a bottleneck: how would anyone be able to evacuate all 10 billion people (see estimated global growth forecast for 2050, Fig. 13.1)? Evacuation of even one percent of this number (100 million) seems to be an almost unrealistic task for logistical, technological, political, and financial reasons. I am far from any attempt to estimate how many people could live in such a shelter; I leave that for experts in logistics, transportation, and other relevant fields. However, it may be expected that the number will be small, given the small area of the Moon and transportation and logistics constraints. It may be assumed that such evacuation policy would interfere with the growing challenges on Earth. If we are going to evacuate as many inhabitants

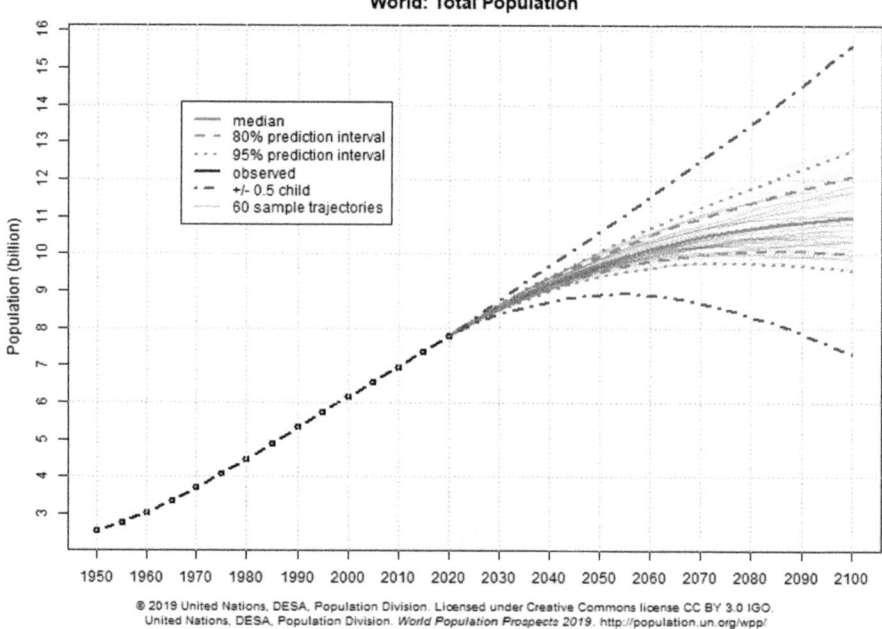

Fig. 13.1 Estimated global growth forecast for humanity up to 2100. credit: united nations, department of economic and social affairs population dynamics. (CC BY 3.0 IGO) https://population.un.org/wpp/Graphs/Probabilistic/POP/TOT/900

as possible to the cosmic colony, it means unequivocally that life on Earth will either no longer be possible or no longer be attractive at all.

If so, it can also be assumed that the technical and financial resources on Earth may not necessarily allow for such an expensive and complicated process as building a shelter in space and a mass interplanetary flight program. It is not entirely clear what a possible conflict of interest in investing in the protection of the Earth's environment and the improvement of climate and living conditions may look like, and, at the same time, investing in the space program. As we argued in our previous works (Szocik, 2019; Szocik et al., 2020), Earth protection and the space program are two separate undertakings, which are not mutually exclusive but serve different purposes and can be seen as mutually complementary. We can and should do everything we can to protect Earth and enable us to continue living on it. However, further life on Earth may no longer be possible for at least part of the population. And even if it were, continuous economic expansion and uninterrupted exploitation of resources may simply require the search for new resources beyond Earth.

As we can see, the concept of a space colony represents a great organizational and logistical challenge, causing, besides obvious technical and economic problems, serious ethical and political challenges. We do not discuss them in this chapter because they form a separate group of problems and as such are suitable for a separate paper. In any case, it is worth keeping in mind that—perhaps a little paradoxically—technological and logistical problems will be the least of the problems when compared with ethical challenges. They include not only selection of future refugees who can be evacuated to a space refuge but also many other challenges, such as the risk of restricting human freedom and perhaps the need to guarantee the right to return to Earth, which may not necessarily be possible for technical, economic, or social reasons (see James Schwartz's contribution to this volume).

A separate group of ethical and bioethical challenges represents the idea of human enhancement by radical means such as implants or genetic editing, which is considered as possibly obligatory for long-term and deep space missions. Such an idea causes obvious ethical concerns, especially if it were to turn out that only genetically modified people have the right to leave Earth. If this were the case, can we allow for a situation where it is mandatory to undergo genetic modification to be granted the right to fly to a space refuge? Some philosophers say no and have strong arguments for this (Schwartz, 2020). Someone could answer that such modification—if it is really necessary to protect against the negative effects of cosmic radiation or microgravity in space—is to protect life or increase efficiency and health. It is therefore not dictated by trivial motives.

However, Schwartz answers that even if it were, then it is necessary to take care of the appropriate mission design so that no modification is necessary. Therefore, the responsibilities are on the side of the organizers and not on the side of the participants—normal citizens. If we assume that radical genetic modification or another form of obligatory improvement is indeed something to be avoided, the organizers of such a mission should exclude the need to modify each member of the mission by significantly shortening the travel time, inventing pharmacological means to reduce or prevent negative effects of reduced gravity and space radiation, and creating artificial gravity and appropriate structures to protect against space radiation during interplanetary travel and in the space colony. If they are not able to, they should postpone the mission in time. However, time is a key factor here. Modification can be a procedure that will be available faster and, as such, can be applied soon. It can also be much cheaper than artificial gravity of a faster interplanetary spaceship. Thus, both time and cost can act as an advantage to apply the modification and be important factors in bioethical analysis.

There are good reasons, and perhaps advantages, in considering the Moon as a space refuge for humanity when compared with a Martian space refuge. Lunar settlement offers some advantages over other possible space locations such as on Mars basically in the field of ethics and bioethics. As it is argued in this chapter, a mission to the Moon possibly will not require substantial human enhancement, which may be mandatory—but also rightly considered highly controversial from an ethical point of view—like it may be in the case of Mars settlement or other further locations.

13.2   Lunar Settlement as a Space Refuge

The idea of a space refuge is not new. But very rarely and, in fact, only recently has it been discussed in academic texts (Szocik, 2019; Szocik et al., 2020). Others only mentioned the concept of a space refuge at the margin of their studies on terrestrial refuges as remedies for global catastrophes on Earth (Baum et al., 2015; Jebari, 2015). Finally, supporters of the mission to the Moon do not mention the concept of a space refuge among the many expected benefits (see, for example, Crawford, 2005). Consequently, for many, talking about a space refuge sounds like science fiction discourse. That impression is strengthened by the fact that this is a popular motif in science fiction novels and films. Nevertheless, the main challenge with the concept of a space refuge—located on the Moon, Mars, or elsewhere—lies in its rationale. By rationale we do not mean here a primary justification in building a space refuge. Such justification is relatively obvious and connected with the global catastrophes that humanity may meet in the future. The rationale for a space refuge is treated here in comparison to the rationale for an alternative terrestrial refuge such as subterranean or aquatic, just to mention the most feasible kinds of terrestrial refuges. Consequently, the challenge lies in arguing for some extra added value that would be offered by a space refuge and that would be impossible to be provided by terrestrial refuges (Stoner, 2017; Szocik, 2019). As we argued elsewhere (Szocik et al., 2020), as long as the initial living conditions in a space refuge may be similar to those on Earth after a disaster, in the long run a space refuge should provide stability and a constant supply. There are good reasons to assume that they may be challenging for a community of survivors on Earth even if based in a safe terrestrial refuge.

Finally, there are good reasons to assume that space refugees—even if recruited from civilians and not from specially trained military personnel—will be expected to satisfy a minimal level of physical performance, efficacy, and health required to survive spaceflight in relatively good conditions. As such, they may be more resistant to hard environmental conditions than civilian survivors left on Earth in terrestrial refuges. We should not ignore that difference because the concept of a space refuge seems to be inextricably linked to prelaunch training and at least minimal physical preparation and medical checking. All these requirements will be beneficial for space refugees. It is worth adding that no analogical checking and preparation can be expected regarding a terrestrial refuge. It is also worth noting that an evacuation scenario would be challenging in regard to both terrestrial refuges and space refuges. In this chapter, we do not analyze possible evacuation scenarios and criteria of selection for terrestrial or space refugees. However, the case of the current policy for vaccination during the COVID-19 pandemic, as well as the many challenges that mass vaccination programs met in different countries, shows that an analogical policy of refugee selection will not be a trivial issue and may have serious logistical constraints.

However, even if for some readers the concept of lunar settlement as a space refuge seems to be too far from reality, there are strong reasons to treat the concept of lunar settlement—even if initially planned only as a research outpost or a starting point for asteroid mining—as a more feasible and rational venture than Mars settlement for technological, medical, and political reasons. It is assumed here that an economic model aimed at space mining, both lunar and asteroid, may be a starting point for further development of lunar settlement. Space mining may also inspire and challenge international discussion about space law, the Outer Space Treaty, and the issue of property rights in space (Jakhu & Buzdugan, 2008) and, as such, influence our vision of outer space as a common heritage. Some possible advantages of lunar settlement over Martian settlement are worth a short presentation.

13.3   Medical Advantages

A mission to the Moon offers evident advantages over missions to Mars for human health and life. First of all, much shorter travel time to the Moon than to Mars means lower and shorter exposure to microgravity. Microgravity causes many hazardous medical effects. The same is especially true about another hazardous factor in space: space radiation. Exposure to space radiation is the highest during the interplanetary trip from Earth to Mars. The shorter the trip, the lower the exposure to harmful radiation in space. While exposure to space radiation during spaceflight is challenging for a mission to Mars, this is not a challenge for lunar missions due to the short distance between Earth and the Moon. Progress in technology may provide better protection against space radiation, for example, by reducing the time of spaceflight between two planets. Finally, permanent lunar settlement may serve as a research field for the proper preparation of astronauts before their later mission to Mars. This refers not only to studying astronauts' physiology and psychology in reduced gravity but also to testing all systems involved, including the life support system (Crawford, 2004).

A separate group of problems involves psychological challenges. Threats such as isolation, distance from Earth, and habitat confinement are expected to be riskier and more challenging during missions to Mars than in lunar settlement. It is true that the degree of confinement is the same on both the Moon and Mars insofar as refugees can survive only in a habitat. However, it may be assumed that the proximity to Earth becomes an important issue. There are good psychological arguments to assume that life on the Moon will not have the same negative effects as expected during a mission to Mars because of the awareness of the possibility of quick evacuation and the feeling of relative proximity to Earth. It can be expected that also the possibility of contact in real time, without the tens of minutes of delay inherent in a Mars base, can play a positive psychological role.

13.4 Ethical and Bioethical Advantages

Due to fewer expected medical constraints during a mission to the Moon (there are good reasons to assume that substantial human enhancement is not required for living in a lunar settlement), some of the ethical and bioethical issues disappear. For example, no bioethical debate is required around human enhancement, which is discussed as permissible or even morally required in the context of a mission to Mars (Szocik, 2020). Consequently, the lunar settlement causes fewer bioethical issues—if any—than the analogical concept of Martian settlement. The above objections discussed by Schwartz (2020) in relation to a mission to Mars disappear in the case of a mission to the Moon, both short-term and long-term. This is an important difference from the point of view of ethics and bioethics, especially the idea of an inclusivist society. Someone might be afraid that it would not be good to base a new human settlement on a discriminatory idea, even if it is currently medically justified. Analogically, someone might want to justify slavery by saying that the current economic situation forces the use of slaves, hoping that in a few decades' time economic progress will make it possible to give up slavery. That sounds ludicrous. The question arises if similar thinking about temporarily obligatory radical human enhancement can be compared to the aforementioned thinking about slavery or whether these are two different ethical situations.

Regardless, it is worth keeping in mind all the threats to human freedom and equality that can already be identified and that may arise during the space colonization program. As Schwartz rightly reports in his contribution to this volume, our main goal should be always the care for quality of life of humans, not just space colonization as such and associated care for cost-effectiveness or optimization. Although the lunar settlement program also brings some ethical risks and challenges, such as the right to migration discussed by Schwartz, which is not obvious even in the case of the Moon, there is no doubt that even if the same ethical issues will occur on the Moon and on Mars, the proximity of the Moon to Earth makes it probably easier to solve them.

We assume that a lack of bioethical obstacles associated with the idea of human enhancement, which should appear naturally when such an idea is discussed, is a great benefit of the concept of lunar settlement. But the Moon also offers other ethical advantages associated with environmental ethics. One commonly discussed topic in space ethics is concern for the pristine space environment. The main risk is contamination by humans but also simply by anything sent from Earth that may contain earthly bacteria and viruses. Some philosophers even go further and question the human right to colonize the universe. In this regard, the Moon offers an advantage over Mars. There is no confirmed existence of life on the Moon, contrary to the expected traces of past and perhaps existing life on Mars. From that scientific and ethical point of view, the concept of lunar settlement is free from this type of ethical hazard. This is a kind of hazard that cannot be ignored. Many authors argue that proper Mars exploration must be preceded by purely scientific missions. Nevertheless, some environmental challenges remain on the Moon, for example, the risk of ice contamination by space traffic on the Moon (Witze, 2021).

13.5   Technological Advantages

Because the Moon is much closer to Earth than Mars, humanity already possesses transportation and life support system technology to reach the Moon and to stay there at least for a short period of time. This physical fact offers another advantage for the lunar settlement. While the Mars launch window is 26 months, 13 missions to the Moon can be sent within the same time (Mendell, 1991). That makes a mission to the Moon not only more feasible and safer from the point of view of human health and life (a risk of emergency evacuations) but also allows for better testing of transport and life support system technologies. Mars, due to its long distance from Earth and narrow launch window, does not provide such possibilities. This is a factor important especially for the concept of a space refuge, where many launches may be required in a short period of time. Let's imagine a scenario of a mass evacuation of Earth due to an expected and coming catastrophe. Let's also assume that such a catastrophe will touch only Earth, and the Moon will remain a safe place. Due to the limited capacity of spaceships, many spaceflights may be required to evacuate at least some portion of the human population. The lack of restrictions caused by a launch window like in the case of Mars makes mass evacuation from Earth to the Moon easier and more feasible than to Mars.

Another technological challenge lies in the feasibility and reliability of different systems required during space missions, including the life support system or communication system. The risk of their failure and long service life under unknown conditions will be a challenge for the concept of Mars settlement. To reduce the risk of failure as much as possible, the concept of lunar settlement as a precursor to Mars settlement may be considered, at least to test our operational skills in an environment that is easier, better known, and closer to Earth (Mendell, 1991; Crawford, 2003; NASA,) . Some authors question the rationality and efficiency of lunar in-situ resource utilization and argue instead for total focus on Martian ISRU (Rapp,). Mars is also larger than the Moon and has some physical advantages over the Moon. Since we have a strong case for both human lunar and Mars missions, it seems that the long-term human space program should include both lunar and Mars missions, rather than focusing on just one location.

13.6   Political and Economic Advantages

There are issues that matter from a political point of view in space policy. Some of them overlap with economic issues, but not all. In regard to economic issues overlapping with political, the main challenge lies in the fact that both kinds of space settlements, lunar and Martian, are open-ended programs that go far beyond annual budget planning. No one can guarantee that the space exploration policy approved in one year will be continued in subsequent decades, but it should be the case of space settlement conceived as a space refuge in the very long term. This is a challenge that Haqq-Misra (2019) calls a "deep altruism". Deep altruism in funding means a situation where sponsors have to give up any benefits. This is almost impossible from a political point of view focused on the benefits of cyclical parliamentary or presidential elections. The concept of lunar settlement is more reasonable, feasible, and, at least apparently, more beneficial—also from the public point of view—than analogical settlement on Mars. In particular, it enables the involvement of the private sector and the development of mining, whether directly on the Moon or at least using the Moon as a base for asteroid mining (Near Earth Objects).

13.7   The Ethics of Quality of Life

The ethics of "quality of life" is usually considered in bioethical discourse as related to earthly matters related to the beginning and end of human life. Such issues as abortion or euthanasia are sometimes discussed in terms of length of life and quality of life. Sometimes it is better not to give someone's life or shorten it than to condemn or force someone to live in suffering. This is a factor that should not be ignored but is often ignored by supporters of the so-called pro-life party. What is known and discussed in relation to abortion and euthanasia can be applied, of course, in a certain sense only, to the context of space settlement and the concept of a space refuge. The key idea is that, while life on the Moon may be less challenging for humans than life in a Mars base, any space refuge will be hazardous and challenging for humans.

Taking into account the conceptual framework offered by an ethics of quality of life, (issues such as daily comfort, level of stress, exposure to hazardous factors in space or on the planetary surface, and fulfillment of basic human rights such as freedom or the right to return to Earth), all should be considered as possible challenges to the idea of the Moon as a space refuge. What is important here is the question of the alternative that humanity has when giving up the concept of a space refuge. We rather take for granted the fact that life in a space refuge must offer a lower standard of living than life on Earth before a disaster. Space philosophers consider the conditions with which we can approve such a refuge. One of them is Koji Tachibana, who discusses the very conditions under which a space refuge for saving humanity can (or cannot) be justified. My point of view coincides with Tachibana's approach. He proposes the space refuge without genetic enhancement, which can be justified from a virtue ethical point of view even if it may require the refugees a lower standard of living than life on Earth before a disaster (Tachibana, 2020).

The only alternative to a space refuge remains the space refuge on Earth. We recommend simultaneously preparing refuges on both Earth (subterranean and aquatic) and in space. Refuges on Earth should be dedicated for as many people as possible due to the above-mentioned technological, economic, and logistical constraints in sending millions upon millions of people to space. Refuges in space should be—and, in principle, they have to be, because of the obvious limitations— dedicated for more carefully selected refuges who will be designated as founders of future new human civilization. The current analogy on Earth for these two types of shelters is the use of mass shelters, such as subway stations, next to a relatively small number of specialized shelters usually intended for politicians. Someone could rightly see that since humanity has no choice, the question of quality of life will disappear. This is only partially true. If the only alternative to a space refuge is an unstable shelter on Earth or mass assisted dying, there are at least two things humanity can do to increase the quality of life in a space refuge. One of them is training and education before sending refugees to a space colony. It is also important to design the space refuge in such a way that it limits the basic rights and comfort of living as little as possible.

13.8 Environmental Ethics and a Prevention Policy

Humanity has perhaps already destroyed Earth, causing irreversible damage that will lead to an existential risk, for example, through global warming and climate change. However, doubts remain about a future that could be impacted by human destructive potential. The idea of a space refuge should be considered in this context in terms of a possible threat not necessarily for the Moon, but as the next existential, anthropogenic risk, that is, humanity's time in a space colony, not on Earth. Despite this and other challenges, the concept of a space refuge in general and lunar settlement in particular is considered here as a duty of humanity to guarantee continuity of our species in the long run. The Moon, next to asteroids, is listed as a source of potential raw materials that may be useful at least for in-situ resource utilization by future settlers. As such, the Moon meets the criteria of a self-sufficient habitat. A little paradoxically, it is not necessarily minerals—however, oxygen, hydrogen, and water resources on the Moon have an obvious value (Anand et al., 2012)—but solar energy that may be the largest and most valuable raw material on the Moon (Crawford, 2015). An availability of resources, which are still waiting to be discovered, is limited. As long as humans will not have any backup planet (it may be assumed that resources on Earth in the near future will be almost completely exhausted), it is possible that the lunar settlement may be the only source of resources.

We do not know if a focus on lunar colonization excludes Mars colonization. Different experts and different futurologists have varying opinions on the subject. However, there are good reasons to assume that humanity due to economic, technological, and/or political reasons will decide not to pursue two very challenging and long-term space colonization projects. In that case, lunar settlers should significantly reduce their exploitation of resources, but only insofar as current scientific and technical possibilities require it. Environmental ethics of lunar settlement should allow for exploitation of resources. In this sense, this is rather consequentialist than deontological ethics, which does not give the resources in space an intrinsic value. Their value is instrumental. However, despite their instrumentality, they are subordinated to human survival. As such, they should be exploited carefully.

As we argued elsewhere (Szocik and Reiss, in press), humans should colonize space to be able to continue to draw energy from nature and to survive and develop. We are skeptical about the possibility of people limiting themselves and abandoning a culture of prosperity for the sake of an ascetic lifestyle. For these reasons, humanity should exploit space to meet its energy needs. But, in contrast to the destructive exploitation of Earth, which led to the near extinction of humanity and many other species, future exploitation of space, including the Moon, Mars, and asteroids, should be limited by ecological rules.

One of the possibilities is the so-called one-eighth principle (Elvis and Milligan, 2019). This idea states that people should exploit not more than one-eighth of the total available resources that are detected in the Solar System. Of course, any other rules can be implemented, which will take into account the resources available to exploit with the current technology. The maximum number of colonists will be the result of available resources and the technological feasibility of exploitation, but also the idea of interplanetary sustainability. Consequently, the adopted prevention policy protocols should make it possible to send only such a number of inhabitants to the colony that will not only make it possible to maintain them, but also not cause super-exploitation that could endanger future generations.

It is worth thinking even longer term and asking what will happen when humanity exploits the Moon and those asteroids that will be available at the current moment of technological development of humanity. The possible answer depends on many factors. In a scenario where no humans survived on Earth, we may expect that pro-environmental awareness may evolve among lunar settlers. However, in a scenario where at least some number of survivors still live on Earth, the answer depends on the status of the lunar colony and the specificity of the space exploration program at some point in the future. Missions realized by private companies will be oriented to the return on investment, and space will work as an extension of the raw materials market for mining. This is the point where bioethical issues may emerge even on the Moon. They may include issues associated with human reproductive rights when the birth policy can be regulated to avoid a risk of overpopulation in the lunar colony. It is likely that in the lunar settlement much more than on Earth, the availability of human rights and the level of personal freedom will be directly connected with consumption behavior and environmental policy.

13.9   Conclusions

The concept of lunar settlement in terms of a space refuge raises ethical issues. Some of them are common to all astronomical objects that humanity can theoretically colonize. Here we focus on only two of them: the Moon and Mars. Slightly simplifying, the Moon has an advantage over Mars in almost all areas that may give rise to ethical or bioethical controversy, except one. Namely, gravity on the Moon is six times lower than on Earth, while on Mars it is about three times. This does not mean that lunar reduced gravity will be harmful and Martian will not. However, it may be expected that the gravity force is the more optimal for humans, the closer it is to Earth's gravity. In terms of other potentially ethically controversial points, the Moon is either completely free of them—as in the case of bioethical issues such as the concept of genetic modification—or they are less problematic and easier to solve. Colonization of the Moon before a mission to Mars offers advantages in areas such as medicine, ethics, and politics. Whether the mission to the Moon will be seen only in terms of a temporary scientific and economic expedition or as a future refuge for humanity, the application of a prevention policy seems essential to avoid super-exploitation and guarantee survival for future generations.

Acknowledgment Konrad Szocik was supported by Polish National Agency for Academic Exchange under Bekker NAWA Programme grant number PPN/BEK/2020/1/00012/DEC/1.

References

Anand, M., Crawford, I. A., Balat-Pichelin, M., Abanades, S., van Westrenen, W., Péraudeau, G., Jaumann, R., & Seboldt, W. (2012). A brief review of chemical and mineralogical resources on the Moon and likely initial in situ resource utilization (ISRU) applications. *Planetary and Space Science, 74*, 42–48.

Baum, S. D., Denkenberger, D. C., & Haqq-Misra, J. (2015). Isolated refuges for surviving global catastrophes. *Futures, 72*, 45–56.

Crawford, I. (2003). Space: Next step is an international moon base. *Nature, 422*, 373–374.

Crawford, I. (2004). The scientific case for renewed human activities on the Moon. *Space Policy, 20*, 91–97.

Crawford, I. A. (2005). Towards an integrated scientific and social case for human space exploration. *Earth, Moon, and Planets, 94*, 245–266.

Crawford, I. A. (2015). Lunar resources: A review. *Progress in Physical Geography, 39*(2), 137–167.

Elvis, M., & Milligan, T. (2019). How much of the Solar System should we leave as wilderness? *Acta Astronautica, 162*, 574–580.

Haqq-Misra, J. (2019). Can deep altruism sustain space settlement? In K. Szocik (Ed.), *The human factor in a mission to mars: Space and society* (pp. 145–155). Springer.

Jakhu, R., & Buzdugan, M. (2008). Development of the natural resources of the moon and other celestial bodies: Economic and legal aspects. *Astropolitics: The International Journal of Space Politics & Policy, 6*(3), 201–250.

Jebari, K. (2015). Existential risks: Exploring a robust risk reduction strategy. *Science and Engineering Ethics, 21*(3), 541–554.

Mendell, W. W. (1991). Lunar base as a precursor to Mars exploration and settlement. NTRS–NASA Technical Reports Server. https://ntrs.nasa.gov/citations/19920038015

NASA. (2019). NASA's plan for sustained lunar exploration and development.

Rapp, D. (2018). *Use of Extraterrestrial XE resources for human space missions to Moon or Mars* (2nd ed.). Springer Praxis Books.

Schwartz, J. S. J. (2020). The accessible universe: On the choice to require bodily modification for space exploration. In K. Szocik (Eds.), *Human enhancements for space missions. Lunar, martian, and future missions to the outer planets. Space and Society.* Springer, Cham

Stoner, I. (2017). Humans should not colonize Mars. *Journal of the American Philosophical Association, 3*(3), 334–353.

Szocik, K. (2019). Should and could humans go to Mars? Yes, but not now and not in the near future. *Futures, 105*, 54–66.

Szocik, K., Norman, Z., & Reiss, M. J. (2020). Ethical challenges in human space missions: A space refuge, scientific value, and human gene editing for space. *Science and Engineering Ethics, 26*, 1209–1227.

Szocik, K., & M. Reiss. (in press). Why only space colonisation may provide sustainable development: Climate ethics and the human future as a multi-planetary species.

Tachibana, K. (2020). Virtue ethics and the value of saving humanity. In K. Szocik (Eds.), *Human enhancements for space missions: Lunar, martian, and future missions to the outer planets* (pp. 169–181). Springer

Witze, A. (2021, January 5). Will increasing traffic to the Moon contaminate its precious ice? *Nature.* https://www.nature.com/articles/d41586-020-03262-9?utm_source=Nature+Briefing& utm_campaign=a6a2937a3b-briefing-dy-20210106&utm_medium=email&utm_term=0_c9df d39373-a6a2937a3b-44694681

Chapter 14
The Emergence of an Environmental Ethos on Luna

Ziba Norman and **Michael J. Reiss**

Abstract Questions of how humanity should undertake and regulate activities on the Moon have been considered for over half a century. In this chapter we outline the various philosophical approaches out of which an environmental ethos may emerge on Luna (the Earth's Moon). We draw on existing thinking within environmental ethics, particularly in regards to wildernesses, and consider the diversity of religious and philosophical frameworks and nascent relevant legal structures. Commercial and competitive considerations militate against stakeholders readily adopting shared norms. Nevertheless, there remains a need for some sort of a Treaty or other form of binding agreement that can be signed, ratified and enacted by all spacefaring nations, rather as the Antarctic Treaty has been by all nations with direct involvement there. This need is becoming more urgent.

14.1 Introduction

Almost immediately after Apollo 11 landed on the Moon on 20 July 1969, questions of how to regulate activities on the Moon, and potentially other celestial bodies, became a subject of focused study and concern. It still took a further 15 years before the Moon Agreement, also known as the Moon Treaty, was ratified, after the minimum required five members states had signed. However, this Treaty, for reasons we discuss below, has proved ineffectual, despite providing a valuable statement of intention.

The regulation of lunar activities can be examined from a number of perspectives– legal, political, scientific and ethical. Our principal focus in this chapter is on ethical considerations. Our argument is that there is much to be learnt from the well-founded discipline of environmental ethics, with the Moon being considered as an example of an extreme environment. We therefore look at wildernesses to see what lessons

Z. Norman (✉) · M. J. Reiss
Institute of Education, University College London, 20 Bedford Way, London WC1H 0AL, UK
e-mail: z.norman@ucl.ac.uk

M. J. Reiss
e-mail: m.reiss@ucl.ac.uk

© The Author(s), under exclusive license to Springer Nature Switzerland AG 2021 221
M. B. Rappaport and K. Szocik (eds.), *The Human Factor in the Settlement of the Moon*,
Space and Society, https://doi.org/10.1007/978-3-030-81388-8_14

might be learnt. We conclude by examining what would be desirable with respect to how humans use the resources of the Moon, and considering how agreement might be reached.

14.2 Approaches to Environmental Ethics

Ethics is the branch of philosophy concerned with how we should decide what is morally wrong and what is morally right. Humans have been guided by ethical considerations since before our species, *Homo sapiens*, evolved. If we look at other animal species, we can discern rules that guide their behaviour. Indeed, in our closest evolutionary relatives, the great apes, we can see evidence of many of the same ethical issues that occupy us. Jane Goodall's pioneering work on chimpanzees, for example, showed how individuals formed alliances to their mutual benefit, behaved as though they had certain moral obligations and routinely attempted to deceive other chimpanzees (van Lawick-Goodall, 1971).

Traditionally, Western normative accounts of human ethics, which are concerned with how we should behave, have examined the deontological approaches of Immanuel Kant and others (including the role of religion), consequentialist thinking (particularly Jeremy Bentham's utilitarianism and John Stuart Mill's refinement of this) and then the more recent turn to virtue ethics (deriving from Aristotle). A different approach is to examine more recent works in anthropology and the social sciences to look at what 'ordinary people' (rather than academic ethicists) think about what is morally right or wrong, and at what people actually do–the branch of ethics that is sometimes termed 'descriptive ethics'. For lunar ethics, there is value in taking both the normative and descriptive approaches seriously.

One major recent study analysed ethnographic accounts of ethics from 60 different societies, around the world (Curry et al., 2019). The authors found seven moral rules: help your family, help your group, return favours, be brave, defer to superiors, divide resources fairly and respect others' property. Examples of most of these rules were found in most societies and there were no examples of societies in which any of these behaviours were considered morally bad. Significantly, these rules were observed with equal frequency across continents: they were not the exclusive preserve of 'the West' or any other region.

However, one notable feature of these rules is that they apply only to humans. Similarly, while religions have some ethical precepts applying to non-humans, the focus in the Abrahamic faiths is overwhelmingly on relations between people, and between people and God. Indeed, in Judaism, Christianity and Sufism, humans are specifically seen as created in the image and likeness of God–*Imago Dei*, placing them above the rest of the created order. However, there have been moves within Abrahamic theology to come to a deeper understanding of how humans should relate to the whole of the creation. In part, such moves have been driven by greater awareness of ecological considerations (e.g., Page, 1996).

Other religions have more consistently given ethical consideration to non-humans. One of Buddhism's central precepts is compassion for all of life and many Buddhists are vegetarian. Hinduism (though it is difficult to generalise as, more than most religions, it is an amalgam of many traditions and philosophies) teaches that all living creatures have a soul, and meat eating is typically either restricted or avoided. Jains not only are vegetarians but avoid eating underground vegetables such as potatoes and onions to prevent harm to soil organisms. The Bahá'í faith emphasises that animals should be treated with kindness, and recognises the need for stewardship of the natural environment.

Secular philosophies have also paid attention to non-humans. Bentham is rightly remembered for his "The question is not, Can they reason?, nor Can they talk? but, Can they suffer? Why should the law refuse its protection to any sensitive being?" (Bentham, 1789, 1970, p. 283n) and utilitarians such as Peter Singer (1975) have developed this train of thought, arguing that much of human attitudes and behaviour towards non-human animals is speciesist. Our focus in this chapter is on still broader issues, namely environmental ethics.

It has long been appreciated that indigenous communities can have and operate with a rich conceptualisation of environmental ethics. There is a danger of romanticisation here—we should remember that in many parts of the world indigenous people have been responsible for environmental degradation, indeed, the extinction of many species, particularly large mammals and flightless birds through overexploitation (e.g., Smith et al., 2018). Nevertheless, though a range of mechanisms (religious traditions, mythical narratives, long-lasting informal agreements), many indigenous people do make more sustainable use of their environmental resources than do many modern people (e.g., Kelbessa, 2005). In the West, too, the eighteenth and nineteenth centuries saw a growth in appreciation of wild places and land more generally (cf. Leopold, 1949).

In the Western tradition, a key issue in environmental ethics has been the distinction between 'instrumental value' and 'intrinsic value'. In essence, the notion of intrinsic value is simply an extension of Kant's categorical imperative—that humans (in Kant's case) and environments (in environmental ethics) should be valued and treated not merely for what they can do for us (the 'ecosystem services' approach) but, at least in part, for what they are in themselves. This approach to valuing the environment developed in a number of countries (Brennan & Lo, 2020). In the USA, Leopold's arguments that land as a whole is worthy of our moral concern was echoed by Stone (1972), who argued that natural objects such as trees should have the same standing in law as corporations, Rolston III (1975), who argued that humans have a moral duty to preserve species, and Taylor (1986) who argued that whole ecosystems are worthy of moral consideration. But it is the Norwegian Arne Næss who is perhaps most associated with this movement, with his call to 'deep ecology', which again espoused the importance of the intrinsic value of nature and rejected the notion of humans as individuals, separate from the rest of the world (Brennan & Lo, 2020). Næss was born in 1912 and in 1939 took up the post of Professor of Philosophy at the University of Oslo. He was an accomplished mountaineer and took part in Green Party politics and in nonviolent environmental action, for instance, chaining himself in 1970 to the rocks at the Mardal waterfall to protest against a projected dam (Krabbe, 2010).

As part of these considerations there is also the issue of the exercise of our human capacities, including our rights, perhaps even obligation to life itself, to carve out our place in the cosmos, and that in doing so we fulfil our potential as a species, 'an ethics of fulfilment'. This may include adapting environments, recognising that stewardship in its most complete sense is not simply about keeping things in balance, but is a dynamic process of which we are a part (Norman, 2020).

14.3 Wilderness on Earth

What we might term 'space ethics' is at an early phase in its development (Szocik et al., 2020) and issues to do with the environment have only begun to be examined. We can draw usefully by looking at ethical attitudes and issues that have arisen in respect of wilderness on Earth.

Wilderness consists of natural environments on Earth that have been relatively undisturbed by human action. Traditionally, wilderness was assumed to be terrestrial but more attention is now being paid to marine wilderness. The rapidly increasing and global reach of human activity means that interest in wilderness has unsurprisingly grown over the last century or so. If we just focus for a moment on the country in which the two of us live, the United Kingdom, there has been a rapid and recent growth in what is now called 'rewilding'. Rewilding is a form of ecological restoration in which humans let nature take its course to a greater extent than is traditional in conservation, where habitat 'management' take a lot of time and effort. So, for example, The Great Fen Project (https://www.greatfen.org.uk) lies between Huntingdon and Peterborough and entails moving land from high-intensity arable use to low-intensity beef and lamb production as a result of the use of cattle and sheep for habitat management. While the main arguments in favour of the project are to do with the enhancement of wildlife diversity, there are expected to be human benefits too through less soil erosion, less flooding and new sources of income such as willow and reed harvesting. Rewilding often goes hand in hand with the reintroduction of species that became endangered or were driven to extinction by human activity and the UK has seen the reintroduction of a number of animals, including sea eagles, red kites, bitterns, pool frogs, natterjack toads, sand lizards, smooth snakes, wild boar, pine martens, the chequered skipper butterfly, the ladybird spider and the Eurasian beaver (Rewilding Britain, 2021).

On a larger scale, the John Muir Wilderness in California covers some 2,350 km^2 (Fig. 14.1). It was established in 1964 as a result of the Wilderness Act and named after the Scottish-American naturalist John Muir (1838–1914), whose early environmental activism helped to preserve the Yosemite Valley and Sequoia National Park. Today, the most comprehensive listing of wilderness and other protected areas is available at https://www.protectedplanet.net/en and updated regularly.

Fig. 14.1 Little Lakes Valley from above Mack Lake in the John Muir Wilderness in California (Jane S. Richardson), CC BY-SA 3.0 https://creativecommons.org/licenses/by-sa/3.0, via Wikimedia Commons). *Source* Taken from https://commons.wikimedia.org/wiki/File:Little_Lakes_Valley_from_above_Mack_Lake.jpg

Perhaps the area on Earth that has the most useful parallels with Luna in terms of the issues raised by the possibility of environmental protection is Antarctica, the only continent on Earth with no indigenous humans. There have been calls for both legal and policy lessons to be learnt for Outer Space from Antarctica (Kerrest, 2011; Race, 2011). In 1959, the Antarctic Treaty came into force, signed by the twelve countries active in the Antarctic at that time (Scientific Committee on Antarctic Research 2020). The original Treaty has subsequently been augmented by various Recommendations, a Protocol and two Conventions. The primary purpose of the current Antarctic Treaty System is to promote scientific research, to hold all territorial claims in abeyance and to ensure that Antarctica continues to be used exclusively for peaceful purposes. As of 2019, there are 53 states party to the treaty, of which 29 (including the original twelve signatories) have voting rights. It is also worth mentioning that the Antarctic Treaty came into existence at a time of increasing international tensions; it was the first arms control agreement during the so-called Cold War between the Soviet Union and the USA and their respective allies.

14.4 Environmental Ethics in Space

Humans seem initially to have treated the issue of waste in space just as we tended to treat the issue of waste in the oceans–simply assuming that it (space / the oceans) was so vast that we would have almost no effect on it. Yet, so-called 'space debris' began to accumulate from the time of the launch of Sputnik 1 on 4 October 1957, which orbited for a few months before falling back into the Earth's atmosphere. The European Space Agency's Space Debris Office estimates, using statistical models, that there are 34,000 artificial objects orbiting the Earth that have a diameter great than 10 cm, 900,000 objects in the 1–10 cm size range and 128 million objects in the 1–10 mm size range (European Space Agency, 2021). Only about 20,000 of these (including the 2500+ operational satellites) are large enough to be tracked; the rest pose hazards that cannot be avoided by altering flight paths.

Sources of space debris include dead spacecraft, lost equipment, garbage bags intentionally jettisoned from space stations, booster rockets and the results of anti-satellite weapons testing. Damage caused by collisions with space debris varies from scratches (which can compromise solar panels, telescopes and cameras) to wholesale destruction. The first major collision between two satellites occurred on 10 February 2009 when the operational 560 kg commercial communications satellite Iridium 33 collided with the derelict 950 kg Russian military communications satellite Kosmos 2251. The relative speed of impact was about 42,000 kms per hour and both satellites instantly broke up, creating thousands more pieces of space debris. In 1978, the NASA scientist Donald Kessler proposed what has come to be known as the Kessler syndrome, in which the abundance of objects (including space debris) reaches the point that collisions become ever more frequent due to a cascade effect resulting from the proliferation of new pieces of debris from collisions. In a worst-case scenario, this could prevent space activities, including the use of satellites, for generations. Perhaps unsurprisingly, there have been calls for some procedure akin to terrestrial environmental impact assessments to be developed for use in space (Kramer, 2020).

In an early examination of the implications of space exploration for environmental ethics, Hartmann (1984) suggested that new discoveries raised the possibility of a long-term shift of mining, refining and manufacturing from the Earth's surface to locations outside Earth's ecosphere, potentially allowing Earth to begin to return toward its natural state. However, he also acknowledged ambivalence amongst environmentalists about such a possibility, with many distrusting such a scenario. Intriguingly he also wrote: "Due to impending resource depletion on Earth, we may have only until the mid-twenty-first century to pursue the promising potential of space exploration to alleviate environmental problems of Earth. Subsequently, there may be too little industrial base to support vigorous exploration and exploitation of resources in space" (Hartmann, 1984, p. 227).

More recently, Elvis and Milligan (2019) ask how much of the Solar System we should leave as wilderness. They point out that humans are poor at estimating how long it takes seriously to deplete environmental resources–a point that seems to be corroborated by the depletion of marine fish stocks and terrestrial reserves of such very different resources as peat, groundwater and rare earth metals. This human limitation is exacerbated by the effects of a typical pattern in which demand for a finite resource grows exponentially. Accordingly, they suggest a rule of thumb in which "as a matter of fixed policy, development should be limited to one eighth, with the remainder set aside" (Elvis & Milligan, 2019, p. 574). This 'one-eight principle' can be criticised as being somewhat arbitrary. However, the lesson of history is that if principles can be written into international Treaties at an early stage, they can be long-lasting and effective.

In the language of environmental ethics that we introduced earlier, Elvis and Milligan's approach provides instrumental reasons for environmental protection in space. Other authors have either argued on intrinsic grounds or combined arguments that draw on instrumental value and intrinsic value. Perhaps the most sustained treatment is provided by Schwartz (2020), who develops and extends arguments developed by his collaborator Milligan (2014). Schwartz' arguments are situated within both an instrumental and an intrinsic valuation of space exploration, with his principal focus being on the generation of scientific knowledge and understanding. One of the rather refreshing aspects of Schwartz (2020) is the way in which he debunks a number of the standard arguments for space exploration. He argues, for example, that there is little evidence that spending money on space flight leads to more STEM (Science, Technology, Engineering and Mathematics) graduates, that space colonies might result in undesirable autocratic or totalitarian governments, and that space settlements are not needed urgently as a way of preserving the human species because there are cheaper ways of doing that (e.g., improving asteroid detection to avoid collisions, averting ecological collapse on Earth).

Schwartz examines the environmental arguments for protecting the Earth from contamination by life from space ('back contamination'–cf. *The Andromeda Strain*) and for protecting other planetary bodies from contamination with life from Earth ('forward contamination'). Most of his focus is on forward contamination and he is sympathetic both to the possibility that non-terrestrial life has intrinsic value and to the argument that even in the absence of non-terrestrial life, there are reasons for protecting space environments. Building on the arguments of Holmes Rolston III, Schwartz concludes that:

> … we should not proceed in discussion about the scope of and rationale for planetary protection under the presumption that we already know what is interesting, valuable, or worth protecting in the space environment … what we discover and what we learn through the exploration of space will affect what we find interesting and what we value … what we need for the moment is a growing rather than an attenuating appreciation of the space environment.

(Schwartz, 2020, pp. 143–4)

14.5   The Particular Case of Luna

At present, the basic legal framework of international space law (including the exploitation of Luna) is provided by the 1967 Outer Space Treaty. There are 110 countries that are party to the Treaty, including China, India, Russia, the USA and almost all the countries in Australasia, Europe and South America. Its main points are that it prohibits the placing of nuclear weapons in space, limits the use of the Moon and all other celestial bodies to peaceful purposes only, establishes that space shall be free for exploration and use by all nations, and holds that no nation may claim sovereignty of outer space or any celestial body (United Nations for Disarmament Affairs, 1967).

However, the 1967 Outer Space Treaty is silent on the issue of commercial activities, such as mining. Ten years of negotiation led to the Moon Treaty of 1979 (United Nations, 1980), which, *inter alia*, treats the Moon and its natural resources, as the 'common heritage of mankind' and says that an international regime would be needed to govern exploitation of the Moon's resources. However, this proved too ambitious. Only 18 states are party to the Treaty, which means they are prepared to support the process involved in further development of the treaty. Crucially, it does not mean they have ratified the treaty, and thus consented to be bound by it. While India is a signatory, it has not ratified it. China, Russia and the USA have not even gone that far, and have given clear indications, including via recent Executive Orders of the US Administration, that they have no intention of travelling down this route. In the case of India, there is a call for them to exit from the Moon Treaty altogether (Chennai, 2020). Aware of the importance of the principles to be established, the United Nations Committee on the Peaceful Uses of Outer Space (COPUOS) held a high-level meeting in 2018 that tried to produce a consensus on a framework of laws for the sustainable development of outer space, but this too failed (O'Brien, 2019).

In the absence of international agreement as to how the Moon's resources might be exploited (or not), it is unsurprising that there are increasing signs of a 'wild West' free-for-all. In 2020, the then President of the USA signed an Executive Order to support mining on the Moon and elsewhere (Wall, 2020). This included the statement that:

> Americans should have the right to engage in commercial exploration, recovery, and use of resources in outer space, consistent with applicable law. Outer space is a legally and physically unique domain of human activity, and the United States does not view it as a global commons (Foust, 2020).

At the time of writing (in the early days of the Biden administration), the USA plans to have a 'manned' (one of the two astronauts will be a woman) trip to the Moon in 2024. There are also plans to extract valuable deposits of water–ice from the lunar South Pole. These could be used to make rocket fuel on the Moon, serving as the foundation for a lunar economy (Rincon, 2020). What are called 'the Artemis Accords'–a set of guidelines for the crewed exploration of the Moon–have been signed only by eight countries (Australia, Canada, Italy, Japan, Luxembourg, the United Arab Emirates, the UK and the US), with Russia stating that it is 'too US-centric' for it to sign (Newman, 2020).

China has an expanding set of Moon activities. The Chinese Lunar Exploration Program is an on-going series of robotic Moon missions that had its first launch, the Chang'E-1 lunar orbiter, in 2007. In 2019, Chang'E-4 provided the first robotic visit to the far side of the Moon (Li et al., 2019). It is difficult to be sure what China's long-term lunar aims are. In November 2017, China signed an agreement with Russia to cooperate on lunar and deep space exploration, and Goswami (2018) has argued that China's long-term aims are to meet its burgeoning economic and energy needs.

In reality, it may be that with existing technologies, mining on the Moon or any other extra-terrestrial body for the benefit of those of us remaining on Earth is simply economically unfeasible (Crawford, 2015)–rather as deep-sea mining was touted for decades before it finally began to take off (probably causing considerable environmental damage in the process–cf. the Deep Sea Mining Campaign http://www.deepseaminingoutofourdepth.org).

However, it is very possible that mining on the Moon might take place for the benefit of astronauts and, eventually, residents there. Moon rocks and dust are typically about 45% oxygen by mass and in 2020, the UK company Metalysis was awarded a European Space Agency contract to develop the technology (electrochemistry) to turn Moon dust and rocks into oxygen, leaving behind aluminium, iron, silicon and other constituents for lunar construction workers to build with. The oxygen itself might be used for breathing or in rocket propulsion systems (Sample, 2020).

Furthermore, the need for an environmental ethics of Moon exploration is stressed by Elvis et al. (2021) who point out that there are only a handful of sites of particular interest on the Moon to those whose focus is relatively high concentrations of resources of special value. *Sinus Medii* is one of two places ideal for siting a lunar elevator (Fig. 14.2). The same point obtains if, for example, you are interested in erecting an astronomical telescope on the Moon. Elvis et al. conclude that "diverse actors pursuing incompatible ends at these sites could soon crowd and interfere with each other, leaving almost all actors worse off. Without proactive measures to prevent these outcomes, lunar actors are likely to experience significant losses of opportunity" (p. 1). At least eight spacecraft from nations including China, India, Japan, Russia and the USA are set to go to the Moon by the end of 2024 (Witze, 2021).

Of course, there are lessons from history on Earth so, whatever the ethics of environmental exploitation, one hopes that pragmatism prevails and governance arrangements are put in place. In addition, it is entirely possible that a new kind of lunar culture will emerge over time, especially if there are a number of different groups present on the Moon at any one time. The various international space stations have shown how international competition can occur even at times of heightened terrestrial tensions between nations. Such cooperation is particularly likely if astronauts who know one another are in danger or require assistance–international cooperation is notable in extreme environments, such as on Antarctica. One could hope that such cooperation might lead the way for more formalised arrangement to come into being. As an aside, some readers may be reassured that there are plans for a decent 4G wireless network on the Moon (upgradable to 5G), with Nokia having been awarded a US$14 m contract to this end in 2020 (France-Presse, 2020). Presumably, some of us will eventually have a coveted @luna email address: @luna.edu or perhaps ending with lu as the first planetary suffix.

Fig. 14.2 *Sinus Medii* is one of two places ideal for siting a lunar elevator, a proposed transportation system between the Moon's surface and a docking port in space (James Stuby based on NASA image, Public domain, via Wikimedia Commons). *Source* Taken from https://commons.wikimedia. org/wiki/File:Sinus_Medii_2093_med.jpg

14.6   Conclusions

It isn't likely that a single set of coherent principles to govern activities on the Moon, with the accompanying development of an environmental ethos, will be established simply by means of the organs of the United Nations or some other existing international body. All indications are that lunar governance and environmental ethos will be a messy affair, led by prudential considerations and come into being, as on our own home planet Earth, piecemeal. We hope the multiple lunar missions planned over the next decade will result in situations that lead to the establishment of an environmental ethos that is not based purely on short-sighted considerations, but will be based on deep ethical reflection drawing on the full diversity of traditions within moral philosophy. Furthermore, if our presence in space is to be sustained, and supportive of life on Earth, then maximum cooperation will be necessary. As situations emerge it is likely there will increasingly be a push for more developed structures. Some sort of development of the Artemis Accords will hopefully result in a Treaty that can be signed, ratified and enacted by all spacefaring nations, rather as the Antarctic Treaty has been by all nations with direct involvement there.

References

Bentham, J. (1789/1970). *An introduction to the principles of morals and legislation, ed. by* J. H Burns and HLA Hart. Oxford: Clarendon Press.

Brennan, A., & Lo, Y. S. (2020). Environmental ethics. In E. N. Zalta (Eds.), *The Stanford encyclopedia of philosophy.* https://plato.stanford.edu/archives/win2020/entries/ethics-env ironmental/

Chennai, M. R. (2020). Why India should exit the Moon Agreement. *The Hindu Business Line.* 20 May. https://www.thehindubusinessline.com/news/science/why-india-should-exit-the-moon-agr eement/article31634373.ece

Curry, S., Mullins, D. A., & Whitehouse, H. (2019). Is It good to cooperate? Testing the theory of morality-as-cooperation in 60 societies. *Current Anthropology, 60*(1), 47–69.

Crawford, I. A. (2015). Lunar resources: A review. *Progress in Physical Geography: Earth and Environment, 39*(2), 137–167.

European Space Agency. (2021). *Space debris by the numbers.* https://www.esa.int/Safety_Security/Space_Debris/Space_debris_by_the_numbers

Elvis, M., & Milligan, T. (2019). How much of the solar system should we leave as wilderness? *Acta Astronautica, 162,* 574–580.

Elvis, M., Krolikowski, A., & Milligan, T. (2021). Concentrated lunar resources: Imminent implications for governance and justice. *Philosophical Transactions of the Royal Society A, 379,* 20190563. https://doi.org/10.1098/rsta.2019.0563

Foust, J. (2020, April 6). White house looks for international support for space resource rights. *SpaceNews.* https://spacenews.com/white-house-looks-for-international-sup port-for-space-resource-rights/

France-Presse, A. (2020, October 20). Talking on the moon: Nasa and Nokia to install 4G on lunar surface. *The Guardian.* https://www.theguardian.com/science/2020/oct/20/talking-on-the-moon-nasa-and-nokia-to-install-4g-on-lunar-surface

Goswami, N. (2018). China in space: Ambitions and possible conflict. *Strategic Studies Quarterly, 12*(1), 74–97.

Hartmann, W. K. (1984). Space exploration and environmental issues. *Environmental Ethics, 6*(3), 227–239.

Kelbessa, W. (2005). The rehabilitation of indigenous environmental ethics in Africa. *Diogenes, 52*(3), 17–34.

Kerrest, A. (2011). Outer space as international space: Lessons from Antarctica. In P. A. Berkman, M. A. Lang, D. W. H. Walton, & O. R. Young (Eds.), *Science diplomacy: Antarctica, science, and the governance of international spaces* (pp. 133–142). Smithsonian Institution Scholarly Press.

Krabbe, E. C. W. (2010). Arne Næss (1912–2009). *Argumentation, 24,* 527–530.

Kramer, W. R. (2020). A framework for extraterrestrial environmental assessment. *Space Policy, 53,* 101385.

Leopold, A. (1949). *A sand county almanac.* Oxford University Press.

Li, C., Wang, C., Wei, Y., & Lin, Y. (2019). China's present and future lunar exploration program. *Science, 365*(6450), 238–239.

Milligan, T. (2014). *Nobody owns the Moon: The ethics of space exploitation.* Jefferson: McFarland & Co

Newman, C. (2020, October 19). Artemis Accords: Why many countries are refusing to sign Moon exploration agreement. *The Conversation*. https://theconversation.com/artemis-accords-why-many-countries-are-refusing-to-sign-moon-exploration-agreement-148134

Norman, Z. (2020, February 11). Conservation v colonisation: The ethics of a human presence in space. APEX talk delivered at University College London. https://mediacentral.ucl.ac.uk/Play/22303 (video recording), http://zibanorman.co.uk/ZibaNorman-Conservation-vs-Colonisation.pdf (text)

O'Brien, D. C. (2019, January 21). Beyond UNISPACE: It's time for the Moon treaty. *The Space Review*. https://www.thespacereview.com/article/3642/1

Page, R. (1996). *God and the web of creation*. SCM.

Race, M. S. (2011). Policies for scientific exploration and environmental protection: Comparison of the Antarctic and Outer Space Treaties. In P. A. Berkman, M. A. Lang, D. W. H. Walton, & O. R. Young (Eds.), *Science diplomacy: Antarctica, science, and the governance of international spaces* (pp. 143–152). Smithsonian Institution Scholarly Press.

Rewilding Britain. (2021). *Reintroductions and bringing back species*. https://www.rewildingbritain.org.uk/explore-rewilding/ecology-of-rewilding/reintroductions-and-bringing-back-species

Rincon, P. (2020, September 22). NASA outlines plan for first woman on Moon by 2024. *BVC News*. https://www.bbc.co.uk/news/science-environment-54246485

Rolston, H. (1975). Is there an ecological ethic? *Ethics, 85*, 93–109.

Sample, I. (2020, November 9). UK firm to turn moon rock into oxygen and building materials. *The Guardian*. https://www.theguardian.com/science/2020/nov/09/uk-firm-to-turn-moon-rock-into-oxygen-and-building-materials

Schwartz, J. S. J. (2020). *The value of science in space exploration*. Oxford University Press.

Scientific Committee on Antarctic Research. (2020). *The Antarctic treaty system*. https://www.scar.org/policy/antarctic-treaty-system/

Singer, P. (1975). *Animal liberation: A new ethics for our treatment of animals*. HarperCollins.

Smith, F. A., Elliott Smith, R. E., Lyons, S. K., & Payne, J. L. (2018). Body size downgrading of mammals over the late quaternary. *Science, 360*(6386), 310–313.

Stone, C. D. (1972). Should trees have standing? *Southern California Law Review, 45*, 450–501.

Szocik, K., Norman, Z., & Reiss, M. J. (2020). Ethical challenges in human space missions: A space refuge, scientific value, and human gene editing for space. *Science and Engineering Ethics, 26*, 1209–1227.

Taylor, P. W. (1986). *Respect for nature: A theory of environmental ethics*. Princeton University Press.

United Nations for Disarmament Affairs. (1967). *Treaty on principles governing the activities of states in the exploration and use of outer space, including the Moon and other celestial bodies*. http://disarmament.un.org/treaties/t/outer_space/text

United Nations. (1980). *Agreement governing the activities of states on the Moon and other celestial bodies*. New York: United Nations. https://treaties.un.org/doc/Treaties/1984/07/19840711%2001-51%20AM/Ch_XXIV_02.pdf

van Lawick-Goodall, J. (1971). *In the shadow of man*. Collins.

Wall, M. (2020, April 6). Trump signs executive order to support moon mining, tap asteroid resources. *Space.com*. https://www.space.com/trump-moon-mining-space-resources-executive-order.html

Witze, A. (2021). Will increasing traffic to the Moon contaminate its precious ice? *Nature, 589*, 180–181.

Chapter 15
Environmental and Occupational Ethics in Early Lunar Populations: Establishing Guidelines for Future Off-World Settlements

Evie Kendal⊙

Abstract The potential for lunar settlements to mitigate the environmental impacts of over-population on Earth, and to provide opportunities for research and mining, carries with it concomitant concerns for occupational safety and environmental protection for the Moon and its future inhabitants. The question of whether the lunar landscape should be preserved for future generations of human visitors, and in particular whether the Apollo landing sites should be considered cultural heritage sites, also speak to shared concerns regarding the appropriate use of celestial objects. While the Outer Space Treaty (United Nations Office for Outer Space Affairs, UNOOSA, 1967 Treaty on principles governing the activities of states in the exploration and use of outer space, including the Moon and other celestial bodies, available at https://www.unoosa.org/oosa/en/ourwork/spacelaw/treaties/introouterspacetreaty.html) makes it clear the Moon is not subject to personal or national appropriation, the boundaries of acceptable commercial and residential use warrant further discussion before settlements begin to form. This will include consideration of workers' rights and safety standards on any lunar mining facilities, citizenship issues for children born in an off-world settlement, and whether people residing in a lunar settlement should be prioritised for any natural resources obtained from the Moon. This chapter will consider these issues using traditional and speculative bioethical frameworks.

15.1 Introduction

The potential for lunar settlements to mitigate the environmental impacts of over-population on Earth, and to provide opportunities for research and mining, carries with it concomitant concerns for occupational safety and environmental protection for the Moon and its future inhabitants. The question of whether the lunar landscape should be preserved, either for its intrinsic value or its commercial potential for future space tourism, also speaks to shared concerns regarding the appropriate use

E. Kendal (✉)
Swinburne University of Technology, John St, Hawthorn, Victoria 3122, Australia
e-mail: ekendal@swin.edu.au

© The Author(s), under exclusive license to Springer Nature Switzerland AG 2021
M. B. Rappaport and K. Szocik (eds.), *The Human Factor in the Settlement of the Moon*,
Space and Society, https://doi.org/10.1007/978-3-030-81388-8_15

of celestial objects. While the Outer Space Treaty (1967) makes it clear the Moon is not subject to personal or national appropriation, the boundaries of acceptable commercial and residential use warrant further ethico-legal discussion before settlements begin to form. This chapter will explore some of the reasons for establishing a human lunar settlement and consider how workers' rights and occupational safety can be promoted in the off-world setting. It will then briefly explore citizenship and employment issues for children born in any future lunar settlement, and whether people residing in a lunar habitat should be prioritised for any natural resources obtained from the Moon, by right of their unique vulnerability, dependency, and moral and physical closeness to the resources and those extracting them.

15.2   Why Go to the Moon?

The first question to ask before determining relevant ethical considerations for Moon settlement is should humanity be going back to the Moon at all? There are many reasons public and private enterprises may value space exploration and lunar settlement beyond scientific curiosity, including that the Moon's surface contains many useful natural resources. The composition of lunar rocks and soil is already well known, with the list of resources that could be extracted from the Moon including oxygen, silicon, hydrogen, helium, aluminium, iron, calcium, and magnesium, to name a few (Ellery, 2020; Hoffstadt, 1994). These resources are useful in natural or refined form, and it is generally believed that having a local source for these elements would greatly contribute to the feasibility of any future human lunar settlement (Crawford, 2015). There are also various rare Earth elements to be found in the lunar regolith, that if recoverable, might increase the commercial viability of mining operations (Ellery, 2020). The discovery of water ice on the Moon has further stimulated discussion of the potential to establish a lucrative lunar economy, with lunar-derived hydrogen and oxygen products having the potential to serve both life support systems and fuel needs (Kornuta et al., 2019). Estimates of the potential pecuniary value of space resources to Earth's economy vary substantially, with Sowers and Dreyer (2019) claiming a "robust commercial economy in cislunar space" could add "trillions of dollars into Earth's economy this century" (235). For most estimates, the cost of exploration and mining, of either the Moon's surface or nearby asteroids and comets, are the limiting factors for establishing a commercially viable Moon base, requiring further research into cost-effective methods of resource extraction and mineral processing to meet any future lunar settlement's resource demands (Cilliers et al., 2020; Pelech et al., 2019).

Beyond its natural resources, the Moon as a travel destination also holds appeal for the unique research opportunities it provides. These include in areas of cosmology and space science generally, but also for specific areas of medical science and astrobiology, such as exploring the impact of fractional gravity on biological systems (ISECG, 2017). As a justification for why humans should return to the Moon though, the latter may require further scrutiny. Some have argued that advancements in aerospace medicine and research into developing off-world life support systems do not make compelling grounds to engage in lunar exploration, as we could avoid the "need" for such research by simply staying on our home planet. As Cockrell (2010) notes:

> It has been pointed out before that much of what has been proposed as important in lunar astrobiology, for example, developing life support systems, is tautologous, in the sense that if we did not go to the Moon we would not need to be using the Moon to test these techniques in the first place (3–4).

However, Cockrell does outline two other essential areas of research that could be conducted on the Moon that might have benefits beyond supporting future lunar missions or settlement: understanding the influence of the Moon on Earth's ecosystem and the emergence of life on the planet, and studying the impact of extremes of temperature, ionising radiation and altered gravity. This could be relevant for other space environments but also potentially for predicting the effects of extreme climate change on Earth. In a 2017 white paper produced by the International Space Exploration Coordination Group (ISECG), it was also noted that a tele-presence on the Moon could provide a useful vantage point for observing such climate change.

Establishing a human lunar settlement might also one day mitigate some of the terrestrial impacts of climate change due to over-population and protect against further environmental degradation due to resource extraction on Earth. It is also possible to imagine a future in which lunar settlements become a haven for economic refugees lacking job opportunities domestically and internationally. This latter point, however, introduces numerous ethical issues that need to be explored as part of any occupational health and safety analysis of lunar working conditions, as off-world employment can be expected to compound the vulnerabilities of employees in relation to the companies that contract their services.

15.3 Worker Rights, Safety, and the Risk of Exploitation

With a view that workers' rights are human rights (OHCHR, 2016), the idea that researchers and mining workers stationed on off-world settlements should enjoy the same rights and protections as their counterparts on Earth seems a reasonable starting point. However, given the complexities of the lunar environment as a potential working location, securing equal protection is not sufficient, due to the unique threats to human health that this environment entails. Working in reduced gravity is associated with bone and muscle loss and the development of osteoporosis and other coordination issues, in addition to being implicated in renal damage, cardiovascular changes and impaired immunity (Blaber et al., 2010; Mann et al., 2019; Uri & Haven, 2005). The isolation and stress of space travel, coupled with the limited entertainment options available to space crews, have also been shown to impact psychological wellbeing, sleep and cognitive performance (Marušič et al., 2014). Furthermore, exposure to high levels of cosmic radiation is associated with fertility and sensory loss. As such, the future lunar workforce can be considered at high risk of occupational injury, necessitating more stringent workplace safety standards to protect workers' physical and mental health. However, it may turn out to be logistically impossible to provide a higher, or even equivalent, level of occupational health and safety for workers off-world, a potentiality that requires detailed cost–benefit analyses as part of any attempt to justify establishing such a workforce in the first place. After all, if it is found to be prohibitively expensive to provide lunar workers with the same level of basic safety as those working on Earth, this becomes a justice issue with off-world employees likely to be unfairly disadvantaged. While exposure to workplace hazards is not uncommon on Earth, and there are compelling ethical reasons to limit or ban some particularly risky employment practices (*see* Resnik, 2019), it is the scale of risk and its pervasive nature for lunar workers that makes this situation unique. Some of these concerns and the concomitant duties they may confer on off-world employers have been discussed in detail elsewhere (*see* Szocik & Wójtowicz, 2019; Kendal, 2020; Szocik et al., 2019), so rather than focusing on potential physiological and psychological risks, this section will consider the social harms lunar employees may face in their industries.

On Earth, paid employment is often highly valued as it carries with it many social benefits beyond the securement of income. This is possibly best recognised by considering the common consequences of unemployment, including poorer health outcomes, loss of identity and self-esteem, and reduced social opportunities (Saunders, 2002). As such, there is a power imbalance already existing between employees and those offering the benefits of employment, that can be further exacerbated in contexts where, for example, continued employment is required to access health or life insurance, as is seen in employer-sponsored US health insurance programs. However, if we consider the lunar workforce, reliance on employer provided resources might extend as far as survival habitats and oxygen supply–literally the air the workers breathe! The unique vulnerability of living off-world renders future lunar workers particularly susceptible to exploitation, especially when considering there may not be alternative employment sources available should an employee wish to leave a certain role or company. For example, a miner on the Moon may feel coerced into working unreasonable hours given they have no other method of supporting themselves if denied access to company resources. Given the geographic constraints for using solar energy to power commercial and residential services on the Moon, it is likely living arrangements will be tightly linked to work location and that large-scale mining operations will be needed just to keep lunar habitats sustainable (Cilliers, Rasera & Hadler, 2020). Most notably this will involve developing reliable systems for high yield oxygen production from lunar regolith (Rapp, 2013). The overlap of domestic and work spaces and dependence of workers on employer provided essential resources might leave lunar employees in a precarious situation should workplace relations deteriorate, leaving them disempowered to demand safety standards are adhered to or working conditions improved, where necessary. A review of whistleblower protection laws and the need for anti-retaliation clauses for mining and mineral companies on Earth paint a concerning picture for mining employees off-world, where regulators may be too far away to serve as a deterrent for unsafe and/or exploitative work practices.

The severity of negative impacts a lunar worker could face if voicing concerns about workplace safety, coupled with the lack of genuine options for alternative employment once residing in the lunar settlement, indicate a need for strong legal protection for all citizens signing a contract to work off-world. Rigorous informed consent procedures will also need to be established to ensure future workers are aware of the occupational hazards of working in the space environment and of their rights as employees to be protected from these hazards as far as is possible (Kendal, 2020). Harsh penalties for any companies failing to discharge their duties to employees or comply with safety standards will need to be established before any crewed commercial enterprise should be allowed to operate on the Moon, in addition to determining which international agencies will be tasked with exacting these punishments. With the current surge of both public and private industry interest in space exploration and lunar resource extraction, including companies such as Elon Musk's SpaceX and Israeli non-profit SpaceIL (Gibney, 2019), now is the time to ensure such ethico-legal protections are firmly in place in advance of any off-world human settlement. However, even these guidelines may be insufficient to manage the unique vulnerability of members of the lunar workforce, especially since it is not only *actual* disadvantage and exploitation that is an ethical issue in this context, but also workers' *perceptions and fears* of retribution that might leave them with reduced capacity to protect their own interests. In other words, it is less about whether a company *would* ever threaten to punish an employee by expelling them from employer supplied housing on a Moon base, to render this situation a potential source of coercion it is enough that they *could*.

When considering all of the above ethical challenges to protecting worker rights and conditions in the off-world setting, a few possible solutions arise. The first is requiring all work contracts to be negotiated on Earth and ensuring that they cannot be adjusted after the worker has arrived in the lunar settlement in any way that can be considered disadvantageous to the employee, e.g. by extending their daily working hours or reducing wages, etc. This would also require the establishment of an independent regulatory body on Earth to adjudicate any disagreements and provide regular auditing services to ensure compliance. Given the limited communication facilities likely to be available on a Moon settlement, methods to anonymously report workplace abuses or securely contact industry liaisons on Earth would also need to be created. Here the unique vulnerability of lunar workers is again apparent, as it is not difficult to imagine a scenario in which employers control communication capabilities in the off-world settlement, making it difficult for workers to contact regulators. The substantial differences in cultural and national standards for worker rights across the globe, mean the proposed regulatory body would need to have diverse international representation and its own negotiated uniform standards. The UN Economic and Social Council represents one possible candidate to help advise on the development of these standards. It could be argued that inconsistencies in labour rights on Earth negate the need for such a standard to be applied off-world, however, improving conditions for more exploited workers globally is already something we ethically ought to be doing. The Kantian principle "ought implies can" is especially meaningful to the space context though, as we *can* choose to establish strict protections before any lunar workforce is formed, thereby avoiding the obstacles to post-hoc renegotiation facing current employees wanting to improve working conditions. It is also possible that international standards developed for off-world settlements could promote the interests of exploited workers on Earth. Whether such protections are pursued on or off-world will depend on political appetite though, as it is unlikely corporations will be motivated to promote these themselves, especially given the considerable expenses involved.

A more substantial commitment aimed at avoiding exploitation of the lunar workforce would involve bestowing on every potential employee a guaranteed "return fare" to Earth at the company's expense. Depending on the scope of the commercial enterprise and its access to transportation, the latter may be prohibitively expensive or impractical. Nevertheless, a right to return to Earth if an employee's situation changes may be the only way to protect against some forms of worker abuse. Another alternative might involve establishing a social welfare system to operate in the lunar settlement, so all residents have guaranteed access to their basic survival needs, regardless of employment status. Such an enterprise could be funded by required contributions from public and private companies operating in the lunar settlement, as it is unlikely the worker base would be sufficiently large to sustain the system through a taxation method, at least in its early stages.

Moving away from mining and considering the situation for lunar scientists and other researchers, the restrictions noted above are likely to similarly protect against exploitation for members of this workforce. However, in the case of medical research there are additional concerns for research participants that need to be addressed. Imagining a scenario in which a Moon-based employer has funded research into the impacts of working in lunar gravity on productivity, employees may feel pressured to participate in such projects for the same reasons outlined above in reference to working conditions. As such, it is proposed that inhabitants of a lunar settlement be considered a vulnerable and confined population for the purposes of research. This will carry certain restrictions on the kinds of research that can be conducted on this population and typically require that the benefits of the research be directly relevant to this group (*see* NHMRC, 2018 or similar national human research guidance documents). Again, this may require that research participation be negotiated before space travel commences to promote autonomous decision-making among research participants. As with all medical research, it is essential to communicate that refusal to enroll in a research trial or study will not lead to prejudicial treatment or refusal of medical care. This is all the more vital in a situation where medical facilities will be necessarily limited (possibly to a single lunar medical facility) and the roles of healthcare provider and medical researcher are likely to contain some overlap.

15.4 Preservation Versus Extractivism and the Issue of Appropriation

Another ethical consideration when developing guidelines for off-world human settlements concerns potential obligations to the space environment, in this example, the Moon. Is there an ethical imperative to protect the Moon from over-mining, or to leave it to continue its development without human interference, a development that some speculate could in many generations' time yield signs of life? Alice Gorman's work on space archaeology also posits that there could be an ethical obligation to preserve as cultural heritage sites places like the Apollo landing sites for future generations of humans to enjoy (Gorman, 2021). At the very least it seems fair to create some restrictions for public and private activities on the Moon aimed at environmental protection. The risk of destroying culturally important landmarks on the lunar surface also warrants attention given what Roger D. Launius labels the "mythical significance" of the Moon landing and romantic connection many people have to the Moon and humanity's history of space exploration (Launius, 2012). Many of the commercial interests involving the Moon are focused on extraction of resources or transport of goods or tourists, all of which come at a cost to the lunar environment (Dawson, 2012). A proposal that an environmental impact report form a necessary component of any lunar-based industry project seems a modest requirement. A bolder protection strategy might also require that industries have a positive obligation to engage the least disruptive methods for conducting their business, even if this involves efficiency sacrifices, and to demonstrate how they will avoid causing unnecessary damage to the lunar environment. Waste management policies will also need to be enforced to prevent the Moon becoming a dumpsite for industrial waste or companies contributing further to the current space debris problem (Gorman, 2021).

At the heart of ethical concerns regarding the extractivist approach to lunar exploration is the issue of resource appropriation. The Outer Space Treaty (1967) makes it clear that the Moon and other celestial objects are not subject to appropriation, in short, they cannot be "owned" by any country or corporation. There is significant disagreement regarding whether engaging in lunar or asteroidal mining constitutes a violation of this principle, and if so, whether this element of the treaty needs to be amended to more closely resemble common views on property rights outside of this context. This is especially the case assuming corporations are being encouraged to engage in the space economy through mining, research and tourism. Legal scholar, Hoffstadt (1994) claims that the origin of the Treaty's wording about space being the "province of all mankind" is typically believed to relate to the Roman concept of *res communis*–a right to "free access and use–but not rights of ownership–to the shared property of the community" (587). The question then becomes whether mining to sustain a lunar settlement should be considered "use" of lunar resources, or appropriation of the same. This is of course particularly relevant when considering non-renewable resources. Hoffstadt (1994) claims that ambiguity regarding exactly what Outer Space Treaty terms like "province", "common heritage" and "equitable sharing" of resources mean in a legal sense, have hindered private investment in lunar mining. Few countries have specified policies regarding private mining in space, but the US and Luxembourg have legislated that companies have a right to retain any space resources they extract (Foster, 2016). This position has been criticised by many other States, with Andreas Losch (2019) citing this as a prime example of why we need to collectively develop an "ethics of planetary sustainability" to prevent "Wild West" attitudes toward space exploration and expansion. This also leads to the final point for ethical consideration in this chapter: who should have priority access to use the Moon's resources and what are the implications for future generations of lunar residents?

15.5   Implications for Moon Residents and Future Generations

Imagining a future in which there is an established human lunar settlement engaging in large-scale mining and lunar research projects that are yielding substantial financial benefits on Earth, it is not difficult to extend this vision further to predict future disagreements about where the resources from this base should be concentrated. In light of the increased occupational risks lunar workers face, there is a compelling argument that the benefits of their activities should be conferred to themselves and their immediate community, rather than distant shareholders on Earth. While the maintenance of habitable residences for lunar workers are both an ethical and practical requirement for continued business, there could also be a case made that wealth generated from the extraction of the Moon's resources should first be dedicated to improving living conditions in the settlement, rather than increasing company profits. In other words, it is not enough that the lunar workforce merely be provided the bare necessities for survival if others are making significant profit from their labour. Matters become even more complicated when considering the potential for a rare element to be in high demand both in the lunar settlement and on Earth. In such a case, whose interests should prevail? There may be more people who stand to benefit by sending the resources to Earth, assuming a cost-effective transportation method is available, but do the lunar residents have a higher claim to the element by nature of their proximity to it and involvement in its recovery?

To answer these hypotheticals, the concept of moral closeness is relevant. In traditional bioethics, this theory relates to the particular "duty to rescue" a nearby person in peril (Miller, 2004). This theory allows for some degree of favouritism in who we provide aid to, preferencing those morally, and in some cases literally, closer to hand. An example would be the strong moral imperative to help a family member in need or care for a child who has fallen over in the street ahead of us, compared to the (arguably) weaker duty to donate to an international charity focused on the needs of strangers very distant to us. Arguments for considering moral closeness in ethical decision-making are in conflict with ethical theories that promote moral impartiality and treating all needs equally. In a debate over resources, moral closeness would justify resources being made available first to those in need in the lunar settlement, due to their social and geographic closeness to those involved in resource extraction. The ethics of closeness prioritises moral duties to those we have existing relationships with, justifying prioritisation of resources to people we have direct experience of, even at the expense of others whose needs might be greater (Myhrvold, 2003). Adopting such an approach to settle resource disputes between a lunar settlement and Earth-based companies would further serve to avoid exploitation of off-world workers, ensuring only those resources in excess of what the settlement requires for maintenance and targeted expansion are available for export.

While the above covers what claims the future children of lunar workers may have to the resources necessary to sustain their existence in the settlement, another consideration for future generations of lunar residents relates to whether they can be considered to have a right to return to Earth, particularly given the limited employment opportunities available off-world. The limits of the UN "Rescue Agreement" (1967), where all parties are required to render available aid to astronauts in distress in their own or unclaimed territory, are also likely to be tested once human settlements start to form off-world. For example, if a lunar worker or their family member suffers an accident on the Moon base what parties might be obligated to provide assistance? It is likely children born to lunar workers will hold the same terrestrial citizenship as their biological parents, however, we will also need clear laws regarding interplanetary repatriation. There is also the issue of establishing multi-generational citizenship protocols and whether a lunar human settlement should be allowed to develop its own national identity. Identifying specific ethical obligations to the children of off-world workers, including whether all humans can be considered to possess an intrinsic right to reside on our home planet, irrespective of place of birth, remains an open area of debate.

15.6   Conclusion

There are many aspects of working in the lunar environment that pose ethical challenges from an occupational health and safety perspective. However, by establishing the rights and responsibilities of all parties in advance of any human lunar settlements forming, it is possible to prevent or reduce various harms to lunar workers, including exploitation, and to strive to protect and preserve the lunar environment. Given the unique vulnerability of this population of workers, and the potential hazards of the space environment for human health and wellbeing, stringent ethico-legal requirements should be developed now to protect the interests of off-world workers in the future.

References

Blaber, E., Marçal, H., & Burns, B. P. (2010). Bioastronautics: The influence of microgravity on astronaut health. *Astrobiology, 10*(5), 463–473.

Cilliers, J. J., Rasera, J. N., & Hadler, K. (2020). Estimating the scale of Space Resource Utilisation (SRU) operations to satisfy lunar oxygen demand. *Planetary and Space Science, 180*, 104749.

Cockrell, C. S. (2010) Astrobiology—What can we do on the moon? *Earth, Moon, Planets, 107*, 3–10.

Crawford, I. A. (2015). Lunar resources: A review. *Progress in Physical Geography, 39*(2), 137–167.

Dawson, L. (2012). *The politics and perils of space exploration: Who will compete? Who will dominate?* Springer.

Ellery, A. (2020). Sustainable in-situ resource utilization on the moon. *Planetary and Space Science, 184*, 104870.

Foster, C. (2016). Excuse me, you're mining my asteroid: Space property rights and the U.S. space resource exploration and utilization act of 2015. *Journal of Law, Technology & Policy*, 407–430.

Gibney, E. (2019). First private Moon lander sparks new lunar space race. *Nature, 566*, 434–436.

Gorman, A. (2021). *Dr Space Junk vs. the universe: Archaeology and the future.* Cambridge: MIT Press.

Hoffstadt, B. M. (1994). Moving the heavens: Lunar mining and the common heritage of mankind in the Moon Treaty. *UCLA Law Review, 42*(2), 575–621.

Kendal, E. (2020). Biological modification as prophylaxis: How extreme environments challenge the treatment/enhancement divide. In K. Szocik (Eds.), *Human enhancements for space missions* (pp. 35–46). Cham: Space and Society, Springer.

Kornuta, D., Abbud-Madrid, A., Atkinson, J., Barr, J., Barnhard, G., & Bienhoff, D., et al. (2019). Commercial lunar propellant architecture: A collaborative study of lunar propellant production. *Reach, 13*, 100026.

Launius, R. D. (2012). Why go to the moon? The many faces of lunar policy. *Acta Astronautica, 70*, 165–175.

Losch, A. (2019). The need for an ethics of planetary sustainability. *International Journal of Astrobiology, 18*(3), 259–266.

Mann, V., Sundaresan, A., Mehta, S. K., Crucian, B., Doursout, M. F., & Devakottai, S. (2019). Effects of microgravity and other space stressors in immunosuppression and viral reactivation with potential nervous system involvement. *Neurology India, 67*(2), S198-203.

Marušič, U., Meeusen, R., Pišot, R., & Kavcic, V. (2014). The brain in micro- and hypergravity: The effects of changing gravity on the brain electrocortical activity. *European Journal of Sport Science, 14*(8), 813–822.

Miller, R. W. (2004). Moral closeness and world community. In D. K. Chatterjee & D. Maclean (Eds.), *The ethics of assistance: Morality and the distant needy* (pp. 101–111). Cambridge: University Press.

Myhrvold, T. (2003). The exclusion of the other: Challenges to the ethics of closeness. *Nursing Philosophy, 4*, 33–43.

National Health and Medical Research Council (NHMRC). (2018). *The national statement on ethical conduct in human research (2007)–Updated 2018*. https://www.nhmrc.gov.au/about-us/publications/national-statement-ethical-conduct-human-research-2007-updated-2018

Pelech, T. M., Roesler, G., & Saydam, S. (2019). Technical evaluation of Off-Earth ice mining scenarios through an opportunity cost approach. *Acta Astronautica , 162*, 388–404.

Rapp, D. (2013). *Use of extraterrestrial resources for human space missions to Moon or Mars* (2nd ed.). Springer.

Resnik, D. B. (2019). Occupational health and the built environment: Ethical issues. In A. C. Mastroianni, J. P. Kahn, & N. E. Kass (Eds.), *Oxford Handbook of Public Health Ethics* (pp. 718–727). Oxford University Press.

Saunders, P. (2002). The direct and indirect effects of unemployment on poverty and inequality. *Australian Journal of Labour Economics, 5*(4), 507–529.

International Space Exploration Coordination Group (ISECG). (2017). *Scientific opportunities enabled by human exploration beyond low-earth orbit: An ISECG white paper*. Noordwijk: European Space Research and Technology (ESTEC).

Sowers, G. F., & Dreyer, C. B. (2019). Ice mining in lunar permanently shadowed regions. *New Space, 7*(4), 235–244.

Szocik, K., Campa, R., Rappaport, M., & Corbally, C. (2019). Changing the paradigm on human enhancements: The special case of modifications to counter bone loss for manned Mars missions. *Space Policy, 48*, 68–75.

Szocik, K., & Wójtowicz, T. (2019). Human enhancement in space missions: From moral controversy to technological duty. *Technology in Society , 59*, 101156.

United Nations Human Rights Office of the Commissioner (OHCHR). (2016). Statement by Maina Kiai, Special Rapporteur on the rights and freedom of peaceful assembly and of association at the 71st session of the general assembly. https://www.ohchr.org/en/newsevents/pages/DisplayNews.aspx?NewsID=20727&LangID=E

United Nations Office for Outer Space Affairs (UNOOSA). (1967). Treaty on principles governing the activities of states in the exploration and use of outer space, including the Moon and other celestial bodies. https://www.unoosa.org/oosa/en/ourwork/spacelaw/treaties/introouterspacetreaty.html

Uri, J. J., & Haven, C. P. (2005). Accomplishments in bioastronautics research aboard International Space Station. *Acta Astronautica, 56*, 883–889.

Chapter 16
Science and Faith Off-Earth

Christopher D. Impey ⓘ

Abstract Colonists living on another world will be physically dislocated from the planet of their birth, they will be living in a highly managed, artificial environment, and they will be under huge pressure to make the experiment a success. Lunar colonists will be making a commitment of years or perhaps a lifetime. The only terrestrial situations to rival the isolation of a lunar colony are for people who spend the austral winter at a research station in Antarctica or for astronauts who live a year or more on the International Space Station. This chapter considers implications of religious faith in a future lunar colony. The colonists will be selected for their particular skills, their propensity to work in a team, and their ability to handle stress. It is possible that funding for a colony might dictate a monolithic population, such as citizens of one country or only scientists. But it is more likely that colonists will embody diversity of gender, nationality, occupation, race, and creed. Establishing a self-contained habitat in a very hostile environment puts a premium on technical skills, so it is also assumed that the population will be skewed towards scientists and engineers. Some level of technical skill will be a sine qua non for the first wave of colonists. In this context, it is interesting to consider the role that faith and religion might play in the lives of future inhabitants of the Moon.

16.1 The Landscape of Science and Faith

Faith is a general term denoting trust or confidence in a person, thing, or concept. What distinguishes it from science is that no evidence or proof is required to sustain it. Lunar colonists will depend on science for their survival, but many may depend on their faith to sustain their motivation to live off-Earth Faith is a subtle concept, with many variations that include existential confidence, knowledge of specific truths as revealed by God, belief in God and in the fact that God exists, and practical commitments that stem from a belief that God exists (Bishop, 2016). Faith can apply to a community and their shared religious practice or to someone with their own individual spiritual

C. D. Impey (✉)
Steward Observatory, University of Arizona, 933 N. Cherry Avenue, Tucson, AZ 85718, USA
e-mail: cimpey@as.arizona.edu

© The Author(s), under exclusive license to Springer Nature Switzerland AG 2021   245
M. B. Rappaport and K. Szocik (eds.), *The Human Factor in the Settlement of the Moon*,
Space and Society, https://doi.org/10.1007/978-3-030-81388-8_16

discipline. Religion is defined as a socio-cultural system of behaviors and practices by a group of people who share certain supernatural and transcendental beliefs. There is no scholarly consensus about what constitutes religion (Nongbri, 2013), but those planning a lunar colony will have to serve religious and spiritual needs of the colonists, just as they will have to serve their needs for exercise and recreation. For off-Earth colonists, the pertinent aspects of religion are the way it guides ethical decisions, the way it can provide a sense of meaning and purpose, and the rituals it requires of practitioners. There may also be a role for, and a need for, experts to guide various religious practices.

Science is intrinsically international. Its methods are commonly accepted across all disciplines and national boundaries, and its success depends on transparency and the free flow of information. English is its common language. Even as developing countries create their own scientific enterprises, the pull of transnationalism eclipses tendencies towards chauvinism (Crawford et al., 1993). Science is a niche profession. Worldwide, there were about 8 million scientific researchers in 2013 (UNESCO, 2015). Religious faith, by contrast, is pervasive and its varieties segregate by geography and culture. Over 80% of the world's population has a religious affiliation, and each of the major religions is dispersed across many countries, although Hindus, Muslims, and Buddhists concentrate in Asia (Pew Research Center, 2012). It is worth remembering that the billion people with no religious affiliation form the third largest group, behind only Christians and Muslims.

The trends in religiosity are important, given that lunar colonists will be drawn largely from a cohort of people who are not yet alive. Religion is in decline globally, especially in the past few decades (Inglehart, 2020). Around the world, survival becomes more reliable as societies develop, and as life expectancy rises and people feel more secure, they tend to become less religious. Worldwide, this population is expected to peak around 2040 (Pew Research Center, 2015). The trend toward being religiously unaffiliated is particularly strong in the United States. At the dawn of the Space Age in the 1950's, only 2% of Americans were non-believers. This fraction saw two surges. The first up to 7–8%, starting at the time of the Apollo Moon landings, and the second to above 25%, starting around the time the Internet was introduced. It has been argued that this is not a coincidence; the rapid growth of science and easy access to online information may have weakened the cohesion of cultural norms like organized religion (Vacker, 2019). It is entirely plausible the people adhering to a major religion will be in the minority for those who choose to live off-Earth.

There are many more people of faith than scientists, but how much overlap is there between the two groups? The only large survey of religious belief among scientists is a 2009 survey of members of the American Association for the Advancement of Science (Pew Research Center, 2009). Scientists are about half as likely as the general public to believe in god or a higher power. These numbers have been stable for almost a century. Nearly half have no religious affiliation, compared with only 17% of the general public. Among the faiths, scientists are less likely to be Protestant or Catholic, and more likely to be Jewish (Fig. 16.1). Lunar colonists will probably be drawn from many countries so international data on faith is relevant. A survey of

Religious Affiliation Among the General Public and Scientists
% who are...

Among general public

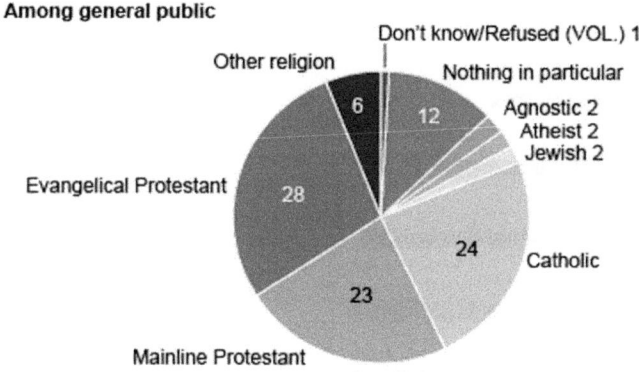

Among scientists

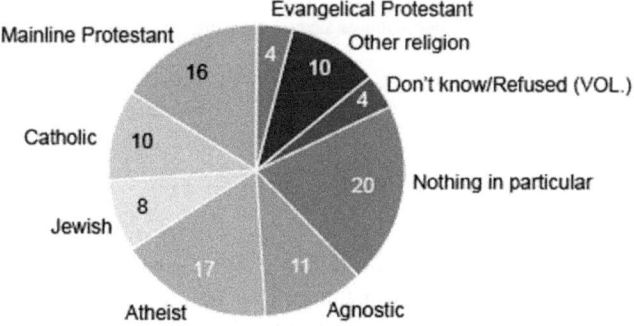

Fig. 16.1   U.S. religious affiliation of scientists and the general public, measured in surveys carried out in 2009 (Credit: Pew Research Center, 2009)

biologists and physicists from eight regions around the world finds them to be more secular that the general population in every region (Ecklund et al., 2016). The margin is around a factor of two in Europe and the United States and less elsewhere in the world (Fig. 16.2). Scientists in this survey did not think that science and religion were in conflict. In a study of the elite scientists who won Nobel Prizes over the last century, just over half were Christians, 24% were Jewish, a much higher fraction than among American scientists in general, and 6% were Atheists and Agnostics, a much lower fraction than among scientists in general (Shalev, 2005).

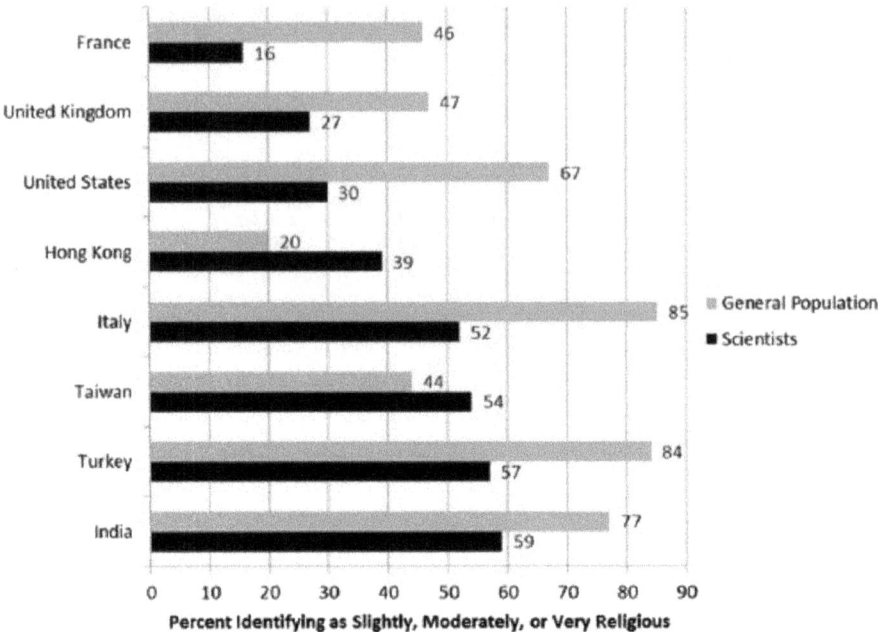

Fig. 16.2 Percentage of physicists and biologists by region identifying as religious compared with the population (Credit: Ecklund et al., 2016)

16.2 The Practice of Religion in Space

The Space Age is sixty years old. Since Yuri Gagarin completed one circuit of the Earth on 12 April 1961, fewer than 600 people have been in Earth orbit, and only 24 have left the Earth and traveled to the Moon. We are currently on the threshold of a major new expansion of the space activity, driven by new government initiatives (Clark, 2020) and burgeoning competition in the private sector (Grady, 2017). Colonies on the Moon and Mars are part of the arc of these initiatives, likely to come to fruition within thirty years (Impey, 2016). Humans in Earth orbit for a limited period of time can tolerate a loss of creature comforts and constraints on their activities. Lunar colonists will be making a commitment of years to an alien habitat and they will understandably want to bring as many aspects of their lifestyle as possible with them. That includes the expression of their spirituality or religion. At its best, religious faith has attributes that could help colonists adapt to their alien environment, by reinforcing virtuous or resistant attitudes, and nourishing hope in the face of adversity.

Space travelers are becoming more diverse. The first few cohorts of astronauts from the United States and the Soviet Union were almost exclusively white and male, drawn from the ranks of test pilots and the military. Gradually, women, ethnic minorities, and people from other countries got their chance to have the unique experience of zero gravity and a global perspective on their planet (Smith et al., 2020). Mercury, Gemini and Apollo astronauts were almost a monoculture, being primarily Protestants from the American Midwest. A tension arose between their free expression of religion, NASA's secular charter, and the separation of Church and State that is enshrined in the U.S. Constitution. As they orbited the Moon on Christmas Eve 1968, the Apollo 8 astronauts read the first ten verses of Genesis. This spurred a lawsuit by Madalyn Murray O'Hair, the founder of American Atheists, who alleged a violation of the First Amendment. The case was dismissed, and over the next six years NASA received eight million letters and petition signatures supporting free religious expression by American astronauts in space (Oliver 2013a). Apollo 15 Commander David Scott left a bible on the lunar rover during an extravehicular activity. These overt displays of Christianity contrasted with Russian ideology, which framed space exploration as a means of proving atheism (Pop, 2009).

As the space age progressed, the variety of religious expression increased. Buzz Aldrin, a Presbyterian, performed communion for himself on the Moon, using materials given to him by his church. He was dissuaded from broadcasting the service back to Earth due to the controversy over the Apollo 8 Genesis reading (Aldrin, 1970). Communion wafers consecrated in a Catholic Mass were used for Holy Communion by three astronauts on the Space Shuttle in 1994, and by one astronaut on the International Space Station in 2013 (Sadowski, 2016). With the fall of the Soviet Union, cosmonauts began to express Orthodox Christianity in space (Pop, 2009). Muslims in space have a particular problem in trying to face Mecca and kneel to pray while traveling in zero gravity at kilometers per second (Lewis, 2013). The National Fatwa Council was forced to modify Islamic rituals for astronauts to resolve these issues (Associated Press 2007). Due to the constraints of space travel, religious artifacts are often transported in miniature. A microfilm Bible was carried to the Moon and subsequently auctioned off, and a microfilm Torah was taken into orbit in 2013 (Patterson, 2011). It is logical to anticipate that Moon colonists will carry their faith and religious practice with them to surface of that forbidding world.

16.3 Illustrative Examples from the Earth

Humans are social animals. Even during the tens of millennia that we were nomadic hunter-gatherers, the mobile unit of under a hundred people maintained a larger web of connections and was embedded in a large and complex society (Bird et al., 2019). Over the history of civilization, people have carried their religion to some of the most remote and isolated places on Earth. Examples include the Havasupai tribe members who live on a canyon floor in Arizona, the 300 inhabitants of Tristan de Cunha in the middle of the Atlantic Ocean, the town of Utquiagvik built on permafrost in northern Alaska, the 500 people crammed onto Migingo Island set in Lake Victoria, and the primitive mining settlement La Rinconada at a height of 16,000 feet in the Peruvian Andes. The United States is unique in its rich history of isolated settlements set up for religious and secular purposes (Sutton, 2003, 2004). Many of these communities were motivated by Utopian ideals as they carved out a new way of living in an inhospitable land (Kesten, 1993, Yuko 2020). The first lunar colonies will require the substantial investment that only governments can provide. However, we can envisage the time when the Moon is host to a variety of settlements, some monocultures of one nation, one religion, or one ideology, and others attempting to be a microcosm of humanity. Some spaces in these settlements will likely be set aside for religious or spiritual practice. A terrestrial analog is the airport chapels that first appeared in the 1950's and eventually spread to most of the world's major airports (Cadge, 2018). These have evolved into multi-faith spaces that can be also found in hospitals, schools, and other public spaces (Bobrowicz, 2018).

The Antarctic offers an analog of the isolation and extreme physical conditions of the Moon (Fig. 16.3). There have been several studies of long-term Antarctic residents and an awareness that this research is directly relevant to the future of space travel (Rothblum, 1990; Taylor, 1987). Residents who winter-over face as many as eight months of darkness and extreme isolation, with serious behavioral and medical consequences (Palinkas 1992). According to an old mariner's adage: "Below 40 degrees south there is no law; below 50 degrees south there is no God." There are, however, seven churches. Most Antarctic research stations have a small, multipurpose room that serves as an ad hoc chapel. All of the dedicated chapels serve Christians, although the Chapel of the Snows at McMurdo Station on Ross Island has also opened its doors to Bahais and Buddhists (Reidel, 2016). The most unusual place of worship is the Ice Cave Catholic Chapel at the Argentinian base on Coats Island. Hindus and Muslims have no specific sites of worship on the continent. Despite its isolation, Antarctica has accreted most of the accoutrements of faith and religion.

These terrestrial analogies miss a vital ingredient of the lunar experience: dislocation from a familiar habitat and the perspective the Moon offers on the Earth.

Fig. 16.3 The U.S. South Pole Station, partly covered in drifting snow; winter residents spend up to eight months in extreme isolation (Credit: Josh Landis, National Science Foundation)

For decades, astronauts have reported an overwhelming emotional response to their experience of space and a feeling of identifying with humankind and the whole planet. This has been called the "overview effect" (Yaden et al., 2016). The genesis of this awareness for most people was the dramatic photograph of Earthrise over the Moon, taken by the Apollo 8 astronauts on the same day that they read from the Bible (Poole, 2010). This image was part of the spur to the modern environmental movement (Henry & Taylor, 2009). There was no explicit change in astronauts' spirituality attached to the overview effect, rather it was a spur to a more humanistic way of thinking (Kanas, 2020). The lunar colonists will perforce experience the confines typical of prisoners and the frugality typical of pioneer settlers. They be able to appreciate daily the beauty and fragility of the home planet.

16.4   East Meets West on the Moon

In general, the issues of religion for a lunar colony are similar to those that have been explored for a Mars colony (Oviedo, 2019; Szocik, 2017). Support for space exploration is not uniform across the religious traditions. One survey in the United States found that Evangelical Protestants had the least knowledge about space, the least appreciation for space exploration, and the lowest support for space policy. Jews, and those practicing Eastern traditions such as Buddhists and Hindus form the core of the religious attentive public for the space activity (Ambrosius, 2015). It has even been argued that advocacy for space exploration has attributes that give it the characteristics of a religion (Launius, 2013). As we consider the role of faith off-Earth, it is important to recall the distinction between religions based on faith such as Christianity and Islam and religions based on practice such as Buddhism and Shintoism (Traphagan, 2020). While an argument can be made for an absolute distinction between faith and science, the tension that is often perceived between religion and science in Western cultures is largely absent in Eastern cultures (Kroesbergen, 2018).

There was a pervasive sense of a Christian mission throughout the ranks of the NASA employees that worked on the Apollo program (Oliver 2013). NASA is a civilian agency so it will eschew any such association in their plans for a lunar colony (Yang, 2019). If we are projecting to decades in the future, the time it will take to establish a colony on the Moon, we should recall that 1.2 billion people worldwide are not affiliated to a major religion. Many of these people claim to be "spiritual" but not in the sense of adhering to any set of tenets (or dogmas). This number is growing, particularly in the United States, where over the past decade the share of Christians has fallen from 77 to 65%, while the proportion of the religiously unaffiliated has grown from 17 to 26% (Pew Research Center, 2019). This trend, combined with the difficulty of maintaining traditional religious rituals and the great distance from the "Mother Church," means that organized religion might not be the norm for lunar colonists. Modes of modern spirituality are variegated. They can manifest as a sense of oneness or closeness to an abstract higher power, a feeling of transcendence, or connection to humanity or nature (Davis et al., 2015). The bonds of organized religion are likely to further loosen, and new religious models emerge, if the colonists live and die on the Moon, and raise families there, forming the first new branch of the human "tree" since we left Africa 50,000 years ago (Fig. 16.4).

When East meets West on the Moon, the biggest impact may come from any Buddhists who live in the colony. Buddhists have a particular perspective on space habitation and the utilization of space resources. For example, 81% of Buddhists think the Moon and other off-Earth locations should be valued and protected, even if they are devoid of life, compared with 21% of a control sample (Capper, 2020). Also, they reject the idea that it is appropriate to globally alter any planetary ecology. These views are consistent with adherence to the principles of non-harm and interconnectedness that underly Buddhist philosophy (Irudayadason, 2013). Buddhists do not have a lock on this way of thinking; the variegated forms of Christianity also

Fig. 16.4 Lunar colonists will have to live underground in highly controlled and artificial environments, creating stresses on social interactions. This artist's impression shows NASA's planned Artemis base (Credit: European Space Agency/Pierre Carril)

include those that strongly support preservation of natural resources. Since Buddhism is more of a philosophy and a way of life than a religion, it can erode barriers of culture and ideology (McMahan & Braun, 2017; Wright, 2017). As an empirical practice, it aligns with science that will drive the creation of the first viable lunar colony (Impey,). Buddhist meditation practice can increase social bonds and reduce prejudice (Kreplin, Farias and Brazil 2018), benefit the immune system (Davidson et al., 2003), and reduce natural brain atrophy (Luders et al., 2015). All these benefits will be valuable as the colonists adapt to physical and psychological challenges of living off-Earth.

References

Aldrin, B. (1970). When Buzz Aldrin took communion on the Moon. *Guideposts*. https://www.gui deposts.org/better-living/life-advice/finding-life-purpose/guideposts-classics-when-buzz-aldrin-took-communion-on-the-moon.

Ambrosius, J. D. (2015). Separation of church and space: Religious influences on public support for U.S space exploration policy. *Space Policy, 32*, 17–31.

Associated Press. (2007). Astronaut to grapple with daily prayer ritual. *NBC News*. https://www.nbcnews.com/id/wbna20894077.

Bird, D. W., Bird, R. B., Codding, B. F., & Zeanah, D. W. (2019). Variability in the organization and size of hunter-gatherer groups: Foragers do not live in small-scale societies. *Journal of Human Evolution, 131*, 96–108.

Bishop, J. C. (2016). Faith, *Stanford Encyclopedia of Philosophy*. https://plato.stanford.edu/entries/faith/.

Bobrowicz, R. (2018). Multi-faith spaces uncover secular premises behind the multi-faith paradigm.*Religions, 9*, 37–44.

Cadge, W. (2018). A brief history of airport chapels. *Smithsonian Magazine*. https://www.smiths onianmag.com/travel/airport-chapels-brief-history-180967765/.

Capper, D. (2020). *Buddhism and space environments*. https://www.buddhismandspace.org/mining-our-moon.

Clark, S. (2020). ESA and NASA Unveil Bold Plans for the Future of Space Exploration. *BBC Science Focus Magazine*. https://www.sciencefocus.com/news/esa-and-nasa-unveil-bold-plans-for-the-future-of-space-exploration/.

Crawford, E., Shinn, T., & Sorlin, S. (eds.). (1993). *Denationalizing science: The contexts of international scientific practice*. Boston, MA: Kluwer Academic Publishers.

Davidson, R. J., Kabat-Zinn, J., Schumacher, J., Rozenkranz, M., Muller, D., Santorelli, S. F., Urbanowski, F., Harrington, A., Bonus, K., & Sheridan, J. F. (2003). Alterations in brain and immune function produced by mindfulness meditation. *Psychosomatic Medicine, 65*, 564–570.

Davis, D. E., Rice, K., Hook, J. N., Van Tongeren, D. R., DeBlaere, C., Choe, E., & Worthington, E. (2015). Development of the sources of spirituality scale. *Journal of Counseling Psychology, 62*, 503–513.

Ecklund, E. H., Johnson, D. R., Scheitle, C. P., Matthews, K. R. W. & Lewis, S. W. (2016). Religion among scientists in an international context: A new study of scientists in eight regions. *Socius: Sociological Research for a Dynamic World, 2*, 1–9.

Grady, N. (2017). Private companies are launching a new space race–here's what to expect. *The Conversation*. https://theconversation.com/private-companies-are-launching-a-new-space-race-heres-what-to-expect-80697.

Henry, H., & Taylor, A. (2009). Re-thinking apollo: Envisaging environmentalism in space. *The Sociological Review, 57*, 190–203.

Impey, C. D. (2016). *Beyond: Our future in space*. New York, NY: W.W. Norton.

Impey, C. D. (2020). What Buddhism and science can teach each other–and us–about the universe. *The Conversation*. https://theconversation.com/what-buddhism-and-science-can-teach-each-other-and-us-about-the-universe-134322.

Inglehart, R. F. (2020). Giving up on god: The global decline of religion. *Foreign Affairs*. https://www.foreignaffairs.com/articles/world/2020-08-11/religion-giving-god.

Irudayadason, N. A. (2013). The wonder called cosmic oneness: Toward astroethics from Hindu and Buddhist wisdom and worldviews. In C. Impey, A. H. Spitz, & W. Stoeger (eds.), *Encountering life in the universe: Ethical foundations and social implications of astrobiology* (pp. 94–119). Tucson, AZ: University of Arizona Press.

Kanas, N. (2020). Spirituality, humanism, and the overview effect during manned space missions. *Acta Astronautica, 166*, 525–528.

Kreplin, U., Farias, M., & Brazil, I. A. (2018). The limited prosocial effects of meditation: A systematic review and meta-analysis. *Scientific Reports*, 8, Number 2403.

Kesten, S. R. (1993). *Utopian episodes: Daily life in experimental colonies dedicated to changing the world*. Syracuse, NY: Syracuse University Press.

Kroesbergen, H. (2018). An absolute distinction between faith and science: Contrast without compartmentalization. *Zygon, 53*, 9–28.

Launius, R. D. (2013). Escaping earth: Human spaceflight as religion. *Astropolitics, 11*, 45–64.

Lewis, C. S. (2013). Muslims in space: Observing religious rites in a new environment. *Astropolitics, 11*, 108–115.

Luders, E., Cherbuin, N., & Kurth, F. (2015). Forever Young(er): Potential age-defying effects of long-term meditation on gray matter atrophy. *Frontiers in Psychology*. https://doi.org/10.3389/fpsyg.2014.01551

McMahan, D. L., & Braun, E. (eds.). (2017). *Meditation, Buddhism, and science*. Oxford, England: Oxford University Press.

Nongbri, B. (2013). *Before religion: A history of a modern concept*. New Haven, CT: Yale University Press.

Oliver, K. (2013). The Apollo 8 genesis reading and religion in the space age. *Astropolitics, 11*, 116–121.

Oliver, K (2013b). *To touch the face of god: The sacred, the profane, and the American space program, 1957–1975*. Baltimore, MD: Johns Hopkins University Press.

Oviedo, L. (2019). Religion for a spatial colony: Asking the right questions, In K. Szocik (eds.), *The human factor in a mission to Mars*. London, England: Springer Nature.

Palinkas, L. A. (1992). Going to extremes: The cultural context of stress, illness and coping in Antarctica. *Social Science and Medicine, 35*, 651–664.

Patterson, T. (2011). *The surprising history of prayer in space*, CNN Religion Blog. https://religion.blogs.cnn.com/2011/07/07/the-surprising-history-of-prayer-in-space/.

Pew Research Center. (2009). Religion and Science in the United States. In *Pew forum on religion and public life*. Washington, DC: Pew Research Center. https://www.pewforum.org/2009/11/05/an-overview-of-religion-and-science-in-the-united-states/.

Pew Research Center. (2012). The global religious landscape. *Pew forum on religion and public life*. Washington, DC: Pew Research Center. https://www.pewforum.org/2012/12/18/global-religious-landscape-exec/.

Pew Research Center. (2015). *The future of world religions: Population growth projections, 2010–2050*. Washington, DC: Pew Research Center. https://www.pewforum.org/2015/04/02/religious-projections-2010-2050/.

Pew Research Center. (2019). *In U.S., decline of Christianity continues at a rapid pace*. Washington, DC: Pew Research Center. https://www.pewforum.org/2019/10/17/in-u-s-decline-of-christianity-continues-at-rapid-pace/.

Poole, R. (2010). *Earthrise: How man first saw the earth*. New Haven, CN: Yale University Press.

Pop, V. (2009). Space and religion in Russia: Cosmonaut worship to orthodox revival. *Astropolitics, 7*, 150–163.

Reidel, K. (2016). Faith in Antarctica: Religion in the land of eternal snow. *Polar News*. https://polar-news.com/antarctic/society/174-faith-in-antarctica-religion-in-the-land-of-eternal-snow.

Rothblum, E. D. (1990). Psychological factors in the Antarctic. *Journal of Psychology, 124*, 253–273.

Sadowski, D. (2016). For catholic astronauts, flying to space doesn't mean giving up faith.*Catholic News Service*. https://www.catholicnews.com/services/englishnews/2016/for-catholic-astronauts-flying-to-space-doesnt-mean-giving-up-the-faith.cfm.

Shalev, B. A. (2005). *100 years of Nobel prizes*. New Delhi, India: Atlantic Publishers.

Smith, M. G., Kelley, M., & Basner, M. (2020). A brief history of spaceflight from 1961 to 2020: An analysis of missions and astronaut demographics. *Acta Astronautica, 175*, 290–299.

Sutton, R. P. (2003). *Communal utopias and the American experience: Religious communities, 1732–2000*. Westport, CN: Praeger.

Sutton, R. P. (2004). *Communal utopias and the American experience: Secular communities, 1824–2000*. Westport, CN: Praeger.

Szocik, K. (2017). Religion in a future Mars colony? *Spaceflight, 59*, 92–97.

Taylor, A. J. W. (1987). *Antarctic psychology*. DSIR Bulletin No. 244, Vilnius, Lithuania: Scientific Information Publishing Center.

Traphagan, J. W. (2020). Religion, science, and space exploration from a non-western perspective. *Religions, 11*, 397–406.

UNESCO. (2015). *UNESCO science report: Towards 2030*. Paris, France: United Nations Educational, Scientific and Cultural Organization.

Vacker, B. (2019). The decline in religious belief in America: The role of NASA and media technologies.*Medium Online Magazine*. https://medium.com/explosion-of-awareness/the-decline-in-religious-belief-in-america-the-role-of-nasa-and-media-technologies-7e8ef3d0b919.

Wright, R. (2017). *Why Buddhism is true: The science and philosophy of meditation and enlightenment*. New York, NY: Simon and Schuster.

Yaden, D. B., Iwry, J., Slack, K. J., Eichstaedt, J. C., Zhao, Y., Vaillant, G. E., & Newburg, A. B. (2016). The overview effect: Awe and self-transcendent experience in space flight. *Psychology of Consciousness: Theory, Research, and Practice, 3*, 1–11.

Yang, J. (2019). Spaceflight and spirituality, a complicated relationship. *Wired Magazine*. https://www.wired.com/story/apollo-11-spaceflight-spirituality-complicated-relationship/.

Yuko, E. (2020). *These forward-thinking Utopias changed design forever. Architectural Digest*. https://www.architecturaldigest.com/story/utopian-communities-the-future.

Part VI
Social Science Perspectives on the Human Lunar Experience

Chapter 17
Cognitive Research and Religious Experience on the Moon

Lluis Oviedo

Abstract The expectation is that the new and very different environment of the Moon will lead to research programs on human cognition, including aspects of religious faith and perception, as well as any increased sensitivities to environmental protection, both on the Moon and on Earth. Since religious experience is not purely subjective, but an embodied and embedded living sense of ourselves and our environment, it can also be anticipated that a strongly different context, as one expects on the Moon, would change the content and quality of any perception of self-transcendence, with direct influence on the way a person feels related to the world and other people. Since this type of awareness has been increasingly related to a sensitivity towards our natural environment, it is possible that a growing intensity or significant change in religious perception on the lunar surface would entail a greater commitment to care for the natural cosmic environment, on our own planet Earth, on the Moon, and beyond.

17.1 Introduction: A Paradigm Shift for Lunar Cognition

The study of human cognition has recently moved in a direction that appears rather odd for what could be expected from scientific advancement–and by this, I mean the achievement of greater simplicity, reduction, and parsimony. Examining the literature published in the last 20 years leaves the impression that we have moved from reducing cognition to few identifiable clues, closely similar to computational systems, and traceable to the brain or neuronal structures, toward much more complex models in which the mind appears embodied and embedded in a thick network of environmental and cultural factors (Newen et al., 2018). The result would be like a field of distinct and competing forces. So, the more we study and try to better understand the human mind, the more it appears hopelessly complex and intricate. Possibly, its mysterious status would apply not just to consciousness and related features, as several scholars

L. Oviedo (✉)
Universita Antonianum, Via Merulana, 124, 00185 Roma, Italy
e-mail: loviedo@antonianum.eu

259

have pointed out, but even to the whole mind, including language, beliefs, hopes, and moral judgement.

The former paradigm is well reflected in the scientific study of religion, which replicates a similar process: from simplicity and reduction, to complexity and integration of distinct levels and dimensions involved in a rich experience that is resistant to one dimension. This is something that could be expected. Developments in the study of mind and cognition should necessarily have repercussions in the study of religion as a mental activity involving–among other things–cognition, itself. Religious experience appears much more embodied, embedded, and context-sensitive than previously conceived when the focus was pointing out cognitive structures that allowed for conceiving supernatural agents (Day, 2007; Barrett, 2010: Angel et.al. 2017). It seemed there were too few structures for what has always been sensed as a much deeper and wider experience, involving social and cultural aspects. In both cases–for cognition, in general, and for religion, specifically–reduction does not help, except that it works as a heuristic strategy to describe several, but invariably, too few factors for a complex process (McCauley, 2013).

Here, I search for a fresher framework when trying to project religion in a future scenario–a Moon settlement established to serve scientific and other objectives, and able to keep a crew of humans alive for a relatively long time while they are involved in different activities. Even if religion appears as a relatively unimportant aspect of such a hugely challenging project (from a sheer technical and practical point of view), this perspective provides a good opportunity to test and to assess the extent to which religious experience is bound to context and a vastly different environment. It will be very interesting, as results unfold.

The hypothetical character of a new lunar setting invites us to represent in the most careful way possible the living conditions that humans would endure and how those conditions could influence different experiences like attitudes, relationships, emotions, and–why not?–religious faith. Obviously, such aspects belong to the so called "human factor." The importance of religious faith is perhaps minor compared to technical, economic, and structural aspects required for a mission's success. However, it would be wrong to ignore those human and social aspects as a reflection of that off-world enterprise. It is a theme already suggested for another hypothetical situation: a deep space expedition or a Mars settlement (Oviedo, 2019) and that can be related to other self-transcending experiences (Yaden et.al. 2017).

17.2   Religion, Cognition, and Spiritual Experience in Space

As noted previously, religion can be studied from several distinct angles, and those approaches allow us to compose a sort of mosaic, perhaps more accurately, a model, analogous to a physical field of vectors, with many tensions, attractors and repellers, and influences–positive and negative–struggling to determine the sense and direction of an experience. In a similar way, human evolution can be best conceived as a combination of at least four processes–genetic, epigenetic, developmental, and cultural–as

Eva Jablonka and Marion Lamb have insightfully described (2005). In their work, the human mind appears as the result of several, determinant variables in a complex process. Religion, too, can be better described as such a complex process and the result of intricate interactions between different features and traits.

The point regarding religious complexity and the need to account for many factors can be stated at the neurological level. It is interesting to recall how the neuroscientist Mario Beauregard put an endpoint to a season of studies trying to discover the "God spot" or to locate the neural networks and the mechanism that was the basis of religious experience (Newberg, D'Aquili & Rause 2002). The truth was that too many areas in the brain were involved in such an experience (Beauregard 2007), and therefore, he suggested that any attempt at reduction would become a scientific failure. This lesson is not yet completely learned. We still perceive an indefatigable effort to explain religion in the simplest terms, perhaps those easiest to control and manipulate at a cognitive or behavioural level, and those that lead to the most easily manageable theories. Religious experience results from many interacting features: internal and external; personal and social; mental and corporeal; cognitive and cultural. There is no purpose in trying to isolate and outline just one or two. A vast literature can confirm the number of factors and how they influence that experience, always in an uncertain process that is always changing and open to new input (Donaldson, 2015; Jones, 2015; Van Eyghen, 2018).

Among these complex factors, external context or environment appears quite interesting, especially a context as different as the Moon. Historical evidence reveals how many important religious experiences in recorded world religions are linked to lonely places, to the desert and to wilderness, away from crowded towns and hectic cities. This may be stating the obvious: religion is very context-dependent, and both solitude and extreme conditions work to enable peak experiences, away from worldly distractions. However, what is less clear and needs more research is the extent to which external conditions can elicit and favour certain types of spiritual awareness, or render people more able to feel a sense of self-transcendence, beyond the usual physical conditions and worries we endure daily. Inspiring landscapes, special weather conditions, or extraordinary natural events that evoke the presence of the supernatural, can all awaken our imagination. They can summon memories that encourage spiritual awareness, even if the awe linked to those special places and events is by no means universal, but at most, amply shared (Van Cappellen & Saraglou 2012).

A good lesson to remember when considering the lunar environment is that religion is configured by its own social and natural environment, and it is rarely disconnected from external features. Now questions arise about the correlations between such special contexts and types of religious or spiritual experiences. For instance, which experience is mostly inspired by the desert? Which one arises in a storm, or afterward when calm is regained? What about inspiring landscapes? Similar questions arise when trying to determine the unique conditions of space travel and an environment like a lunar settlement, and then, how they could influence perceptions of self-transcendence broadly understood, comprising specific religious or theistic experiences, or more fuzzy spiritual ones.

The kind of experience that can be described as an "out-of-this-world" vision has already been the subject of many testimonials that astronauts have left after long periods in the orbital space station, or even by those who were treading on the moon or circling in space around our satellite, the Moon. A broad choice of published books is available, telling us about those experiences, and how what they sensed while away from the earth changed their way of living and gave new meaning to their lives. Some testimonials are self-help books. Good examples for an essential list include the following: Sally Ride and Susan Okie, *To Space and Back* (1986); Michael Collins, *Flying to the Moon: An Astronaut's Story* (1994); Chris Hadfield, *An Astronaut's Guide to Life on Earth* (2013); and Scott Kelly, *Endurance: My Year in Space, A Lifetime of Discovery* (2017). A quick search of these books reveals that their authors were rather uninterested in religion, God, or spirituality. However, what is undeniable is that for all of them, the experience of being out in space and far from our planet changed their lives and caused them to appreciate life and to see the world, its problems, and others differently.

It is remarkable that the earlier stories of the first astronauts landing in the Moon offer many more religious or spiritual insights. Some essays try to explore those experiences. One very relevant to this issue is the book by Frank White, *The Overview Effect: Space Exploration and Human Evolution* (1987), which proposes the thesis that looking at the Earth from outside, and, as a whole, creates feelings of awe and even a new consciousness, a greater concern for our planet and its inhabitants, and in several cases a spiritual or religious experience. A second book specifically explores the religious experiences and dimensions in the early stage of space exploration, including the missions to the moon: Kendrick Oliver, *To Touch the Face of God: The Sacred, the Profane, and the American Space Program, 1957–1975* (2013). This book gathers many testimonials from astronauts, especially some who landed in the Moon, and who were witness to deep religious or spiritual experiences in that strange, faraway, and awesome landscape. A summary by Daniel Oberhaus, in an article published in *Wired,* is particularly revealing:

> Mitchell, for one, reported a feeling of "universal connectedness," and Apollo 15 astronaut James Irwin said he felt God all around him. Indeed, Irwin was so overcome by his experience on the lunar surface that he asked his colleague David Scott if they could hold a religious service atop some nearby hills before they departed from the moon. Scott ultimately shot down this request, so Irwin made do by quoting Psalm 121: "I will lift up mine eyes unto the hills, from whence cometh my help." This experience, Irwin later reported, was "the beginning of some sort of deep change taking place inside me" marked by a profound belief in the power of God (Oberhaus, 2019).

Oberhaus claims in the same article that, "In the 30 years since the term was coined, astronauts have repeatedly reported experiencing the *overview effect* [emphasis added] from the International Space Station." He even suggests that the effect could have important consequences helping to raise a greater awareness concerning the big issues our planet faces now, like climate change. He reports on attempts to render that awakening experience more available through new technical means–like space travel simulations–in order to build a new consciousness connected with religious faith and ethical concern.

The general impression when considering this literature on astronaut testimonials is that the spiritual or religious dimension seems to be much more linked to an early stage in space and lunar exploration, something that may be fading away in later stages. One is tempted to say that a general context of greater religiosity was paramount in the sixties and early seventies in the US (although this surely was not true for everyone), and that it could be determinant in that perception. If this is true, then a later secularization and loss of religious interest could render the *overview effect* much less common. If this is the case, then it may be less the exceptional conditions of space and the Moon, and more the cultural context, as well as one's general sensitivity to those perceptions. It remains a topic for research what would be determinant in enabling religious or spiritual experiences in those extreme conditions.

Indeed, later astronaut testimonials like those quoted in books by Hadfield (2013) and Kelly (2017) explicitly declare themselves to be non-religious, ignoring the sort of spiritual experiences others report from outer space. However, they were orbiting around the Earth (even if for a long time), and they did not go as far as astronauts on expeditions to the Moon. It is not clear whether they could feel the *overview effect*, or if they felt it in a completely secular way with no hint of self-transcendence. This is the way it appears from their reports. Kelly writes that, "this vantage point [seeing Earth from the space] has never created any particular spiritual insight for me" (2018). This "drier view" could discourage great expectations raised by White and Oberhaus concerning the special effects of an out-of-this-world experience, or even lunar travel, when it becomes routine and less risky. Possibly what can be called The Luhmann Law–with reference to the great German sociologist Niklas Luhmann (1977)–could be verified. It might be stated this way: If religion is a way to deal with greater risk or indeterminacy, then spiritual perceptions might be enabled in dangerous or unusual circumstances. In other words, transcendence could be seen as a coping strategy in the face of greater risk and uncertainty, and, once those risk levels lower, we would expect less religious interest. Religion would fade away as soon as the living conditions become more reassuring. Once more, this is a topic for further investigation (Szocik & Van Eyghen, 2021).

17.3 Anticipating Types of Religious Experience in a Lunar Station

Now the reflection moves from the empirical level, developed from reliable testimonials, towards the hypothetical, when trying to anticipate what religious experience in a future Moon settlement might be like. We can build our expectations based on previous experiences of people living in extreme conditions of isolation, precariousness, and high risk, where things might not work and life itself could be dangerously compromised. However, this is not the point. Travel to, and life conditions on the Moon will become, with enough time, more routine and reassuring. The point is that living in a long-term lunar settlement would probably entail some level of spiritual insight, following White's clues about the overview effect, or something similar that emerges as a type of perspective or insight. This anticipation is less speculative than one might think, since similar types and levels of insight and perception have been often achieved among humans in new and unusual environments. Considering the available data, it becomes more prudent to rely less on a kind of hypothetical "conversion story", as could be triggered by distant travel and extreme life conditions, and to point more to "continuity stories" or even to some "confirmation bias". Religious conversions do not happen when people visit some faraway shrine or a moving sanctuary, but when people awake an interest in visiting such places or paying tribute to a distant temple.

The key to all this lies in the beliefs and the believing process people hold when they are involved in an important space mission, with a huge sense of responsibility and dedication (Connors & Halligan, 2015; Angel et.al. 2017). Traveling to the Moon and staying there for some time could possibly confirm to religious people their faith, and even nourish rich mystical experiences. Conversely, that long trip would probably not raise any spiritual awareness in those who lack religious faith or any spiritual interest. The so called "Mathew Law" could apply here. According to a quote in that Gospel, "For to everyone who has will more be given, and he will have an abundance. But from the one who has not, even what he has will be taken away" (Mt 14,29). Or, simply put: Those who believe would believe still more, and those who do not believe would confirm their non-religious feelings.

The centrality of beliefs and believing is better perceived through the distinct experiences of different people in the same contexts and circumstances, where some react in a way that entails a religious answer, and others react in a way that, at most, involves some awe, but would be understood in sheer secular ways–nothing transcending. It could be that an experience on the Moon would prolong and confirm the previous experience of those leaving our planet for a long-term mission. More secularized times and cultural contexts do not allow for spiritual views, which always require some transcending framework or set of religious beliefs that nourish and make sense of those new perceptions.

However, it cannot be excluded that conversions and revelations happen in extremely special conditions, and even that a new spiritual awareness can be born in that case. Let us re-state our initial generalization about context and religious cognition: The model we have in mind is of a field–in the sense used in physics–in which components and forces interact in a complex manner. This probably applies better to our case scenario. Context is important, but it works only in conjunction with the internally held beliefs, as well as the emotions they can trigger, and the social networks in which we live. The question now is whether being religious, or better, holding religious beliefs, make a difference for those who travel to the Moon and have to build a settlement there.

Having religious beliefs could be a less relevant factor compared with the skills, endurance, and the physical and psychological strength needed in that distant setting.

The issue at stake in this case, is which kind of beliefs become most helpful and functional in dealing with the Moon's extreme environment and coping with risk and uncertainty. This is an issue I have raised in former studies dealing with religion in a hypothetical Mars colony and in deep space travel (Oviedo, 2019, 2020). Similar reflections apply in this case, which is closer to the Earth, and possibly involves less risk because of a closer distance from Earth. After all, astronauts would be risking their lives for participating in the mission. Surely some religious forms provide higher levels of endurance and the virtues needed to live through difficult times, as has been shown in past examples of ascetic men and women. However, religion is not the only way to provide those virtues and capabilities.

The interesting point is, again, how a lunar experience could influence the religious beliefs of those who hold them, as well as those who believe in a milder way. An experimental approach to this issue could build on some hypothesis that would need more evidence and testing. The question is whether people of religious faith would live the lunar experience as a confirmation of their faith; as a source of doubt that possibly could change several aspects of it; or, as in a crisis, give rise to a different system of beliefs, enriching a previous religious sense or awareness.

We already have instruments and methodology to conduct such research. The *Creditions* program, born in Graz, Austria, and the different applications that such an approach has taken, encourages our plans for a further application to observe, document, and assess how beliefs are formed, kept, affected, and even dismissed by new perceptions, especially the environment (Seitz & Angel, 2014). The methodology is based upon a known system that assesses cognition, emotion, and culture, and derives sets of beliefs that need, time after time, to integrate new inputs to modulate and to re-stabilise the entire system. Using this instrumentation to study beliefs and believing on the Moon could reveal to what extent religious beliefs are held as stable entities, resistant to change, or whether they adapt to changing environments before (or after) challenging situations.

The expectation is that such a research program could contribute toward better understanding an aspect of human cognition–religious faith and perception–which is absolutely unavoidable in any human community or settlement. If past is predictor of future, it has played and will play a great role in every human settlement and many activities in them. It connects religious and secular dimensions of human life. In the end, it must do so, because of the need to resort to beliefs to guide one's own life and to take important decisions that, at least in a lunar context, could mean life or death for the base crew. What is at question in this case is which beliefs become more functional and more useful in a new context, and how we can nourish them? We refer to both religious and non-religious, for in a lunar mission they must work together to spell success.

17.4   Religion and the Moon as a Sustainable Planetary Body

The previous analysis suggests a very open panorama when trying to come to terms with the lunar experience from a religious point of view, even if all humans share a similar cognitive architecture constraining religious cognition. The Moon would become–from this perspective–a large experimental setting in which the cognitive and experiential theses advanced here could be tested. Of special interest will be what has been described analogously as "a tensor field", i.e., a field combining different forces, to mediate and enable the expression of religious experience. These forces or "tensors" could include: previous religious training or experience; one's opinions, beliefs, and attitudes; personal disposition toward the mission and the lunar setting; current meaning systems (new and old); and objective measures of the new physical context and social conditions provided by a moon base and the Moon, itself, as an exceptional environment.

Another aspect needs to be considered in this hypothetical panorama we examine. As White points out in his book, the "overview effect" is not just spiritual but includes a greater consciousness regarding our planet Earth and humanity as a whole. The point is that the ability to observe the Earth from afar, even as a distant blue ball (a spectacle reserved to those who settle on the Moon) has often raised the level of concern for our planet Earth and the life it nourishes, and hence, for its sustainability. According to some testimonials, that exclusive experience belongs to those able to see the Earth from a distance and reframe minor worries or petty issues, and therefore focus on the common questions for humanity and its survival (White, 1987).

Possibly, and still moving in the hypothetic field, looking for future empirical evidence, the described feeling of an "Earth concern" could be shared by religious and non-religious crew alike, or, it could be an independent factor, separate from religious intensity, as many studies show here on Earth. Research shows that environmental concern often does not depend on religious beliefs (Hope & Jones, 2014; Chuvieco et.al. 2016). If so, what we can anticipate and what would be a great hope regarding the lunar experience, is that everybody could converge in that special setting–beyond particular beliefs, religious or not–in the conviction that it is urgent to behave in a way to preserve our planetary ecosystem, to overcome differences among human groups, and to find out ways to ensure a more peaceful life and a greater equality among all humans. This would naturally be extended to the environment of the Moon, itself, as a new home.

This does *not* mean that the Moon would necessarily become a place where humans can at last overcome their differences and achieve perpetual peace, even beyond religious contrasts and conflicts. The point is that such an environment could help to mature or assist in the evolution of the religious mind, so that this new perspective would encourage even newer views related to the current needs and conditions of humanity. In no way would that perspective entail the "overcoming of traditional religion", to be replaced by new forms more fitting in the present conditions. It suggests a healthy and helpful cultural and religious evolution resulting from new experiences on the Moon and awakened by its distant gaze of Earth. Like every previous discovery in human history, and each scientific and technical development, it will shape the religious mind (in many cases assisting a greater awareness and giving rise to more adaptive versions). We could therefore expect that a similar process would take place because of the establishment of a lunar settlement.

This proposed perspective is by no means something spontaneous or that could be expected just from a kind of natural evolution that brings human betterment. Things can still go wrong: the space race to the Moon and Mars could exacerbate rivalries and nationalistic sentiments (see Chaps. 10 and 11 in this volume). The universalistic and environmental concern we anticipate could instead become a particularistic and self-serving project. Since things can go wrong, and nobody can ensure the scenario we long for, in other words, because so much risk is involved in space exploration, religious faith, including its healing and redemptive aspects, will still need to be considered when planning our great endeavour of establishing a human settlement on our satellite the Moon.

References

Angel, H. F., Oviedo, L., Paloutzian, R. F., Runehov, A. L., & Seitz, R. J. (2017)*Processes of believing: The acquisition, maintenance, and change in creditions. Springer*

Barrett, N. F. (2010). Toward an alternative evolutionary theory of religion: Looking past computational evolutionary psychology to a wider field of possibilities. *Journal of the American Academy of Religion, 78*(3), 583–621.

Beauregard, M., & O'Leary, D. (2007). *The spiritual brain: A neuroscientist's case for the existence of the soul.* Harper One.

Chuvieco, E., Burgui, M., & Gallego-Álvarez, I. (2016). Impacts of religious beliefs on environmental indicators: Is christianity more aggressive than other religions? *Worldviews, 20*(3), 251–271. https://doi.org/10.2307/26552264

Collins, M. (1994). *Flying to the Moon: An Astronaut's Story.* Farrar, Strauss & Giroux.

Connors, M. H., & Halligan, P. W. (2015). A cognitive account of belief: A tentative roadmap. *Frontiers in Psychology, 5,* 1588.

Day, M. (2007). Let's Be realistic: Evolutionary complexity, epistemic probabilism, and the cognitive science of religion. *Harvard Theological Review, 100*(1), 47–64.

Donaldson, S. (2015). *Dimensions of faith: Understanding faith through the lens of science and religion.* Lutterworth Press.

Hadfield, Ch. (2013). *An astronaut guide to life on Earth.* Little Brown.

Hope, A. L. B., & Jones, C. R. (2014). The impact of religious faith on attitudes to environmental issues and Carbon Capture and Storage (CCS) technologies: A mixed methods study. *Technology in Society, 38,* 48–59. https://doi.org/10.1016/j.techsoc.2014.02.003

Jablonka, E., & Lamb, M. (2005). *Evolution in four dimensions: Genetic, epigenetic, behavioral, and symbolic variation in the history of life.* MIT Press.

Jones, J. W. (2015). *Can science explain religion?: The cognitive science debate Oxford.* Oxford University Press.

Kelly, K. (2017). *Endurance: My year in space, a lifetime of discovery.* New York: Knopf–Doubleday

Luhmann, N. (1977). *Funktion der Religion,* Frankfurt a.M: Suhrkamp.

McCauley, R. (2013). Explanatory pluralism and the cognitive science of religion: Why scholars in religious studies should stop worrying about reductionism. In D. Xygalatas & W. W. McCorkle Jr (Eds.), *Mental Culture* (pp. 11–32). London: Routledge

Newberg, A. B., D'Aquili, E. G., & Rause, V. (2002). *Why god won't go away: Brain science and the biology of belief*. Ballantine Books.

Newen, A., De Bruin, L., & Gallagher, Sh. (2018). *The Oxford handbook of 4E cognition*. Oxford University Press.

Oberhaus, D. (2019, July 16). Spaceflight and spirituality, a Complicated Relationship, *Wired*. https://www.wired.com/story/apollo-11-spaceflight-spirituality-complicated-relationship/

Oliver, K. (2013). *To touch the face of god: The sacred, the profane, and the american space program, 1957–1975 (2013)*. Johns Hopkins University Press.

Oviedo, L. (2019). Religion for a Mars colony: Raising the right questions. In K. Szocik (Ed.), *The human factor in a mission to Mars* (pp. 217–231). Springer.

Oviedo, L. (2020). Religion as human enhancer: Prospects for deep spatial travel. In K. Szocik (Eds.), *Human enhancements for space missions: Lunar, martian, and future missions to the outer planets* (pp. 279–288). Dordrecht: Springer

Ride, S., & Okie, S. (1986). *To space and back*. Harper Collins.

Seitz, R. J., & Angel, H. F. (2014). Psychology of religion and spirituality: Meaning-making and processes of believing. *Religion, Brain & Behavior.* https://doi.org/10.1080/2153599X.2014.891249

Szocik, K., & Van Eyghen, H. (2021). *Revising cognitive and evolutionary science of religion*. Springer.

Van Cappellen, P., & Saroglou, V. (2012). Awe activates religious and spiritual feelings and behavioral intentions. *Psychology of Religion and Spirituality, 4*(3), 223–236. https://doi.org/10.1037/a0025986

Van Eyghen, H. (2018). What cognitive science of religion can learn from John Dewey." *Contemporary Pragmatism 15*(3), 387–406.

White, F. (1987). *The overview effect: Space exploration and human evolution*. Houghton Mifflin.

Yaden, D. B., Haidt, J., Hood Jr, R. W., Vago, D. R., & Newberg, A. B. (2017, May 1). The varieties of self-transcendent experience. *Review of General Psychology,21*(2), 143–160. http://dx.doi.org/ https://doi.org/10.1037/gpr0000102

Chapter 18
Religion in a Lunar Settlement: An Anthropological Assessment

Gerald F. Murray

Abstract This chapter anticipates the probable impact that lunar settlement will have on the cultural domain known as "religion". We examine a hypothetical settlement that neither forbids religion nor promotes a new religion but that accommodates existing religious practices of settlers. We begin with an operational definition of religions as systems with distinct cognitive, behavioral, and organizational components: spirit-beliefs, rituals, and specialists ("clergy"). Each component will be affected differently by the lunar environment. After discussing existing research on astronaut religion and cosmology, we identify religious practices that may or may not remain stable on Luna. We end with recommendations for institutional accommodation to different settler religions. For some, the protracted physical confinement and restricted movement required by objective dangers in the lunar environment will predictably produce higher levels of subjective psychic stress among settlers. Existing religions have serenity-enhancing meditative practices, now largely under-utilized, which, if encouraged along with selected non-religious meditative practices, could mitigate the internal stresses generated by the lunar environment.

18.1   Introduction

The article discusses the probable impact which life in a lunar settlement is likely to have on the beliefs and practices of the cultural domain labeled "religion". On a lunar settlement it is assumed here that religion will not be forbidden, nor will a new religion be designed and promulgated. Several causal factors will govern what happens religiously on the Moon. (1) Characteristics of the religions brought from earth; (2) policies of the sponsoring country regarding religion; and (3) features of the lunar environment that will facilitate or impede (usually the latter) the full replication of terrestrial religions.

G. F. Murray (✉)
Department of Anthropology, University of Florida, Gainesville, FL 32611, USA
e-mail: murray@ufl.edu

18.2   Conceptualization and Information Sources

The paper will have four sections: (1) an operational definition of the term "religion" to be used in these pages; (2) examination of clusters of existing data on settlers and astronauts to empirically ground predictions; (3) an identification of religious beliefs and practices that can or cannot be replicated easily on the Moon; and (4) a proposal for institutional accommodation of religious practices in a lunar settlement.

The proposed operational definition covers three distinct systemic elements—cognitive, behavioral, and organizational—empirically present in human religions. Religion is here defined as *a belief-behavior system whose members (a) believe in the existence of invisible spirits, (b) engage in behaviors ("rituals") to interact with these spirits; (c) and do so under the guidance of practitioner-specialists.* This is not *the* anthropological definition; no such agreed-on definition exists. It is similar to that found in (Winkelman & Baker, 2010; Crapo, 2003). Other elements are present in some religions (e.g. sacred scriptures, food taboos, ethical codes, codes of sexual behavior). But for conciseness we will examine the three universal core elements as they might unfold on the Moon.

Each element requires its own definition. "Spirits" here are invisible beings conceived of as conscious, active agents: God or gods, angels, demons, souls of the dead, ghosts, and many others. They usually have names and are usually viewed as having some power to affect humans and in turn can be influenced (or warded off) by humans. A "ritual" here is a behavior (collective or private) directed toward the spirits. A procession for the Virgin of Guadalupe would here be considered a ritual. A 4th of July parade would not, though colloquial English might loosely label it so. Fasting to atone for one's sins is a religious ritual. Fasting to lose weight is not. We can avoid futile nitpicking about the "real-essential-meaning of *religion, ritual, spirit* and other polysemic terms by simply operationalizing criteria for using the terms here.

Two final paradigmatic points. We will discuss "religious systems". Different core components—spirits, rituals, specialists—logically relate to each other. Buddhist monks don't celebrate Catholic Masses to worship Allah. Muslim imams don't officiate at Yom Kippur rituals to venerate Jesus. There is logical consistency among spirits, rituals, and specialists within systems. Disaggregation of specific systemic components facilitates prediction of what can or cannot be transferred to the Moon.

Finally, we consider it essential to distinguish systemic *components* from systemic *functions.* The components are the elements of a religious system. The functions are the purposes people pursue or the unintended effects which occur. Example: defined componentially, a "knife" is a tool with a short blade and handle. We can safely throw in the function "used to cut" since agreement exists on its core function, *No such agreement exists on the functions of religion*—its intended purposes or unintended consequences. Analysts often inject their own favorite (or despised) function—e.g. social solidarity, sentiments of awe, interclass exploitation—as the "real" function of religion. Transient functions should be a matter of research, not built into the definition of religion. To anticipate: many of the *core components* of terrestrial religions can remain stable on the Moon. Their *functions* may have to shift in response to certain features of the lunar environment.

18.3   Empirical Precedents: Earthbound Colonists and Astronauts

Despite some demurral in certain Evangelical creationist circles, which view extraterrestrial exploration as "driven by man's rebellion against God … to prove evolution" (Weibel, 2017), contemporary religious systems across the board are largely supportive of space travel. Explicit institutional support has been historically strong in Catholicism, which initiated astronomical studies in the sixteenth century. (They were brought to China by Jesuit Matteo Ricci.) The Specola Vaticana (Vatican Observatory) under Jesuit management near Rome opened a branch observatory in Arizona's mountains to avoid Rome's light-pollution interference. This U.S. observatory has actively collaborated with NASA (Drake, 2008). A Carmelite parish in Houston near NASA activated canonical procedures to commission Catholic astronauts as eucharistic ministers for communion services while in orbit, sent a scapular of Our Lady of Mt. Carmel into multiple orbits, and hosted an ecumenical prayer service for the 1986 Challenger victims (Carmelites, 2007). With less explicit institutional involvement, other religious systems have also encouraged space travel. There is even a popular Protestant genre of science fiction involving space travel.

Though religious authorities on Earth approve of space travel, what will happen to religion on the Moon? Three information sources will help empirically ground predictions: (a) the historically documented religious behavior of settlers on planet Earth; (b) research into the impact of space travel on astronaut cosmology and religion; (c) an identification of the lunar environmental features that are likely to exert causal impacts on terrestrial religions.

Item (a) is general knowledge. Most colonists on Earth neither jettison their religion nor invent new ones; they bring pre-existing religions with them and often proselytize the locals, as did Spanish, French, and Arab conqueror-settlers. (British Protestant settlers had little interest in proselytizing 17th-century Amerindians. 19th and 20th century Jewish immigrants to Israel had no interest in converting Palestinian Muslims to Judaism.) Caution must be exercised in extrapolating knowledge of terrestrial settlers to a lunar colony. Space travelers, with their advanced engineering or aeronautic, degrees, belong to an atypical subset of any population. Nonetheless we know empirically that many of them profess religious beliefs. Following terrestrial precedents, we can therefore reasonably predict that they will bring these beliefs with them to the Moon, but without proselytization. There is no native population to proselytize. And given prevailing American MYOB cultural norms, Settler-X is unlikely to badger Settler-Y to adopt the true religion—i.e. Settler-X's religion. A further basis for possible extrapolation to the Moon are reports and formal studies about the impact of space travel on cosmology and religiosity among astronauts.

Several astronaut-authors discuss their religion (Irwin & Emerson, 1973; Williams, 2010; Jones, 2006).

Some have engaged in religious practices while in space. A group of astronauts read passages from Genesis back to Earth in 1968. (Following that, a lawsuit against NASA was unsuccessfully launched by freedom-from religion activists) (Dexter, 2016). Catholic astronauts had a eucharistic service in outer space (Drake, 2008). A Jewish astronaut wished her fellow Jews a happy Hanukkah from orbit and showed them her socks with a menorah and Star of David (Times of Israel, 2019). An Israeli brought kosher food into orbit (Halily, 2008). One fasted for Yom Kippur. In short, many (but not all) astronauts have expressed involvement in a religious tradition while in space. (Soviet and later Russian cosmonauts have displayed no such inclinations.)

Systematic data exist on this matter. In his Overview Effect, White collected 100+ pages of statements from 30 astronauts, all but a few of them Americans, concerning the impact of space travel (most of it orbiting Earth) on their worldview (White, 2014). Major religious patterns can be summarized.

- There is religious heterogeneity among astronauts: Protestantism, Catholicism, Judaism, Mormonism, and Islam in descending frequency. (The latter two had one each.) One non-religious astronaut later explored Buddhism.
- Powerful cognitive insights and emotional reactions were *humanistic rather than religious*. Several Astronauts report sudden insights into the geographical unity of the planet, the artificiality of borders between countries, the fundamental unity of the human species. *But virtually none of the reactions was religious in character*. Few astronauts mentioned God or other spirits. When the word "spirit" came up, it usually referred glowingly to the "human spirit" transcending barriers of gravity.
- Pre-existing religious beliefs and feelings are largely validated rather than challenged. A case in point: An astronaut was asked if space travel would affect human evolution. Her answer: "I'm a Christian. I believe we were created not evolved." (White, 2014, 272). (Five recent Popes have encouraged evolutionary studies. Their Christian status is apparently in doubt.) Astronauts rely on scientific texts for engineering, aeronautical, or medicinal purposes. But at least some of them feel no incongruity in relying on scriptural texts for information on human origins. One Evangelical astronaut back on Earth searched for the remains of Noah's Ark [3]. One study that tests White's "Overview Effect" indicates that views of planet Earth and the oneness of humanity are significantly affected by space travel whereas religious beliefs remain statistically unchanged (Kanas, 2020) (Other studies on the impact of space travel on world-view are (Yaden et al., 2016; Suedfeld & Weiszbeck, 2004; Suedfeld et al., 2010; Ihle et al., 2006).

In short, the information available suggests that (a) the level and diversity of religiosity among U.S. astronauts parallels that of the general population and (b) that religious beliefs are largely unaffected by travel to outer space. A few became more religious after returning to earth. None report an abandonment of religious belief.

So what's going to happen on the Moon? Several working assumptions will guide answers. (1) The lunar settlements will be established by individual countries, which will have different religious policies. Religion on a Saudi settlement will be treated differently from a Chinese settlement. We will here discuss a hypothetical U.S. colony. (2) Rather than discussing "religion" in the abstract, we will instead examine the religions that are likely to have most the lunar adherents in a U.S. settlement. The procedure for analysis will be applicable, *mutatis mutandis*, to other religions.

18.4   Spirit Beings on the Moon

Let's begin with the invisible spirit beings. They will do fine on the Moon. The Elohim/Adonai/Ein-Sof of Judaism, the Trinitarian God of Christianity, and the Allah of Islam are viewed as omnipresent and omniscient. Their lunar followers will comfortably invoke them. The angels, present in all Abrahamic traditions, will also fare well. Catholics can continue to pray to St. Michael the Archangel and to their personal guardian angels for protection. Even Jews, whose prayers are generally directed to God alone, poetically welcome angels every Friday evening at the beginning of the Sabbath meal: Shalom aleichem, malachei hashalom, malachei Elyon. (Greetings to you, angels of peace, angels of the Most High.) There's no reason why angels couldn't receive a lunar welcome. Protestantism also accepts the existence of biblically recorded angels but does not generally greet or invoke them. In Islamic belief, the Quran was given to the Prophet by the angel Jibril.

Belief in survival of the souls of the dead will also survive on the Moon. All Abrahamic religions posit a "differentiated afterlife"—i.e. post-mortem reward or punishment. The Catholic and Islamic afterlife is tripartite: Heaven, temporary Purgatory, and eternal Hell fire. Judaism posits two post-mortem destinations: Paradise or temporary punishment in Gehinnom. (Rabbinic sources indicate that post-mortem punishment for Jewish souls lasts at most for a year, after which they ascend to paradise, Gan Eden—the "Garden of Eden". There is no eternal Jewish Hell.) Traditional Protestant conceptions of the afterlife consign souls immediately and irrevocably to only one of two places: Eternal Heaven or eternal Hell. Nothing about lunar life is likely to shake these differing afterlife beliefs. In short, the cognitive component of religious systems, spirit beliefs, will comfortably be transplanted on the Moon.

18.5   Rituals on the Moon

That may be less true of the behavioral component of religion, rituals to interact with the spirits. There could be problems in the domain of rituals. Certain ritual objects can easily be transported to, and used on, the Moon with no problem: Muslim prayer beads, Jewish mezuzas (sacred texts attached to doorposts), Catholic rosaries. But collective rituals could be more problematic. Though some public, collective rituals may also take place in non-religious venues, they usually take

place in churches, synagogues, mosques, temples, ashrams, etc. Specialized buildings are not strictly required. Jewish Torah services, Catholic Masses, Pentecostal
glossolalia, and Islamic salat can validly occur in private homes. But congrega-
tional rituals, including rites of passage like baptisms, weddings, confirmations, and
bar-mitzvahs that signal life-cycle transitions, usually occur in specialized worship
structures with religion-specific architecture and décor. This will be virtually impossible in a confined, insulated lunar settlement whose residents practice five or six
different religions. Even if space were available, but the settlers could vote, we could
predict a preference for a movie theater, a Starbucks, an Outback, and a gym rather
than five holy buildings. One holy building would probably do.

And it would have to be watered-down generic, with no sectarian artefacts or
décor. Many Protestants and Jews will refuse to enter a Catholic church with statues
of saints or the Virgin Mary. Some recoil at the presence of a Catholic crucifix—a
cross with an icon of the dead body of Jesus attached. Traditionalist Catholics would
chafe at a denuded generic worship site watered down to placate *sola scriptura*
Evangelical sensitivities.

The intra-religious sectarian problems would be even greater among Jews on the
Moon. (In the U.S., among affiliated Jews, about 38% are Reform, 33% Conservative, 22% Orthodox, and 7% other (Demographics of Judaism). Higher birthrates and
religious endogamy may eventually create an Orthodox majority (Forward, 2018).
Current disagreements could be problematic for lunar Judaism. Judaism is unusual,
perhaps unique, in requiring the presence of at least ten Jews for a full Torah service.
The Orthodox require a minyan of ten men, place a *mechitsah* (physical barrier)
between men and women, and call only men to the Torah (aliyah). The liberal Reform
and Conservative streams count women in a minyan, have mixed seating, and call
women to the Torah. The Orthodox will militantly refuse, on the Moon as on Earth,
to participate in mixed-gender services. Liberal-stream Jews will predictably recip-
rocate the boycott, refusing to pray where women receive no aliyot and are not even
counted in a minyan. But even liberal streams disagree among themselves about
"who is a Jew" for synagogue purposes. Conservatives follow tradition requiring a
Jewish mother (or valid conversion) to count someone as a Jew. For the Reform,
either a Jewish mother or father suffices for temple participation.

Thus, even if there were 20 Jews on the Moon,—10 males 10 females—distributed
proportionally among the different streams, it might take a *nes gadol*, a miracle, to
organize a minyan which ten of the lunar Jews would attend. Jewish solidarity as a
people crosscuts internal religious diversity. But the solidarity is ethnic, not religious,
in character. There's a Jewish joke. A rescue team finally found a Jew who had been
shipwrecked on an island years earlier. They noticed he had built two synagogues.
"Why did you need two synagogues!!?" "This is the synagogue I pray in. That's the
one I wouldn't get caught dead in."

Quite apart from sectarian issues, the lunar environment itself will create havoc to
many rituals, particularly calendrical rituals. A common Abrahamic sequence is six
days of work followed by a religious sabbath, based on the Genesis creation account.
It's extremely strict among traditional Jews. Friday pre-sunset initiates a full day of
strict observances. (For Islam, Friday is a day of mosque attendance but not rest.)

There'd be a problem on the Moon for Christians and Jews. the Moon takes about 29.5 Earth-days to rotate on its axis. A lunar sunrise-to-sunrise "day", lasts about an earthly month. A six-day work week on the Moon followed by a sabbath rest would entail working for six Earth'months straight and resting the following month. Sabbath would be more like a month-long vacation than a 24-h rest period. A radical adaptation would of course have to be made.

It can be managed. Astronauts on the International Space Station have the opposite dilemma: sunrise occurs every 90 min. One Jewish astronaut wanted to observe the Yom Kippur fast from sunset to sunset. A 90-min fast seemed insufficiently penitential. When should fasting begin and end? The general ISS solution is a work—sleep schedule based on GMT+0—a compromise between Russian and American time zones. A similar solution is feasible on the Moon. The lunar "day" would be disassociated from the lunar sunrise and sunset and calibrated to some terrestrial time zone. Religious Jews or Muslims might prefer Jerusalem or Mecca time. Catholics wouldn't particularly fight for Vatican time. In an American settlement, calibration with Houston time might win out, to facilitate communications with NASA staff.

18.6   Religious Specialists

Let's examine the third element of religious systems: *specialists.* Rituals requiring ordained clergy may be hard to replicate on the Moon. Catholicism would be particularly challenged by the requirement of an ordained priest for three central sacramental functions: (a) the transubstantiation of bread and wine at Mass into what is believed to be the Real Presence of the body and blood of Christ, (b) the conferring of sacramental absolution after confession of sins, and (c) Anointing of the Sick (formerly called Extreme Unction). Sixteenth century Protestant groups all eliminated priestly absolution, and most eliminated the real-presence belief. As far as rites of passage (such as baptism and weddings) Protestantism as a whole requires ordained ministers. But for ordinary weekly liturgies, Protestants could conceivably have their standard Sunday services on the Moon, with bible reading, hymn singing, prayers, and possibly even preaching, without an ordained minister.

In contrast, without a priest lunar Catholics would have no face-to-face access to two major rituals: Mass and sacramental absolution. Canonical accommodations exist for the absence of Mass: a shortened ritual of scripture reading followed by distribution of previously consecrated hosts, bypassing the Offertory and Canon. Other accommodations to priestlessness are an act of "spiritual communion" when the Eucharist is unavailable and an act of "perfect contrition" pending availability of sacramental absolution. A casual outside observer may wonder: what's the problem? Why doesn't the Pope just authorize Catholic astronauts to consecrate the Eucharist and forgive sins? Short answer: Popes can't do that. Not even an Argentinian Jesuit Pope.

Judaism would have no parallel specialist dilemma. In the distant past, kohanim (Jewish priests) were required for obligatory daily animal sacrifices, which only they could do, and only in the Jerusalem Temple. When the Romans destroyed the Temple in 70 CE, the kohanim were suddenly unemployed, stripped of their major raison d'être. The rabbis became the sole authority in Judaism. In contemporary Rabbinic Judaism, ordained rabbis are not required for synagogue rituals. Ten ordinary Jews can hold a Torah service as long as one of them can read Hebrew out loud. (Understanding of the Hebrew is desirable but not required.) The main barrier to a lunar Torah service, at least among American Jews, would be earlier-discussed internal sectarian divisions, not the absence of a rabbi.

Other religious rituals would require lunar adaptations. Parallel challenges, most of them solvable, would also confront lunar Hinduism and Buddhism. There is no space for exhaustive treatment. The point here is methodological. The use of a systems paradigm that disaggregates spirits, rituals, and specialists provides an ethnographic framework for bringing discussion of "religion on the Moon" down from the stratosphere. It permits discussion of specific dilemmas in specific systems.

18.7 Programming the Lunar Adaptations to Terrestrial Religious Systems

So what stance, if any, should settlement authorities take toward religion? First, some realism. During the initial lunar settlement period of technical challenges and dangers, the accommodation of religious preferences will be justifiably low on the list of institutional priorities. The International Space Station, for example, has a mini-gym (Howell, 2018) but not a mini-chapel. American astronauts slept before reaching ISS on the Dragon in 2020. They had pre-programmed an alarm with wake-up music. Their choice was not Amazing Grace or Gregorian chant, but Black Sabbath (Bartels, 2020). That's not Jewish Kabbalistic music; it's a heavy metal British rock group. Their musical choice provides a healthy antidote against over-projecting religious concerns into the minds of astronauts or lunar settlers.

It can simply be pointed out that there are precedents in the U.S. for lunar accommodation to diverse religions: (a) military chaplaincies and (b) non-denominational chapels in public hospitals and airports. Several factors complicate the chaplain issue. On a hypothetical settlement of 1,000 individuals that reflects U.S. religious distribution, 490 would be Protestant, 210 would be Catholic. Mormons and Jews would have 20 settlers each. Muslims, Buddhists, and Hindus would have 10 each. 230 would be unaffiliated. How do you choose chaplains for the Moon? Occupational slots will be severely limited. If given a choice, the lunar settlers themselves might vote for a physician or a dentist rather than a minister, priest, or rabbi.

There's another chaplain option. Beside ritual roles, chaplains also counsel in times of stress or sorrow. This could be done via Earth-to-Moon tele-counseling. Settler-A's message would take about 1.3 s to reach Chaplain-B on Earth. The response would be equally rapid. There are limitations to tele-solutions. Under current canonical rules, for example, Catholic chaplains could not tele-consecrate the Eucharist or give tele-absolution. But religious tele-counseling would be possible across the religious spectrum. Tele-counseling actually has advantages in a settlement with, for example, six religions. Six resident chaplains would be out of the question. But tele-counseling gives access to chaplains of any religion. (Tele-counseling and tele-medicine will probably become permanent options in a post-pandemic world.)

A second support option builds on the "non-denominational chapels" in public hospitals and airports. A building or space on the Moon could be set apart for religious or quasi-religious gatherings. It should not be labeled a "chapel", a religious term which could trigger alarms and provoke litigation from increasingly militant U.S. freedom-from-religion activists. Call it instead the Meditation Center. Except for scheduled liturgical or educational events, it would be a place of silent recollection, whether individual or group.

Arguably, its most powerful function should indeed be the encouragement of meditation practices that now exist both in secular and religious circles and can legally be supported with public funding. It is now common to hear assertions to the effect: "I don't believe in religion, but I do believe in spirituality" The definition of "spirituality" is even more elusive than that of "religion". But it has something to do with enhanced consciousness and meditative states. Ever since the 1970s, meditation and altered states of consciousness (ASC) have moved from the arcane periphery into the secular mainstream. Transcendental Meditation a-la-Beatles came into vogue in the 60s but raised secular eyebrows with its Hindu guru, secretly assigned mantras, and a required initiation ritual. In contrast, Benson's "Relaxation Response" (Benson, 1976) and Kabat-Zinns "Mindfulness Meditation" (Kabat-Zinn, 2016) have both been promoted by secular academics. Subsequent EEG and fMRI research has documented how these and other methods affect the brain. Going beyond neurological correlates, the Mayo Clinic, for example, reports positive medical effects of meditation: alleviation of stress, anxiety, pain, depression, insomnia, and hypertension (Mayo Clinic Staff, 2018).

Meditative and other mind-control skills might be particularly urgent on the Moon. The incidence of negative mental turbulence and chronic stress will be more acute among long-term lunar settlers than among astronauts in brief orbit around Earth. Earth orbiting provides astronauts with spectacular, constantly shifting euphoria-generating views of sunrises, sunsets, continents, islands, mountains, rivers, forests, and oceans. As noted above, it provokes mind-changing humanistic insights. Life on the Moon will differ. When the novelty wears off, long-term confinement within a protective shelter on the sterile lunar environment—two weeks in darkness followed by two weeks of unabated scorching sunlight—is more conducive to depressing boredom, interpersonal tensions within a crowded settlement, and intrapersonal anxiety and stress (cf. Vakoch, 2019). There are no forests for momentary escape, no streams to sit by, no birdsongs to listen to. In past centuries, had travel been possible, the Moon might have been chosen as a penal colony. Currently it will certainly be seen as a dangerous and temporary hardship post. It is unlikely that anyone would take their children to live there.

Objection: wouldn't it attract adventurous spirits? It will indeed, particularly adventure tourists with enough spare cash for a round trip to the Moon. But adventurers who reach Everest or the North Pole rush home to receive accolades; they don't try to settle there. Lunar adventurers will likewise not want to live and die there. For those living there on some long-term assignment, meditative techniques to gently disengage from inevitable bouts of stress and anxiety generated by the lunar environment could be every bit as important as daily physical exercise. In that light, the practice among Westerners of daily meditation, whether secular or religious, is probably more common among those with higher levels of education. In that sense there may be an enhanced readiness among highly educated lunar travelers to engage in meditative practices as an antidote to the special stresses of the lunar environment.

Where does religion fit in? Long antedating secular meditation, world religions developed ancient techniques for altered states of consciousness (ASC) under the rubrics of "meditation", "contemplation," "spirituality", "mysticism". Buddhism and Hinduism are rated high on that dimension by Westerners who, unfamiliar with mystical traditions of their own religions, rush to the East for enlightenment. Catholic meditative traditions began with 3rd-century Desert Fathers continuing later with figures like John of the Cross, Teresa of Avila, Ignatius of Loyola, the anonymous Cloud of Unknowing, and more recently Therese of Lisieux and Thomas Merton (cf. Fleming, 1978). Cistercian monks now promote "centering prayer", which incorporates certain mind-emptying techniques common in Buddhist meditation (Keatng, 2009). More recent Protestant ventures into mysticism have also been documented (Foster, 2017). Jewish meditation derives from centuries-old Kabbalistic traditions associated with sixteenth century names like Abraham Abulafia, Isaac Luria and, in later centuries, with the Baal Shem Tov and Shneur Zalman (Kaplan, 1989; Kaplan & Sutton, 1990). Islamic mysticism and ASC are associated with the Sufi movement (Schimmel, 2011).

The use of religious meditation by those so inclined would constitute the "functional readjustment" mentioned earlier. First: an analytic caution, mystical experience of the numinous and the sacred is often mis-classified as a central component of religion. It is not. Increasingly empty churches suggest that religious boredom, especially among the young, is more frequent than happy-clappy hymn-singing highs. Enhanced consciousness with awe-filled insights is better analyzed as *one of the multiple functions* that religious systems can and have occasionally played.

It's a useful positive function that can produce measurable cerebral impacts. Neuroscientists have studied Carmelite nuns during meditations (Beauregard and Paquette, 2021), and (with the encouragement of the Dalai Lama) Buddhist monks (Josipovic, 2013). One study combines fMRI and EEG to explore neural correlates of ecstatic meditation (Hagerty et al., 2013). Another documents advantages of meditation training on attention (MacLean et al., 2010). Yet another discusses the differential neural correlates of three different meditation methods (Josipovic et al., 2012). Parallel to such studies of religious meditation are above-mentioned scientific studies comparing major secular traditions. Also of relevance is the research into altered consciousness by the Institute of Noetic Sciences founded by former astronaut Edgar Mitchell. (The Institute's website is https://noetic.org/.) An example of their research into meditation is (Delorme, 2019).

In short, scientific research exists, not only on neural correlates but also on positive health impacts of meditation. From a mental health perspective, the threatening lunar environment justifies giving a Meditation Center equal priority with a lunar gym. No particular methods should be mandated. Users could explore a menu of methods, secular or religious. The Center would be available as well for the scheduling of conventional religious services, but information on meditation would be among its core offerings.

In short, the prediction made here is for neither the disappearance of religion on the Moon nor the emergence of a new religion, but for an evolutionary *shift in the function* of the religions which earthlings will bring to the Moon. Many historically earlier functions of religion—explanations of nature, diagnosis of illness, sociopolitical control—are largely irrelevant on the Moon. This opens the way to an evolutionary shift toward a focus on the potential intrapsychic functions of religion linked to latent capacities of the human brain. Like musical, artistic, and mathematical potentials, however, the genetic capacity for ASC will remain untapped unless a cultural system values and activates it.

The notion of "activation of inborn capacities" is important. A recent study by Rappaport and Corbally explores the evolution of innate "religious capacity" in humans (Rappaport & Corbally, 2019). This is a refreshing departure from simplistic and anthropologically uninformed Marxist demonization of *all* religion as an "opiate" of ignorant exploited masses and from chic middle-class Freudian dismissals of religion as collective delusion driven by sexual repression—religion as a mental illness to be cured, not a capacity to be developed. When used harmfully, religion can indeed do harm. But when linked to meditation and mental health, it can be a source of enhanced awareness, intrapersonal serenity and interpersonal harmony. The lunar environment, by forcing a de-emphasis of certain conventional clergy-led ritual modes, and by its heightened probability of increased stress levels, opens the door (but in no way guarantees) functionally reoriented religiosity.

For this, there is no need of a new religion, but for a functional refocusing of those that exist. All meditators would aim for a still mind and emotional serenity. Religious meditators could do so with the symbols of their religious tradition: the silent presence of the Ein Sof of the Jewish Kabbalah or Islamic *taffakur* or *dhikr* with beads. Christians can meditate in the silent presence of the Trinity or of the incarnate Son or (among Catholics) in the comforting and loving presence of his Virgin Mother. For some meditators, inner silence and serenity are themselves the goals. For religious meditators they are often pathways to something beyond, believed to require special support at advanced stages—"ruach haqodesh" in Judaism, "infused contemplation" in Catholic tradition, the "pull" of Allah felt by Sufi mystics. A lunar Meditation Center could simultaneously accommodate and encourage explorers of all pathways.

To conclude: these pages have proposed a systemic paradigm that permits identification of some barriers against the full replication of terrestrial religious systems on the Moon. The bad news about barriers leads to some potentially good news about functional alternatives. Not everyone is constitutionally wired (or personally inclined) to become a mystic or reach Nirvana, just as not all can be star athletes or mathematical geniuses. But all can strive for physical fitness via sustained disciplined effort, and all can learn to multiply and divide. Likewise, all humans have the cerebrally based genetic capacity (often untapped) for increasingly higher levels of inner clarity and emotional serenity. The confined, threatening environment of a lunar settlement may help in this regard by creating a special urgency for such inner pursuits. It will not happen spontaneously. Meditative procedures could and should enjoy institutional recognition and encouragement in the lunar setting.

A brief concluding caveat is in order. As is true of physical fitness, little will be achieved by occasional dabbling. Progress in inner growth, whether secular or religious, requires sustained, disciplined practice. Strategies for exploring inner space can and should be mastered on planet Earth before venturing into outer space. If a lunar settler population is unfamiliar with, or disinterested in, enhanced consciousness, the lunar Meditation Center will end up largely underutilized, with embarrassingly fewer users than the lunar Starbucks or gym next door.

Acknowledgements I wish to thank the following for helpful comments on earlier drafts of this article: Maria Alvarez, Chris Corbally, Thomas Dorney, Douglas Fraiser, Paul Magnarella, Michael Moran, Alex Rödlach, Michael Wolf, and Daniel Yelenik.

References

Bartels, M. (2020). SpaceX's first crew Dragon….at space station today. Space.com, May 31. https://www.space.com/spacex-demo-2-astronauts-space-station-docking-webcast.html.
Beauregard, M., & Paquette, V. (2021). Neural correlates of a mystical experience in Carmelite nuns.
Benson, H. (1976). *The relaxation response.* Avon.
Carmelites. (2007). Carmel breaks the bonds of earth. *Catholic Exchange*, October 29, 2007. https://catholicexchange.com/carmel-breaks-the-bonds-of-earth.
Crapo, R. H. (2003). *Anthropology of religion: The unity and diversity of religions.* McGraw Hill.
Delorme, A. (2019). When the meditating mind wanders. *Current Opinion in Psychology, 28,* 133–137.
Demographics of Judaism. Georgetown University: Berkeley Center for Religion, Peace, & World Affairs. https://berkleycenter.georgetown.edu/essays/demographics-of-judaism.
Dexter, P. (2016). Scripture from space. https://pointofview.net/viewpoints/scripture-from-space/.
Drake, T. (2008). NASA Catholics mark 50 years. *National Catholic Register*, September 30, 2008. https://www.ncregister.com/site/article/nasa_catholics_mark_50_years.
Fleming, D. (Ed.). (1978). *The fire and the cloud: An anthology of catholic spirituality.* Paulist Press.
Forward. (2018). Dramatic orthodox growth is transforming the American Jewish community. https://forward.com/news/402663/orthodox-will-dominate-american-jewry-in-coming-decades-as-population/.
Foster, R. (2017). *Streams of living waters: Celebrating the great traditions of Christian faith.* Hodder & Stoughton.

Hagerty, M. R., Isaacs, J., Brasington, L., Shupe, L., Fetz, E. E., & Cramer, S. C. (2013). Case study of ecstatic meditation: fMRI and EEG evidence of self-stimulating a reward system. *Neural Plasticity*.

Halily, Y. (2008). The kosher space shuttle. Ynetnews.com. April 6, 2008.

Howell, E. (2018). International space station: Facts, history, & tracking. Space.com. https://www.space.com/16748-international-space-station.html.

Ihle, E. C., Ritsher, J. B., & Kanas, N. (2006). Positive psychological outcomes of spaceflight: An empirical study. *Aviation Space Environmental Medicine, 77*, 93–101.

Irwin, J., & Emerson, W. A. (1973). *To rule the night: The discovery voyage of astronaut Jim Irwin*. A. J. Holman Co.

Jones, T. D. (2006). *Sky walking: An astronaut's memoir*. HarperCollins.

Josipovic, Z., Dinstein, I., Weber, J., & Heeger, D. J. (2012). Influence of meditation on anti-correlated networks in the brain. *Frontiers in Human Neuroscience*.

Josipovic, Z. (2013). Neural correlates of nondual awareness in meditation. *Annals of the New York Academy of Sciences, 1307*, 1–10.

Kabat-Zinn, J. (2016). *Mindfulness for beginners: Reclaiming the present moment and your life*. Sounds True (publisher).

Kanas, N. (2020). Spirituality, humanism, and the overview effect during manned space missions. *Acta Astronautica, 166*, 525–528.

Kaplan, A. (1989). *Meditation and Kabbalah*. Weiser Books.

Kaplan, A., & Sutton, A. (1990). *Inner space: Introduction to Kabbalah, meditation, and prophecy*. Moznaim Publishing Corp.

Keatng, T. (2009). *Intimacy with god: An introduction to centering prayer* (3rd edn.). Crossroad.

MacLean, K., Ferrer E., et al. (2010). Intensive meditation training improves perceptual discrimination and sustained attention. *Psychological Science, 21*(6), 829–839.

Mayo Clinic Staff. (2018). Consumer health: Mindfulness exercises. Mayo Clinic. https://www.mayoclinic.org/healthy-lifestyle/consumer-health/in-depth/mindfulness-exercises/art-20046356.

Rappaport, M., & Corbally, Ch. J. (2019). *The emergence of religion in human evolution*. Routledge Studies in Neurotheology, Cognitive Science and Religion.

Schimmel, A. (2011). *Mystical dimensions of Islam*. U. of N. Carolina Press.

Suedfeld, P., & Weiszbeck, T. (2004). The impact of outer space on inner space. *Aviation, Space and Environmental Medicine, 75*(7 Suppl), C6–C9.

Suedfeld, P., Legkaia, K., & Brcic, J. (2010). Changes in the hierarchy of value references associated with flying in space. *Journal of Personality, 78*, 1–25.

Times of Israel. (2019). NASA astronaut proclaims 'Happy Hanukkah' from space. *Times of Israel*. December 23, 2019.

Vakoch, D. (2019). Astronauts open up about depression and isolation in space. *Psychology Today*, Febuary 10, 2019. https://www.psychologytoday.com/us/blog/home-in-the-cosmos/201902/astronauts-open-about-depression-and-isolation-in-space.

Weibel, D. (2017). Space exploration as religious experience. *The Space Review*, August 21, 2017. https://thespacereview.com/article/3310/1/.

White, F. (2014). *The overview effect: Space exploration and human evolution* (3rd edn.). American Institute of Aeronautics and Astronautics.

Williams, J. N. (2010). *The work of his hands: A view of god's creation from space*. Concordia Publishing.

Winkelman, M., & Baker, J. (2010). *Supernatural as natural: A biocultural approach to religion*. Pearson Prentice Hall.

Yaden, D. B., Iwry, J., Slack, K. J., Eichstaedt, J. C., Zhao, Y., Vaillant, G. E., & Newberg, A. B. (2016). The overview effect: Awe and and self-transcendent experience in space flight. *Psychology of Consciousness: Theory, Research, and Practice, 3*(1), 1–11.

Chapter 19
The Sociology of Lunar Settlement

Riccardo Campa

Abstract The possibility that, in a not too distant future, a human settlement could be established on the Moon becomes increasingly real. Sociological knowledge invites us not to underestimate social problems that could arise in a dangerous lunar environment. Issues of social order and effective collaboration among settlers from different cultures may be particularly sensitive. Physical and physiological stressors caused by the lunar environment can amplify the psychological and sociological stressors that characterize any work environment. Where even air is an essential and rare good, sabotage by individuals or non-integrated groups can be lethal for the entire community. The emergence of unwanted phenomena such as anomie and "free riding" is possible. Recommendations from the sociological literature on space exploration—particularly with regard to selection and training of space crews—may still prove useful in preventing conflicts when the human presence on the Moon is limited to a base of a few dozen astronauts and technicians. This study suggests that selection should take into account not only the professional skills and mental status of the settlers but also their degree of "sociological imagination."

19.1 Background

In recent years several national space agencies have expressed the intention to establish a permanent lunar base. On May 14th, 2019, NASA Administrator Jim Bridenstine announced that the long prepared program to land the first woman and next man on the Moon by 2024 would be named Artemis, after the Greek goddess of the Moon, sister of the solar deity Apollo. More details about the program have been made public in September 2020. As the document *Artemis Plan* reveals, NASA and its partners will develop an Artemis Base Camp at the lunar South Pole to support longer expeditions on the lunar surface. Among the elements that will compose the planned Base Camp, the document mentions power systems, a habitable mobility platform (pressurized rover), a lunar terrain vehicle (LTV, or unpressurized rover), a

R. Campa (✉)
Institute of Sociology, Jagiellonian University in Cracow, Ul. Grodzka 52, 31-044 Cracow, Poland
e-mail: riccardo.campa@uj.edu.pl

lunar foundation habitation module, and in-situ resource utilization systems (NASA, 2020).

NASA also announced that the Artemis Base Camp will be built in collaboration with international partners such as the European Space Agency, the Japan Aerospace Exploration Agency, the Canadian Space Agency, the Italian Space Agency, the Australian Space Agency, the UK Space Agency, the United Arab Emirates Space Agency, the State Space Agency of Ukraine, and the Brazilian Space Agency. Last but not least, the Artemis project also involves private partners, in particular US spaceflight companies contracted by NASA. The establishment of the infrastructure is intended to enable a sustained lunar surface presence, which, in turn, would be propaedeutic to the next ambitious project of NASA and its partners, namely a crewed mission to Mars.

The Chinese National Space Administration (CNSA) has also announced the intention to build a permanent base near the Moon South Pole and perform a manned lunar exploration mission in about ten years. The head of the CNSA, Zhang Kejian, announced the plan on April 24th, 2019, as reported by the state news agency Xinhua (Letzter, 2019). The successful lander-ascender combination of China's Chang'e 5 robotic lunar probe, which drilled for and packed rocks and soil from two meters beneath the lunar surface, proved that the Chinese space program is both ambitious and serious (Lei, 2020).

In 2016, we learned from the Russian News Agency Tass that Russia also plans to build a base on the Moon in the period from 2030 to 2035. The executive director of manned space programs of the Roscosmos State Space Corporation, Sergey Krikalev, specified that cosmonauts will land on the Moon by 2030, and in the following five years they will assembly the lunar base, which will include a landing and launch area, an orbiting satellite, a solar power station, a telecommunication system, a technological station, a scientific laboratory, and long-range research rover (TASS, 2016). However, the role of Roscosmos is still unclear. NASA's *Artemis Plan* mentions Russia as a potential partner but, after the successful Chinese mission, the head of Roscosmos Dmitry Rogozin declared that Russia is no more interested in participating in the American project and rather oriented in joining the Chinese one (Sheetz & Dzhanova, 2020).

Overall, barring unforeseen circumstances, in the next years, we will see the foundation of two or, perhaps, three lunar permanent bases on the Moon, very close to each other. These developments pave the way for a permanent human colony on the moon, a scenario that deserves to be analyzed from a sociological perspective.

19.2   Approaches, Theories, and Methods

The "Sociology of outer space" is as old as the Space Age, dating back at least to 1957. In the year of the first Sputnik mission, and one year before the foundation of NASA, the *Journal of Jet Propulsion* reported about two discussion panels dedicated to "Space sociology" in the frame of a conference organized by the American Rocket

Society (ARS, 1957). The human factor of space exploration was discussed along with engineering solutions of space carriers' propulsion. The term "sociology" was used in a very broad sense that included the legal, physiological, and psychosocial issues of spaceflight. It is worth noticing that another denomination for the study of space societies, namely "Astrosociology", has gained popularity over time (Ross, 1964, 1976; Tough, 1998; Pass, 2006a; Harris, 2009).

The sociological literature produced in the last sixty years focuses both on the sociology of space exploration (e.g. Bainbridge, 1976; Bluth, 1983; Lundquist et al., 2011) and the sociology of space colonization (e.g. Bluth, 1979, 1981; Pass, 2007; Rudoff, 1996). Still, the first type of literature largely surpasses the second one. This is quite understandable. Max Weber (1949: 51) has stressed "the logical *(prinzip-ielle)* distinction between 'existential knowledge', i.e., knowledge of what 'is', and 'normative knowledge', i.e., knowledge of what 'should be'." While the first type of knowledge is, at least ideally, objective, the second type of knowledge is intrinsically subjective. It can never be the task of empirical science to formulate binding norms and ideals. Sociology is an empirical science that studies social facts, for what they are, whether we like them or not. If the sociologist's mission is put in these terms, there seems to be very little room for the sociology of lunar settlement. Since space colonies do not yet exist, there is no way to interview the settlers and empirically assess how human life unfolds on the Moon. However, this is somewhat too a drastic and limiting view of the sociological work. Weber (1949: 52) also clarifies that "the question of the appropriateness of the means for achieving a given end is undoubtedly accessible to scientific analysis." In other words, whether it is right or not right to engage in space colonization is not a scientific problem. It is a question that concerns the political system and public opinion (Bainbridge, 1991, 2015; Etzioni, 1964). Yet, once the decision is made, the sociologist can be called upon to give sociotechnical advice. The sociologist will draw on the knowledge accumulated in the past, even in different contexts, to venture into the possible futures. After all, futures studies themselves are to a considerable extent an offspring of sociology (Son, 2015). It is not surprising, then, that, despite the colonization of space is still in an embryonic stage, the literature on the sociology of space colonization is not lacking.

In this article, we will formulate some recommendations concerning predictable issues of the lunar settlement project, by building on the already existing litera-ture and discussing neglected or little considered problems. Since this article will be possibly read also by non-sociologists, a premise on the general state of the discipline is necessary. Sociology is a multiparadigmatic discipline. This means that sociological research and theory takes many different forms. Different schools of thought, with their specific denominations (structural-functionalism, critical soci-ology, cultural sociology, social interactionism, pragmatic sociology, analytical soci-ology, etc.) adopt different theoretical perspectives, methods, and vocabulary to a point that sociologists of different orientations may find it difficult to understand each other.

I will not dwell much to situate this study into a specific paradigm, as much of the reciprocal criticism of these sociological movements is anything but a struggle for the allocation of academic resources. When one is committed to a multidisciplinary project like space exploration and colonization, one has to give up intradisciplinary identity struggles. The only way to achieve the result is to focus on "social problems" rather than "sociological problems," as the first can be understood also by non-sociologists.

The different sociological schools, besides dwelling on theoretical issues, have pointed the finger at different social problems. We will transversally pick from the substantive achievements of the many paradigms based on a single criterion, namely their utility for *the social engineering of lunar settlements*. Given this goal, not all social problems deserve to be discussed. For instance, it is unlikely that extreme poverty can occur on the Moon. Unemployed homeless people can survive in Los Angeles or Rome, not in such a hostile environment where even air is an expensive commodity (Ashkenazi, 1992). There is no way that space agencies and corporations can permit the emergence of extreme poverty on the lunar surface. Only people with a job contract will have a ticket to the Moon. Those that lose the job will be likely sent back to Earth (Heppenheimer, 1985).

This criterion applies also to the selection of theories. What is no more fashionable on Earth could still be useful in Space. For instance, contemporary sociologists tend to criticize structural-functionalism (Merton, 1968; Parsons, 1951) because this approach is too much focused on order, harmony, equilibrium, stability, and cooperation, seen as the preconditions for the persistence of a social system. So-called critical sociologists notice that social conflict is not necessarily dysfunctional, as it may bring positive social changes. In other words, functionalism has often been accused to be intrinsically "conservative" and therefore incompatible with a "progressive" view of society. This perspective may be fruitful on Earth but it is rather sterile in Space. We cannot change a society that still does not exist. If we aim at building a space society capable of surviving, persisting, reproducing in a very hostile environment, focusing on such issues as social order, manifest or latent functions, and dysfunctions makes much more sense than celebrating the cathartic role of social conflict. That is why our main focus will be on the possible sources of social disorder, seen as an unwanted side effect of lunar settlements, and the possible solutions to prevent this scenario.

19.3   Possible Sources of Dysfunctional Conflicts and Social Disorder

Three decades ago, Michael Ashkenazi (1992: 367) lamented that "discussions of space colonization address sociological issues, if at all, as if life in space were merely a continuation of life on Earth," and added that "this is highly unlikely, given the conditions of life off Earth." This is a trap in which not a few analysts keep falling.

We should never forget that in a space settlement, unlike on Earth, *social order is a matter of life and death*. Because of the amplified consequences of psychosocial stressors, governance is of no less importance than engineering systems design. Stress is strictly related to interpersonal conflict, being both a cause and an outcome of the latter. In other words, a vicious circle where stress generates conflict, and the latter generates more stress, is a deleterious mechanism often observed in closed human groups such as families or companies. This mechanism is of major concern in space for two main reasons. The first is that a conflict in space may not only be detrimental to the life of some individuals, as it happens sometimes on Earth, but could be deadly for the whole settlement. Back in 1977, a team of space scientists including sociologist Gordon Sutton underlined that "the small size of the settlement, combined with a rather precarious manufactured environment, may emphasize a concern for internal security." In a space settlement, whether an orbital station or a lunar base, "any individual or small group could, in prospect, undertake to destroy the entire colony by opening the habitat to surrounding space, by disrupting the power supply, or by other actions which have few corresponding forms in Earth-based settings" (Johnson & Holbrow, 1977: 27). As Ashkenazi (1992: 374) put it, "space colonies cannot afford to have violent revolutionaries, whether they are Bakunists or not." If "poor disaffected revolutionaries may be a danger to public order and public safety on Earth; in a space colony they endanger every living thing and the existence of the biome as a whole."

The second reason why the issue of social order needs extra-attention is that the space environment may provide not only physical and biological stressors unknown on Earth but also additional sources of psychological and sociological stress. Further-more, the two types of stress are strictly related. Back at the beginning of the Space Age, Dunlap (1966: 441) noticed that "the presence of environmental stressors, such as weightlessness, ionizing radiation, and atmospheric contaminants, can lower the threshold of tolerance to psychological and sociological stressors." His focus was on psychosocial stressors such as sensory deprivation, isolation and confine-ment, and small group dynamics. Drawing knowledge from research on submarine cruises, polar expeditions, prisoner-of-war and survivor-of-disaster accounts, labo-ratory studies of sensory deprivation and social isolation, pilot experiences, and space cabin simulator studies he underlined that confinement can generate anxiety, fatigue, irritability, and hostility in an individual. Since then, much research has been done on actual space missions to unveil the psychological and sociological problems experienced by astronauts and cosmonauts living in space for a long time. Kanas and Fedderson (1971) remarked that "in isolation, interpersonal conflict becomes exaggerated, and there is less chance to go outside to blow off steam, or escape from the difficulties of adjustment." More recently, Russian scientists analyzed the communication process between space crews and outside monitoring personnel in Mission Control. Two isolated crews were studied over a period of 135 and 90 days. The presence of psychological closing and information filtration in the crews over time was observed. One month after the beginning of the mission the total inten-sity of communication dropped. The researchers concluded that "the communication between confined groups and outside monitoring personnel is affected by psycholog-ical closing and information filtration and by the make-up of the teams that comprise the monitoring groups" (Gushin et al., 1997). We are in presence of a tendency of the crewmembers to become more 'egocentric', which has been named 'auton-omization'. The burden of communication was increasingly left on the shoulders of the Commander. These results have implications not only in relation to astronauts' psychological health but also to sociological issues such as governance and social order. Indeed, "egoistic behavior in such environmental conditions would be ultimately self-defeating" (Ashkenazi, 1992: 368).

As a report of the NASA Office of Inspector General underlines, "during a mission crew cohesion may be affected, long-term sleep loss can lead to hypertension, diabetes, obesity, heart attack, stroke, and psychiatric disorders such as depression or severe anxiety may occur." These psychological issues may have social order implications. "Although conflict among crew members has been relatively infrequent during ISS missions, these issues may take on more significance with longer duration missions and in more closely confined spaces" (NASA, 2015).

Even if these studies and reports mainly focus on long-term space missions, rather than permanent space settlements, their utility for our research is patent. Indeed, we may predict that such types of problems may emerge also on the Moon, at least until the terraforming of the satellite takes place. The problem of the effective planning of the living place, to minimize physiological discomfort, boredom, claustrophobia, frustration, psychological issues, dysfunctional conflicts, and social disorder demands the inclusion of a diversity of professionals in designing the Moon Base and investigating human factors in a broad frame. Space psychologist Jesper Jorgensen (2010: 258) states clearly that "we need historians, sociologists, psychologists, artists, doctors, engineers, information technology specialists and all the others who can give their learned views to the mission design procedure."

In particular, sociology can be useful to assess issues such as anomie and deviance from the perspective of group dynamics and collective behavior theories. Since these issues depend on the group size, it is convenient to envision three different phases of lunar colonization marked by growing demographical magnitude.

19.3.1 Base Phase

Initially, there will be only a lunar base inhabited by a small number of individuals, mainly astronauts, scientists, engineers, and logistic workers. In this phase, a certain degree of anomic response is a possibility. As Dunlap (1966: 443) stressed, "group norms are to a certain extent formed, reinforced, and maintained by the larger society of which the group is a part" and "lengthy isolation from this larger social context can lead to shifts in group standards and values toward those of the individual group members and of expediency." Ignited by the atypical situation, the following phenomena can occur: status leveling, weakening of the group hierarchical structure, social withdrawal, desocialization, and reduction of group cohesiveness. These are conditions that may stress and threaten group structure and solidarity. It must be clear, however, that there is no linear correlation between group size and structural stability. Much depends on the quality and quantity of the livable space in which the group is confined. Empirical observations and experiments performed on the International Space Station since 1998, when the first components were delivered to orbit, have shown that "even changes implemented to improve conditions could have unintended, negative consequences" (Stuster, 2010). Instructive cases are the shifts to three-person crews with Expedition 14, to multi-person crews with Expedition 16, and to permanent six-person crews with Expedition 20. On the one hand, larger crews have resulted in greater science productivity and have enabled a better arrangement of tedious tasks such as inventories and logistics duties. On the other hand, "the doubling of crew size inevitably has had a cascading series of negative effects, such as increased competition for exercise equipment, loss of privacy, and more opportunities for interpersonal conflict, to name a few" (Stuster, 2010).

19.3.2 Village Phase

Let us assume that potential health problems of lunar settlers are prevented thanks to artificial gravity or human enhancements (Szocik et al., 2019), and the community can numerically expand to reach the dimension of a village. The commonly assumed number of settlers in orbital or planetary colonies is ten thousand (Johnson & Holbrow, 1977; Russell, 1978), but we can admit that a few hundred settlers make already a village. In theory, at this stage, the isolation problem should end. "As group size increases, the members become more organized and efficient in the way data are presented. Membership polarizes with greater tendency for leaders and followers to emerge. The larger the group, the greater the tendency for leader–follower relationships to emerge…" (Kanas & Fedderson, 1971: 43). Still, we should consider that a lunar village would not be a small-scale, kinship, and neighborhood-based community, namely a *Gemeinschaft* in the sense given to this word by German sociologist Ferdinand Tönnies (2001). As the settlers will be selected based on their professional competencies to work on industrial or scientific projects, it is unlikely that they will share a common ethnicity, cultural background, and religion. Therefore, a certain degree of anomic response, triggered by long-term confinement in a close environment, cannot be excluded. In particular, if the sense of community is weak, we may expect the emergence of social disorder as a consequence of lack of cooperation. As Elster (1992: 15) clarifies, "collective action theory identifies the free-rider problem as the main obstacle to cooperation" and "bargaining theory suggests that the main problem is failure to agree on the division of the benefits from cooperation." Roughly speaking, when the number of settlers will grow, a certain percentage of them will be prone to take a free ride. If someone contributes less than others to the maintaining of the colony, conflict may arise. Besides, if prejudices of any type lead to unfounded generalizations and incorrect identification of the free riders, conflict fatherly increases.

19.3.3 City Phase

Let us assume that at one point in the future the terraforming of the Moon becomes a reality and the urban tissue can grow to the level of a city with many thousands of inhabitants. At this stage, the psychosocial pathologies deriving from confinement in small spaces would be reduced. To put it in simple words, if individuals or groups start arguing or fighting for any reason, they could eventually sober up by going for a walk outside. Also, the existential threat for the whole settlement deriving from the self-destructive behavior of deviant colonists would vanish. However, this reduction of tension would be compensated by the typical anomic and criminogenic tendencies of big urban agglomerates. "American sociologists and psychiatrists have often characterized cities as sites of social disintegration conducive to insanity" (Pols, 2003). On the contrary, small-town rural life has been presented as ideally suited for nurturing mental health. Much depends on how these hypothetic lunar cities would come into existence. If they develop in an ordered way responding to an abundance of highly qualified jobs offered by government agencies or private corporations, for instance involved in extraterrestrial mining, we may surmise that these problems would not reach a critical point. Different would be the situation if humans would massively emigrate from a planet Earth in agony due to wars, global warming, pollution, pandemics, geological catastrophes, or other existential risks, in a desperate attempt to survive on a terraformed Moon. In this case, a ticket to the Moon and a lunar dwelling could be purchased by anybody, including affluent mobsters and terrorists. Consequently, other typical social phenomena monitored by sociologists, such as social stratification, poverty, inequality, class struggle, organized crime, etc., would become relevant issues.

The discussion could proceed by taking into consideration the emergence of more than one city on the lunar soil, perhaps founded by different nations, with different political systems and cultural backgrounds. But the hypothetical relations between these cities could perhaps be better analyzed from the perspective of political science. Any scenario would be worthy of attention, from tensions between cities to the birth of a lunar federation to the prospect of a lunar colony's revolt against rule from Earth–a scenario already envisioned by science fiction writer Robert A. Heinlein (1966). We will not include these potential developments in our scenario analysis, as they were already explored in other studies (Russell, 1978).

19.4 Possible Solutions to Minimize Dysfunctional Conflicts and Social Disorder

In this section, we will tentatively propose sociotechnical solutions to potential dysfunctional conflicts that are likely to occur in a lunar settlement. Economist von Hayek has emphasized that there are two types of social order. On the one hand, one may have a made, exogenous, artificial, constructed order. This is the case of *organization*s. On the other hand, one may have a grown, self-generating, endogenous, spontaneous order. This is the case of the *market economy*. As Friedrich Hayek (1973: 36) remarks, "classical Greek was more fortunate in possessing distinct single words for the two kinds of order, namely *taxis* for a made order, such as, for example, an order of battle, and *kosmos* for a grown order, meaning originally 'a right order in a state or a community'." Clearly, when talking about human phenomena, the distinction between *natural* and *artificial* can only have a heuristic purpose, rather than a clear ontological fundament. According to Hayek (1973: 286), virtually any government's acts of interference in the catallaxy create disorder. Advocates of social policies notoriously think that this conclusion is too ideologically laden. Yet, there is no need to subscribe to it in order to recognize the utility of concepts such as *taxis* and *kosmos*. To complete the picture by taking a more neutral stance, we can recognize that, if there are two types of order, there must correspondingly be two types of disorder, one artificially created by the wrong decisions of the political authority and one that grows out from the wrong decisions of free individuals (which can by no means be always rational and optimally informed). In this perspective, Hayek's conclusion is just a special case among many possible ones. Indeed, the social disorder could also be generated by top-down regulations that, in order to enhance productivity, disproportionally reward competition instead of cooperation. Excess of competition may result in conflict and, therefore, disorder. The outcome of a specific situation depends also on variables such as individual personality traits and cultural background. If we do not have a simple algorithm to predict order or disorder occurrences *ex ante facto*, we have useful concepts to analyze situations *ex post facto*, so that at least one may see and fix situations in their making. The solutions would be different for the three above-mentioned phases.

19.4.1 Base Phase

National space agencies already dispose of sufficient knowledge on interpersonal, behavioral, and cultural issues to ensure social order in the first phase of the lunar settlement. Selection and training of the personnel, as designed for long-term space missions, should work well also for the moon base (Nicholas & Foushee, 1990; Santy, 1994; McFadden et al., 1994; Tomi, 2007). It has been often noted that the leader–follower relationship is particularly important to group stability. Still, it is debated if this stability can be ensured by military-type discipline or by a certain degree of structural flexibility. Indeed, authorities may design a specific hierarchical structure for the lunar settlement (*taxis*), but the settlers could informally and spontaneously assign the effectual leadership to individuals with lower rank (*kosmos*). Depending on personality traits and cultural background, this situation could create tensions and conflict, or could just work out well based on commonly and tacitly accepted assumptions. On Earth, such type of situations could persist for years without jeopardizing the goals of an organization. Yet, the presence in space of extraordinary psychosocial stressors invites avoiding the crystallization of potentially unstable structures. On

Earth, the members of an organization have other social roles (such as parenthood, club membership, religious functions, political activism, etc.) that cannot be easily performed in space. A lack of gratification coming from extra-work activities could exacerbate interpersonal conflicts. Communication with control personnel must not be predominant, as it is itself problematic (Kanas et al., 2000, 2001, 2007). Ensuring constant communication between lunar settlers and their terrestrial friends should reduce the danger of anomic response. Besides, the periodic turnover of personnel in the moon base—which is anyway needed to prevent health issues—should also serve to re-establish the perfect coincidence of the formal and material constitution of the settlement. This can be done either by formalizing the spontaneous order if it seems to work well given the mission goals, or providing new leaders or a new set of roles in the social system.

19.4.2   Village Phase

Roscosmos director Sergey Krikalev said that "there will be no settlement like a village in which people live, tending cows, for a rather long time yet, most likely" (TASS, 2016). He could be right but a sociological analysis of this scenario seems still useful. Since the two or three bases will be built near the South Pole of the Moon we cannot exclude that either a single village will emerge from the merging of the expanded bases or two/three separate villages will develop each growing around a base. Given the geopolitical tensions and the conflicting economic interests existing on Earth, we surmise that the second option is more plausible. Each agency will try to counteract anomic tendencies by ensuring continuous communications and alimenting in the settlers the feeling of being an integral part of the larger society. However, a tendency to an autonomous development is unavoidable. Let us assume that there will be three villages tied respectively to the USA, China, and Russia. We can surmise that the settlements will not slavishly follow the mother countries if the latter raise the level of international competition and conflict. Living in an extremely hostile environment, American, Chinese, and Russian moon settlers know that in presence of existential threats they may receive immediate help only from the neighbors rather than the relatively far mother countries. In other words, it is unlikely that inter-village competition to study or exploit natural resources could easily turn into open conflict, regardless of the pressure from the national agencies or the associated private industries.

As regards the need to minimize intra-village (or interpersonal) conflicts, a more sophisticated selection of personnel should be implemented than the one adopted for usual space missions. Back in the 1980s, Bluth noticed that "a Lunar Settlement … is not a spacecraft. Rather, it is a facility—a permanent 'place in space' to be staffed by interdisciplinary teams of people who come from many countries, cultures, and communities who will be living together, building things, and pursuing knowledge and information as teams" (Bluth, 1988: 665). Two decades later, astrosociologist Jim Pass (2006b) normatively asserted that the "population should be heterogeneous, and with that, we must prepare for the conflict that occurs within a population characterized by diversity in its many forms (e.g., based on social class, race, ethnicity, sex, power, prestige, age, religious affiliation); we should not strive for homogeneity in order to avoid possible discrimination but should attempt to curtail it as much as possible."

These observations seem predictively correct. Indeed, the space community that could develop as an offspring of the Artemis Base Camp will likely be a multicultural community. However, this will not necessarily be the case as regards the Chinese moon village.

If we take a sociological (and not purely ethical) perspective, it is important to make explicit the *rationale* behind a choice for a heterogeneous or homogeneous space community. Political correctness is not a rationale in the techno-scientific (or Weberian) meaning of the word. Put it simply, is there any good reason to export the problems we face on Earth to other worlds? Intuitively, a homogeneous community is more likely to avoid destructive conflicts and there is no reason to take risks without an adequate pay-off. Available sociological literature confirms that cultural homogeneity and heterogeneity are not irrelevant variables in space (Boyd et al., 2009; Gushin et al., 2001). According to Ashkenazi (1992), to ensure stability and success in such an extraordinarily difficult environment as space, not only ethnic or religious conflicts but also excessive economic competitions should be avoided. That is why he proposes to adopt a kibbutz-like community as the basic model for a space colony.

Still, the multicultural community model can also be defended from the point of view of instrumental rationality. "A small settlement in space, of less than 100,000 people, would necessarily require continuing support from Earth. There is little possibility that such a settlement can be sustained without a steady and sizable movement of materials and information between Earth and the colony" (Johnson & Holbrow, 1977: 27). A multicultural nation is more likely to economically support a multicultural moon settlement, also in loss conditions, as the variegated members of the larger society would always find an individual or a group in the settlement identifiable as "one of us."

This, however, does not imply that the selection should be inaccurate or based uniquely on the assessment of professional skills and mental stability. Apart from obvious needs like planning a balanced community in terms of gender and sexual orientations to minimize sexual frustration, other fundamental psychosocial factors should be taken into account. The main recommendation deriving from this analysis is to see the problem of selection from a *metacultural* rather than a merely *cultural* point of view. *How* people believe is much more relevant than *what* people believe or think to know. What one does not want in a fragile space community is the presence of dogmatic, fanatic, egocentric people, regardless of their worldview or substantial beliefs. Psychologists know well how to identify sociopaths and individuals affected by narcissistic personality disorder. But here we are suggesting that the selection should also leave out people that are not necessarily diagnosed with mental issues. Selfish people are still capable of living a normal or even successful life on Earth, but they could easily put at risk the stability of a fragile lunar community. On the contrary, philosophical sensibility and sociological imagination would be fundamental factors of stability in space. As Wright Mills (1959: 7) defined it, sociological imagination is "the capacity to shift from one perspective to another" and "to range from the most impersonal and remote transformations to the most intimate features of the human self–and to see the relations between the two." People with sociological imagination are able to transcend the cognitive limitations imposed by their own biographical vicissitudes and understand the latter as a result of historical changes of a greater scope. They are capable of seeing situations from the point of view of others, because they have acquired "a new way of thinking" and experienced "a transvaluation of values" (Wright Mills, 1959: 7).

In particular, selectors should hesitate to recruit candidates that fail to understand the difference between *believing* and *knowing*. We start believing when we recognize we do not or cannot know something with certainty, when we are not direct witnesses of an event and must therefore place a certain degree of faith or trust in the source, when we can still reasonably doubt a claim. Believing is legitimate, as we cannot have direct experience of everything. However, not everybody understands that metaphysical ideas, being beyond ultimate empirical proof, cannot be discussed as one discusses whether it is raining or not. The minimization of religious conflict in space is important because the livable area in the moon settlement will be predictably limited. If believers are tolerant enough, they may agree about sharing a unique temple for all the different religious symbols, functions, and ceremonies. An equally tolerant attitude should be required from non-religious settlers. When it comes to *knowledge*, it would be important to make sure that settlers are far from the two extremes of total skepticism, which may result in inaction, and stubborn dogmatism, which may result in "science wars." A certain degree of epistemological fallibilism (and consequent cognitive flexibility) is of fundamental importance when the challenge is to adapt to a largely unknown environment. Selectors should ask the potential settlers: (1) if they were ever wrong about something in the past; (2) if they publicly admitted to be mistaken; (3) which probability they attach to the possibility that their current knowledge is to some extent unfounded; (4) how ignorant they feel about the mysteries of the universe. Failing to admit that they ever changed their mind about something, or that they could still be in principle wrong or ignorant about something, should sound like an alarm bell.

Selecting people with sociological imagination is important not only to prevent dysfunctions such as free riding and conflicts resulting from lack of cooperation but also to prevent conflicts deriving from excessive rigidity and lack of empathy. Elster (1992: 15) rightly observes "that collective action failures often occur because bargaining breaks down. Often, it would be absurd to ask everybody to contribute equally to a public good. Some need it more than others, or can better afford to contribute." To put it simply, a woman may have problems that a man does not have and vice versa, or a young person may do things that an older one cannot do, etc. The capacity of putting oneself in someone else's shoes is fundamental to decide in specific situations who should contribute and who should be allowed to take a free ride.

19.4.3   City Phase

We will not spend many words to assess this possible phase of lunar settlement for two main reasons. Firstly, a lunar city inhabited by hundreds of thousands of humans appears non-feasible in the next few decades, because of microgravity, radiations, extreme temperatures, meteorites, health issues, and other well-known problems. If the human presence on the Moon will be profitable from an economic point of view, it is likely that industrial activities will be automatized as much as possible (Campa et al., 2019).

Secondly, a hypothetical terraformed celestial body, being the Moon or Mars, with rivers, cultivated fields, breathable air, bearable weather conditions, and inhabited by millions of humans and animals, etc., would exhibit features so similar to life on Earth that would not necessitate peculiar sociotechnical instruments to ensure social order. If terraforming were to succeed, space communities would become economically independent from mother Earth and their population would start reproducing in situ rather than being a priori selected. Very interesting configurations from a sociological point of view could come into existence, but a futurological exploration of such possibilities would take us too far away from more pressing problems. True, SpaceX founder Elon Musk said that he plans to send one million people to Mars by 2050 by launching three Starship rockets every day (McFall-Johnsen & Mosher, 2020). Hard to believe that this ambitious plan will become reality, but we would be happy to be proven wrong by the facts.

19.5   Concluding Note

To conclude, it seems worthwhile to outline potential areas for future research. At any stage of the lunar settlement, sociologists can be useful not only by helping to select the most suitable settlers, but also to monitor how social life develops in space. When the settlement will start expanding, perhaps, sociological research could be incorporated into space agencies' scientific programs along with other scientific experiments. Sociologists, armed with questionnaires and audio–video recorders, by mean of surveys or more or less structured in-depth interviews, could collect and record valuable information from space settlers and make sense of it by means of sociological theories. This can be done from Earth by using telecommunication technologies but, if one wants to get 'under the skin' of the lunar settlement's design problem, it could be worth resorting to ethnographic research. If astrosociologists were given a chance to observe and/or interact with lunar settlers in their real-life environment, more relevant behavioral information would be retrieved. Participant observation and qualitative analysis could help to fix organizational dysfunctions before they degenerate into existential threats for the space community.

References

ARS. (1957). Space sociology, astronautics sessions highlight ARS spring meeting. *Journal of Jet Propulsion., 7*(3), 320–324.

Ashkenazi, M. (1992). Some alternatives in the sociology of space colonization: The Kibbutz as a space colony. *Acta Astronautica, 26*(5), 367–375.

Bainbridge, W. S. (1976). *The space flight revolution: A sociological study.* Wiley.

Bainbridge, W. S. (2015). *The meaning and value of spaceflight. Public perception.* Springer.

Bainbridge, W. S. (1991). *Goals in space: American values and the future of technology.* State University of New York Press.

Bluth, B. J. (1988). Lunar settlements: A socio-economic outlook. *Acta Astronautica, 17*(7), 659–667.

Bluth, B. J. (1979). Constructing space communities. A critical look at the paradigms. In R. Johnson et al., (Eds.), *The future of the U.S. space program. Advances in the astronautical sciences* (Vol. 38). Univelt, Inc.

Bluth, B. J. (1981). Sociological aspects of permanent manned occupancy of space. *AIAA Student Journal, 48*, Fall, 11–15.

Bluth, B. J. (1983). Sociology and space development. In T. S. Cheston (Eds.), *Space Social Science.* Retrieved January 2nd, 2021, from https://er.jsc.nasa.gov/SEH/sociology.html

Boyd, J. E., Kanas, N. A., Salnitskiy, V. P., Gushin, V. I., Saylor, S. A., Weiss, D. S., & Marmar, C. R. (2009). Cultural differences in crew members and mission control personnel during two space station programs. *Aviation, Space, and Environmental Medicine, 80*(6), 532–540. https://doi.org/10.3357/asem.2430.2009

Campa, R., Szocik, K., & Braddock, M. (2019). Why space colonization will be fully automated. *Technological Forecasting & Social Change, 143*, 162–171. https://doi.org/10.1016/j.techfore.2019.03.021.

Dunlap, R. D. (1966). Psychology and the crew on mars missions. Paper presented at the American institute of aeronautics and astronautics, and american astronautical society stepping stones to mars meeting (Baltimore, Md.), Mar. 28–30, 441–445.

Elster, J. (1992). *The cement of society.* Cambridge University Press.

Etzioni, A. (1964). *The moon-doggle: Domestic and international implications of the space race.* Doubleday.

Gushin, V. I., Zaprisa, N. S., Kolinitchenko, T. B., Efimov, V. A., Smirnova, T. M., Vinokhodova, A. G., & Kanas, N. (1997). Content analysis of the crew communication with external communicants under prolonged isolation. *Aviation, Space, and Environmental Medicine, 68*, 1093–1098.

Gushin, V. I., Pustynnikova, J. M., & Smirnova, T. M. (2001). Interrelations between the small isolated groups with homogeneous and heterogeneous composition. *Human Performance in Extreme Environments, 6*, 26–33.

Harris, P. R. (2009). *Space enterprise: Living and working offworld in the 21st century.* Springer and Praxis.

Hayek, F. A. (1973). *Law, legislation and liberty* (Vol. 1). The University of Chicago Press.

Heinlein, R. A. (1966). *The Moon is a harsh mistress.* G. P. Putnam's Sons.

Heppenheimer, T. A., et al. (1985). Resources and recollections of space colonization. In C. Holbrow (Ed.), *Space colonization: Technology and the liberal arts* (pp. 129–140). American Institute of Physics.

Johnson, R., & Holbrow, C. (1977). *Space settlements: A design study.* Government Printing Office.

Jorgensen, J. (2010). Humans: The strongest and the weakest joint in the chain. In H. Benaroya (Eds.), *Lunar settlements.* CRC Press

Kanas, N., Salnitskiy, V., Grund, E. M., Gushin, V., Weiss, D. S., Kozerenko, O., Sled, A., & Marmar, C. R. (2000). Interpersonal and cultural issues involving crews and ground personnel during Shuttle/Mir space missions. *Aviation, Space, and Environmental Medicine, 71*(9), A11–A16.

Kanas, N., Salnitskiy, V., Weiss, D. S., Grund, E. M., Gushin, V., Kozerenko, O., Sled, A., Bostrom, A., & Marmar, C. R. (2001). Crew member and ground personnel interactions over time during Shuttle/Mir space missions. *Aviation, Space, and Environmental Medicine, 72*(5), 453–461.

Kanas, N. A., Salnitskiy, V. P., Boyd, J. E., Gushin, V. I., Weiss, D. S., Saylor, S. A., Kozerenko, O. P., & Marmar, C. R. (2007). Crew member and mission control personnel interactions during international station missions. *Aviation, Space, and Environmental Medicine, 78*(6), 601–607.

Kanas, N. A., & Fedderson, W. E. (1971). Behavioral, psychiatric, and sociological problems of long-duration space missions, NASA TM 58067, October.

Lei, Z. (2020). Chang'e 5 lunar probe gathering moon samples, December 3rd, 2020. Retrieved January 2nd, 2021, from http://english.www.gov.cn/news/topnews/202012/03/content_WS5fc82 8d1c6d0f725769412cd.html

Letzter, R. (2019). China plans to build a moon base near the lunar south pole, 27 April 2019. Retrieved January 2nd, 2021, from https://www.space.com/china-moon-base-10-years.html

Lundquist, C. A., Tarter, D., & Coleman, A. (2011). Identifying sociological factors for the success of space exploration. *Physics Procedia, 20*, 331–337.

McFadden, T. J., Helmreich, R. L., Rose, R. M., & Fogg, L. F. (1994). Predicting astronaut effectiveness: A multivariate approach. *Aviation, Space, and Environmental Medicine, 65*(10), 904–909.

McFall-Johnsen, M., & Mosher, D. (2020, January 17). Elon Musk says he plans to send 1 million people to mars by 2050 by launching 3 Starship rockets every day and creating 'a lot of jobs' on the red planet. *Business Insider*. Retrieved January 2nd, 2021, from https://www.businessinsider.com/elon-musk-plans-1-million-people-to-mars-by-2050-2020-1?IR=T

Merton, R. K. (1968). *Social theory and social structure*. The Free Press.

NASA. (2015). Nasa's efforts to manage health and human performance risks for space exploration. *Office of Inspector General, Report No. IG-16–003*. October, 29.

NASA. (2020). Artemis plan. NASA's lunar exploration program overview. *National Aeronautics and Space Administration*. Retrieved January 2nd, 2021, from https://www.nasa.gov/sites/def ault/files/atoms/files/artemis_plan-20200921.pdf

Nicholas, J. M., & Foushee, H. C. (1990). Organization, selection, and training of crews for extended spaceflight: Findings from analogs and implications. *Journal of Spacecraft and Rockets, 27*(5), 451–456.

Parsons, T. (1951). *The social system*. Routledge & Kegan Paul.

Pass, J. (2006). Viewpoint: Astrosociology as the missing perspective. *Astropolitics, 4*, 85–99.

Pass, J. (2006). The astrosociology of space colonies: Or the social construction of societies in space. *AIP Conference Proceedings, 813*, 1153. https://doi.org/10.1063/1.2169297

Pass, J. (2007). Moon bases as initial "Space Society" trials: Utilizing astrosociology to make space settlements livable. *AIP Conference Proceedings, 880*(1), 806. https://doi.org/10.1063/1.2437520.

Pols, H. (2003). Anomie in the metropolis: The city in american sociology and psychiatry. *Osiris, 18*, 2nd series, 194–211. Retrieved December 27th, 2020, from http://www.jstor.org/stable/365 5292

Ross, H. E. (1964). A contribution to astrosociology. *Spaceflight, 6*(4), July, 120–124.

Ross, H. E. (1976). Guidelines to astrosociology. *Spaceflight, 18*(4), April, 135.

Rudoff, A. (1996). *Societies in space*. Peter Lang Publishing.

Russell, A. (1978). Human societies in interplanetary space: Toward a fructification of the utopian tradition. *Technological Forecasting & Social Change, 12*, 353–364.

Santy, P. A. (1994). *Choosing the right stuff: The psychological selection of astronauts and cosmonauts*. Praeger.

Sheetz, M., & Dzhanova, Y. (2020). Top Russian space official dismisses NASA's moon plans, considering a lunar base with China instead, July 15th, 2020. Retrieved January 2nd, 2021, from https://www.cnbc.com/2020/07/15/russia-space-chief-dmitry-rogozin-dismis ses-nasas-moon-program-considering-china-lunar-base.html

Son, H. (2015). The history of western futures studies: An exploration of the intellectual traditions and three-phase periodization. *Futures, 66*, 120–137. https://doi.org/10.1016/j.futures.2014.12.013

Stuster, J. (2010). *Behavioral issues associated with long-duration space expeditions: Review and analysis of astronaut journals. Experiment 01-E104 (Journals): Final report.* NASA/TM-2010–216130. Houston, Texas: NASA/Johnson Space Center

Szocik, K., Campa, R., Rappaport, M. B., & Corbally, C. (2019). Changing the paradigm on human enhancements: The special case of modifications to counter bone loss for manned mars missions. *Space Policy, 48*, 68–75.

TASS. (2016). Russia plans to build lunar base in 2030–2035—space corporation, April 5th, 2016. Retrieved January 2, 2021, from https://tass.com/science/867452

Tomi, L., Kealey, D., Lange, M., Stefanowska, P., & Doyle, V. (2007). Cross-cultural training requirements for long-duration space missions: Results of a survey of international space station astronauts and ground support personnel. Paper delivered at the Human Interactions in Space Symposium, May 21st, 2007, Beijing, China

Tönnies, F. (2001). *Community and civil society.* Cambridge University Press.

Tough, A. (1998). Positive consequences of SETI before detection. *Acta Astronautica, 42*(10–12), 745–748.

Weber, M. (1949). *The methodology of the social sciences.* The Free Press.

Wright Mills, C. (1959). *The sociological imagination.* Oxford University Press.

Chapter 20
Projections for Lunar Culture, Living, and Working: How Will We Be Different?

Christopher J. Corbally and Margaret Boone Rappaport

Abstract Living on the Moon will require even more innovation than living on Mars. While Luna's proximity to Earth is some advantage, gravity is only half of Martian gravity or one-sixth of Earth's. Much is uncertain but there are studies that point toward types of supplies and infrastructures such as human dwellings that are suited to communal living and privacy, both of which will be needed in lunar communities. Within these broad parameters, structures and activities can encourage the values of sustainability through art, recreation, and a cuisine that draws on hydroponic farm products. A new culture will develop, one that remains sensitive to the lunar environment if it is ever to achieve lasting success. In this chapter, we address four aspects of living and working: habitation, recreation, art, and cuisine—all in lower gravity, and often enclosed in rooms where oxygen will be in limited supply, where waste management will be a challenge, and where, if we enjoy our family's home cooked recipes, they may come from a replicator.

20.1 Contexts for Lunar Cultural Emergence

This chapter takes the reader farther into the future than most chapters in this book, and for that reason, we are last in the line-up. We shall be making empirically grounded projections based on comparable findings derived from today's scientific and engineering knowledge base and the methodological tools of futures analysts (e.g., trend analysis, statistical modeling, Delphi techniques, simulation and games, and normative forecasting techniques). Indeed, the application of logical reasoning to our existing knowledge on lunar models for human living increases our confidence in the material and social lunar culture that we envision here. Projections like the

C. J. Corbally (✉)
Vatican Observatory Research Group, University of Arizona, Tucson, AZ 85721, USA
e-mail: corbally@as.arizona.edu

M. B. Rappaport
The Human Sentience Project, LLC, Tucson, AZ 85704, USA
e-mail: msbrappaport@aol.com

ones in this chapter are crucial for the planning of the second phase of long-term human presence on the Moon.

In the first phase, as outlined in the US National Aeronautics and Space Administration's Artemis program (Bukley, 2020; NASA, 2020a), the pioneers will have established a foothold, as it were, beyond the first human footprints there. Basic challenges to human life in the demanding lunar environment will have been mitigated, and the site for a permanent moonbase will have been determined. Next will come the phase that will set the pattern for how people living on the Moon will become different from those on Earth, and the construction of the first Moon Village as a nucleus for expanding into a lunar city (ESA, 2020; Parks, 2019).

We anticipate that the culture of this second phase will be dominated by one thing: work. There will be no room for the unemployed, or even for the retired, at first. In the early phases of lunar settlement, there will no doubt be politicians, celebrities, and the wealthy who visit. In the second stage, probably towards the end of constructing the Moon Village, provision will be made for tourists. After all, it will be important to engage everyone on Earth in the first lunar settlement, in order to gain their goodwill. Still, the activities of (1) constructing buildings and infrastructure will dominate, while remembering our need for (2) periods of recreation (a need imposed by work), (3) satisfying cuisine, and (4) creative art.

We cannot anticipate one feature of the emerging lunar culture: its roots. We cannot forecast the proportions of humans who will come from the cultures of Africa, East Asia, Europe, and Latin America, nor can we foresee how the cultures will combine and blend in a process called, "syncretism." We can anticipate that the English language will predominate, because that is the language of most international conferences, especially scientific meetings, at the present and for the foreseeable future. It is also the language for the communication that takes place by international transportation specialists, especially pilots and flight controllers. Therefore, the culture we foresee here will be flavored by the ethnic roots of the people who emigrate to the Moon, but those roots and their combinations are not yet known.

The main trends in lunar culture, living, and working that we shall consider in this chapter relate to construction and architecture, recreation, cuisine, and art. In them we shall find similarities to the same cultural sectors on Earth, with some notable differences. Analyzing the differences will help the preparation for even greater challenges on Mars. Undeniably, lunar colonies can be more dependent on Earth than those of Mars, which lies at a much greater distance from its home planet. It is not just the possibility of more ready supplies that will provide a greater sense of being "close to home." On the Moon, the Earth gives us a true "eyeful" compared to the Earth's tiny orb from the perspective of humans on Mars. Still, the challenges of surviving and thriving on the two are in many ways similar, and dominated by an atmosphere that is absent (Moon) or poisonous (Mars).

Konrad Szocik and colleagues examine, at least from a Western perspective, the factors that the "Martian human" will face in establishing and building a sustainable human colony on Mars (2020). These factors will involve (1) the existential threat of planet conditions, (2) the human psyche, (3) technological advancement, (4) a "business model" including reciprocity, and (5) a "test program" encouraging future growth.

Studies of pioneering in the Americas, Australia, and in post-World War Germany and Japan correctly point to the future importance of establishing a coherent and organized system of values and beliefs appropriate to Mars. The same will be true on the Moon, or any other off-world human colony (cf. Chap. 14 on the anticipation of the Moon's environmental ethos, in this volume). These values and beliefs will be reflected in the culture created through the architecture, recreation, cuisine, and art of the colony. This is a major reason to look at these aspects in detail.

20.2 Architecture and Living

On the Moon humans will not need to live totally buried, and yet, in the face of an environment quite devoid of atmosphere we shall need to be surrounded by suitably pressurized air and adequately protected from harsh thermal conditions, cosmic rays, and micro-meteorites. Currently, these challenges have two main solutions. There have been other designs in the past, and these two may change in the future.

Solution 1. The European Space Agency (ESA) and the architectural firm, Skidmore, Owings and Merrill (SOM), along with the Massachusetts Institute of Technology's Department of Astronautics, are designing tall, four-level structures that provide living accommodations and laboratories, and which can be connected (Fig. 20.1). Each structure has an inflatable shell with a rigid frame, and all materials will be transported from Earth. The need for transport of radiation shielding from Earth limits its quantity and quality, and so each contemporaneous work group must return to Earth after a year on the Moon (ESA, 2020).

Fig. 20.1 ESA-SOM Moon Village Mockup (Image © SOM | Slashcube GmbH)

Fig. 20.2 ESA-NASA Moon Village under construction (Image © ESA/Foster + Partners)

Solution 2. The other solution is a domed structure in development by NASA and ESA (ESA, 2014). The domes can also be linked to form a village-like network (Fig. 20.2). Industrial buildings, which are of considerably larger size than units for habitation and work, will also form a part of an expanding, multi-purpose complex. Radiation protection for Solution 2 will be far greater than for Solution 1. Therefore, even a lifetime could be spent safely on the Moon (Parks, 2019). Locating work and habitation quarters below ground or below greenhouses can also offer some of the same protection (cf. Chap. 5 by Cazalis in this volume,).

The added protection offered by Solution 2 will be provided by two key factors: regolith from the lunar surface, itself (cf. Haviland, Chap. 3, in this volume), and the use of additive manufacturing (AddMan), more popularly known as 3D printing (Benvenuti et al., 2013; Kading & Straub, 2015; Labeaga-Martínez et al., 2017). The framework for each building in Solution 2 will be a rigid shell, which must be transported initially from Earth along with a binding agent or "ink" used by the 3D printing process, which will consolidate the thick regolith covering for all the buildings.

20.2.1 AddMan Construction and the Environment

Let us examine in detail the construction elements used in AddMan, since they will determine much of living and working space on the Moon. Their use on the Moon will be an important test, because they may also be used elsewhere in the solar system if proven successful on the Moon. They could be used on Mars, the asteroids, Ceres (technically a dwarf planet in the asteroid belt), Europa, or elsewhere in human exploration of the solar system.

A consequence of space exploration is a re-orienting to new shapes. Humans in industrial societies, who are so familiar with living in rectangular buildings, will need to become content with the predominantly spherical spaces produced by inflatable shells. They are like many peoples in more traditional societies all over the world such as the nomads of the Central Asian plateau who live in round "yurts". House forms even extending into recorded history have used a round shape, such as the fifteenth century trullo houses that remain in Alberobello, Italy. In the modern era, we will be aided by designers of interiors and furniture who conceive a variety of forms for these round, domed lunar spaces. They will be interesting, like existing prototypes on Earth, and not just those for submariners, who must accustom themselves to small, curved spaces, too.

So, an optimization for lunar architecture, whether in Solution 1 or 2, would be a pressurized inflatable shell. For Solution 2, the strengthening and all-protective regolith that covers semi-rigid shells will be laid down, layer after layer, by a 3D printing rover that can gather the regolith and then crawl up over shells and deposit the mixture in the shape of an arch (an inverse caternary shape). The regolith will naturally flow over the shell, making a covering of uniform thickness (Benvenuti, 2013, Fig. 12). It needs to be sufficient to protect against meteorites traveling at close to nine times the speed of a bullet shot from a rifle. The shell must also shield the interior from solar winds, solar flares, and galactic cosmic rays (Benvenuti, 2013, 292–3). This implies rather deep windows to let in light, so they might be best conceived as skylights. For the two weeks of a lunar night, available light would become "earthlight" reflected off our home planet Earth, rather than sunlight directly from our star, the Sun.

There are distinct environmental advantages in using thick lunar regolith in the form of a dome for protection of human inhabitants or workers. The building material is readily available and does not need to be transported from Earth. Furthermore, as the mole, the human, and other animals on Earth demonstrate, the most efficient shape for thermal control and conservation of heat is a sphere. It has the maximum volume for minimum surface area, so it will ultimately help to preserve fuel for heating and cooling (Allison, 2015). AddMan uses an abundant natural resource, lunar regolith, in an effective and non-polluting way. It fits well into William Kramer's "A Framework for Extraterrestrial Environmental Assessment" (2020). Just as AddMan was first developed on Earth and will be taken to the Moon and beyond, so too feedback from its application on the Moon should influence its use back on Earth, where it is already important for remote military posts (Jagoda, 2020). With further global warming, it could be used widely to control temperatures of internal spaces for humans in the future.

20.2.2   Solutions for Resource Limitations, Lower Gravity, and Dust

For a limited time, we can manage to live in tight quarters like the International Space Station (ISS) or in tiny bedrooms of astronomical observatories. However, each individual on the Moon will need 90 cubic meters of personal space (Drake, 1998). For couples, that could be doubled. Skylights will likely appear in common areas, rather than personal quarters. Window materials will be specially designed to protect humans from radiation, and conserve heat, thereby limiting the use of energy. Skylights will be a limited resource in any lunar building because they will, at first, be brought from Earth. It may also be more convenient for private quarters to be dark for sleeping in the two-week lunar day. Lunar residents will get used to electronic panels in each room that compensate for the lack of windows by providing well-lit views of the lunar landscape or even views of Earth. Many indoor spaces on Earth suffer from lack of daylight, but the difference in lunar life will be the duration of the longer day. Conservation of resources will not only limit skylights, but many other items that were once thought of as necessities.

Antarctic living illustrates some of the differences that we shall find on the Moon. Indeed, one person who lived for two months on an Antarctic icefield during an expedition to gather meteorites has remarked that it felt like being in an "alien world." Similarly, the experiences of people at the Antarctic stations, which have months-long periods of continuous light or darkness, help improve our understanding of what life on the Moon may be like. For instance, the greenhouses described in Braddock's Chap. 6 of this volume will surely not be too distant. They illustrate well that they or similar environments could provide a pleasant space for recreation and exercise for early inhabitants of Earth's Moon.

No windows will be open since there is no lunar atmosphere—that is, until large moon domes are built to fully encompass plants and "outdoor" spaces, as well as living quarters. This is not dissimilar to some work areas on Earth. However, on the Moon, the air supply will always be artificial, despite the recycling accomplished by greenhouses, and the air will be carefully controlled. That control will allow green plants in living areas, which will help both aesthetically and by turning carbon dioxide into breathable oxygen. Still, no natural fires for warmth or barbequing will be possible. The Moon inhabitants will rely on interior decorators and engineers to devise cheerful and warming alternatives, perhaps dial-able in large wall size panels.

Thanks to the Apollo program, we have seen the effects of gravity on the human gait that is one-sixth that on Earth. The interior layout, furniture, and decor will all allow for hopping or skipping as an optional gait, at least between buildings. Inside, weighted or magnetic booties or slip-ons may help. One question that the authors have pondered is whether the family dog (*Canis familiaris*) will be able to cope with the one-sixth gravity of the Moon, or if other pets will eventually prove less problematic.

In the initial phases of the settlement of Earth's Moon, individual houses will be a luxury for very few. Space, light, warmth, and conviviality will be precious resources for most lunar inhabitants who serve in research, administrative, security, military, agricultural, catering, and construction positions, so the layout of a single lunar unit will be similar and confined to sleep and private time, whatever their occupational specialty. Sitting, dining, kitchen, and utility rooms will all be in common. Somewhere nearby there will surely be a café, serving coffee, tea, and snacks, perhaps from a replicator, and a gym, with equipment tested in the ISS. The authors have pondered the appearance of an early bar serving alcoholic beverages, and they wonder how rules and regulations would be crafted to allow substance use—or disallow it. A non-denominational Meditation Center will be essential for many people from a variety of faiths, or no faith, since meditation can be pursued for personal health rather than for faith reasons. As Murray well argues in Chap. 18 of this volume, regular meditation will lower the stress produced in a "hardship" location that, in past centuries, "might have been chosen as a penal colony."

We have not yet mentioned lunar dust, which has the potential for inducing great stress on human comfort, cleanliness, health, and sanity. Airlocks will help, but as the Apollo astronauts found, lunar dust is highly invasive and abrasive (Dolgin, 2019). People cope with dust in Southern Arizona, a home to both authors, where suggestions may emerge for curbing lunar dust. However, few environments on Earth include dust as abrasive as glass shards. Coping with corrosive dust may be one of the single most difficult aspects of lunar living, and "dust abatement" may be raised to a high level of technological precision, and importance for human well-being. It may also exclude certain humans from serving there, because of health reasons.

20.2.3 Location, Location, Location

Given all the challenges we describe, one might wonder why humans want to go to the Moon at all rather than just sending robots (cf. Campa et al., 2019). Even novelist Kim Stanley Robinson, in his *Red Moon* (2018), complains, that, "The moon has nothing people can make money from" (in Adee, 2019). At first, this appears true, because the Moon seems so barren in comparison to Earth. However, the Moon's immense value will surely emerge primarily because of its location.

The Moon will become a testing field for mining projects headed to the asteroids, a waypoint for spaceships to Mars, a launching point for short and long expeditions, a transfer point, and a break-in-bulk point for all types of commerce to and from Earth. Early in its settlement, humans will go to the Moon for science, and for a beautiful and incomparable view of Earth, humanity's home planet. The Moon's transportation and fueling facilities will establish its importance as a good launching pad for spaceships to the rest of the solar system. This is apart from the strategic military importance of the Moon and indeed of all cis-lunar space, as outlined by Stewart and Rappaport in Chap. 11 of this volume.

20.3 Recreation

An old adage goes this way: "All work and no play make Jack a dull boy." We antic-ipate that work will dominate life on the Moon, the International Space Station, and other off-world venues, for the foreseeable future—except for the requisite periods of recreation that all hard-working people need. To an extent, this domination of life by work will arise because of the extraordinary expense of travel to and residence on the Moon. Everyone will want to make the best of the time they have there, especially if it is limited by excessive exposure to radiation, which will be hard to avoid.

Eventually, there will be tourists who lounge and gawk at Earth in the Moon's sky, and at unearthly sites that Moon residents view as ordinary. For ideas about recreation, we might turn again to Kim Stanley Robinson, who treats the sociology of sports against a backdrop of the colonization and rebellion of Mars in its early years. Robinson's series—*Red Mars* (1992), *Green Mars* (1993), and *Blue Mars* (1996)—provide a range of recreational activities pursued by the cast of characters. If one has ever imagined hang gliding on Earth, just imagine how invigorating it would be in the much lower gravity of Mars. One would soar not like a bird, but like a nymph! However, on the Moon, there is no air and there is only one-sixth the gravity of the Earth, so "air gliding" of any type would require a personal power pack, as on one's back or at one's sides. There are various designs.

20.3.1 The Great Outdoors

Charles Duke, Apollo 16 astronaut, exclaimed on landing, "This is one of the most beautiful deserts ever to be seen" (Kohler & Stoukalov, 2019). Hiking in the lunar desert will be a strong recreational option. Amid the desert landscape, there are myriad craters of varying sizes and heights, whose exploration will be easier in one-sixth gravitation than in equivalent spots on Earth. The mountainous terrain on the Moon, called "highlands," will be prime destinations with the kind of hikers' huts or mountain hotels that walkers and climbers find in the Austrian Alps, which were well-known at one time to one of the authors. The other author knows the same in the form of Civilian Conservation Corps huts on the Blue Ridge Mountain trails. On the Moon, the views will be even better from the vantage points of huts in the lunar mountains.

Outdoor pursuits at nighttime will provide greater opportunities than on Earth. Just as the Moon reflects sunlight back onto the Earth, the Earth also reflects sunlight onto the Moon. Humans can experience this by looking for "earthshine" when the Moon is a thin crescent in the evening sky (Qiu et al., 2003). Then, they can not only see a part of the Moon brightly lit by the Sun, but also dimly view the rest of the Moon's sphere lit by the reflection of sunlight off the Earth. That is earthshine. A simple calculation using the mean albedos of the Earth and Moon, together with their angular diameters, show that earthlight on the Moon is about 40 times more than moonlight on Earth. Outdoor sports should benefit from this all-night and all-day illumination. However, once more, this illustrates the need for shade in the sleeping places of humans in domes. No skylights will be needed in the sleeping quarters.

Outdoor sports activity presupposes spacesuits that are considerably less restrictive than those worn by Apollo astronauts in their lunar surface activities. NASA is developing an xEMU spacesuit for the Artemis generation of astronauts. This is a significant advancement in flexibility, while still showing a bulky backpack (NASA, 2020b). The requirements for an improved version are for a type of suit (1) that is skintight to keep pressure on body tissues and blood, (2) that resists the abrasive regolith and rock well, (3) that maintains warmth or coolness with moisture control, and (4) that provides reliable communication and breathable air. This is a tall order, but improvements will surely be made, and then we can bound off on our lunar treks.

The need for reliable communication, whether indoors or outdoors, will be essential for the safety of individuals in the harsh lunar environment. Artemis is already addressing this through a LunaNet architecture for linking robotic landers, rovers, and astronauts. It will support positioning, navigation, and timing services (NASA, 2020a, 26). Humans will still have their mobile phones and 5G, or something equivalent, on the Moon.

20.3.2 Sports and Indoor Recreation

Outdoor sports will benefit from lightweight spacesuits. It will be interesting to see how traditional games adapt, since new forms of lunar football, baseball, cricket, and baseball may arise. The pitches may have to be quite a lot larger than on Earth. Astronaut Alan Shepard reckoned that the infamous golf ball he hit during Apollo 14's landing went some six times farther than it would have done on Earth with the same stroke.

Indoor sports like basketball and squash will probably adopt differently weighted or configured balls. Such games will be important supplements to the lunar gym, and together they will be essential for maintaining the health, well-being, and work output for the first settlers.

As for other games, one can speculate that the standard deck of 52 playing cards might survive even further into the digital age, and so on the Moon. Playing cards on weightless space journeys will definitely be a challenge.

20.4   Cuisine

20.4.1   Food, Greenhouses, and Nutrition

Humans already have over fifty years of experience in eating on the Moon. According to "one who's been there":

> [B]acon cubes were among the meals stored in the lunar module. And it worked out that meal A, the first scheduled meal to be eaten on the Moon, consisted of bacon squares, peaches, sugar cookie cubes, pineapple grapefruit drink and coffee. They ate history's first meal on Moon slightly ahead of schedule after landing at the Sea of Tranquility (Schultz, 2014).

One cannot help but envision that this meal will be immortalized in some future feast and celebration. After all, that is what humans do. They celebrate the past in today's rituals. It reminds us that it is the human species that is settling the Moon, consciously, with great difficulty and effort, and they will prevail in that dangerous environment, as on Earth.

There has been a long gap since the Apollo program, but eating on the Space Shuttle and the ISS has filled that period, at least as far as freeze-dried meals go (Uri, 2020). For the Artemis program, prepackaged foods will suffice, but for the phase we are considering here, meals must come from a largely self-generated food supply on the Moon, itself. The hydroponic greenhouses, well outlined in Cazalis' Chap. 5 of this volume, have already been cited as a crucial source for this supply. Satisfying human nutritional requirements with all the needed amino acids can turn into a multitude of projects.

Vegan diets will be relatively easy to accommodate, but the AddMan process crucial for constructing buildings will also allow to make a passable New York steak (Kelly et al., 2019). Similarly, religious diets can be satisfied on special occasions by AddMan. Several years ago, it was reported that, "Real-life Star Trek 'replicator' prepares meal in 30 seconds" (Nelson, 2020; Reuters, 2015). Granted, there are additional innovations needed before replicators will be in operation on the Moon, but each of us has reason to hope that our families' favorite recipes will be available, along with lunar creations specially designed for lunar feast days. Those special dishes are going to be important for festive occasions that future lunar populations celebrate.

20.4.2   New Astronomical Calendar, Festivities, Celebration, and Conservation

Celebrations are important for humans, whether in anticipation or at their actual occurrence. They are valued for their conviviality or simply for relaxation at home, and a meal is often the highlight of a celebration. Lunar settlers will no doubt invent their own annual calendar of festivities. Some will continue with religious observances, others will be personal like birthdays, and some may be "seasonal."

The term, "seasonal", will have a different meaning on the Moon. It is a planetary body whose spin axis is virtually perpendicular to its orbital path around the Sun, at just 1.5°. The Earth's axis is at a tilt of 23.4°, and therefore, the Earth's tilt causes the Sun to appear a good deal north and south of the equator, giving us summers and winters, with their varying day/night durations and seasonal temperatures. During the lunar year, which is only 347 earth days long, the degree and a half variation of the Sun above and below the Moon's equator will mute the day/night cycle of the seasons. Besides, the Moon lacks an atmosphere, so the high variations of its surface temperature are due only to the extremes of day-to-night radiation from the Sun, rather than its tilt from the Sun. However, near the poles, where the first settlements will probably be located, there will be significant duration in the length of both days and nights, throughout the seasons, just as at Earth's poles. Daylight will be 173 days, as night will be, with 18 days of sunrise and sunset between them. Early settlements are most likely to be located at the poles because of the probable existence of water in the form of sub-surface ice, and because some mountains and craters located there rise higher into the sunlight. Solar power will be available from panels that change with the sun's angle.

It will be a whole new world.

With this new astronomical seasonality as a backdrop, meal preparation will sadly not include flame-charred steaks, at least cooked over burning coals. Electric convection ovens and induction-powered "burners" on a stove top will work fine, along with microwave ovens. Any cooking equipment on the Moon will have to cope with low gravity, which has some annoying effects on meal preparation. Low gravity reduces convection in liquids that we want to boil, and it spatters hot oil and other droplets further. Experimentation and modeling have long been underway to find solutions (Leber, 2014). The invention of spacefaring diets of a requisite variety and nutrition is vast field of study (cf. Vicens et al., 2008) and even a new business endeavor.

20.4.3 Meal Management for Congregate Habitation

Much more careful design is involved in any cafeteria or mess hall on the Moon, as compared to Earth. Requirements include storage on the incoming side and waste management on the outgoing, with preparation, cooking, and sanitation in between. All this must happen with minimal energy use and wastage of materials and food, and a maximum of recycling and hygiene. Restrictions point toward communal rather than individual preparation and dining.

Meal management will hopefully be integrated into a multipurpose environmental plan for all Moon domes and facilities. Settlers will need skilled catering managers, logisticians, and conservation specialists on the Moon, and they all must interact with industrial agronomists who manage the greenhouses (Chaps. 5 and 6 in this volume).

20.5   Art

20.5.1   Art Gives Humans Sustenance

Art, including music, is a component of most human lives, and it is given varying importance depending on one's culture and society. Art and music should rightly be treated as distinct from recreation, although they often have interwoven roles, for example, in cuisine, festivities, education, and religious observances. The elements of art are also involved in architecture so that, hopefully, the Moon's domed structures will decorate human lives, excite the senses, and make lunar living a pleasure. An occasional splurge on an "architectural feature" should be welcomed, because human emotions respond so directly to beautiful surroundings—natural and manmade. Humans will make the Moon beautiful and interesting, eventually, because that is what they do everywhere else. Arts, crafts, song, and dance are documented in varying degrees and expressions everywhere, until well back into prehistory. It is a trait of our species.

20.5.2   Art and Lunar Sustainability

Hans Dieleman (2008) has pointed to the importance of artists as "change agents," and they could be on the Moon—even part-time artists (Jorgensen, 2010; Ono, 2010). We anticipate that, from the time of the first settlements, sustainability will be a lunar theme, an ethos (see Chap. 14 in this volume), a practical management tool, and a fundamental necessity in a harsh environment where resources are limited. If the lunar colony is to change enough to become self-reliant within a sustained environment, there will be a need to build in flexibility to all lunar projects. Unsurprisingly, this is a prime feature of spaceships, too: back-up systems, designed with potential failure in mind.

Some years later, Dieleman (2017) emphasized that artists and designers as change agents do not just emerge by accident. He explains how arts-based education for sustainability is a key in the transformation process toward enchanting and transdisciplinary sustainability (2017, Fig. 1). A shift away from disenchanting, and toward "enchanting sustainability" will be needed in a new, and in some ways, fragile lunar environment. Most contemporary approaches fail to realize environmental sustainability because they are poor in engaging people in the transformation process. Surprisingly, a lunar colony, as in its use of AddMan, can be a model for what should happen on Earth.

Let us remember that the landscape on the Moon is almost completely monochromatic, although future lunar inhabitants may learn to distinguish shades of gray that were never perceived before. Perhaps a "Lunar monochrome" art will evolve, like the Korean monochrome movement developed out of roots in the minimalist, almost subversive Dansaekhwa art that started in the 1950s (Lynch, 2020). Monochrome is not to be disparaged. After all, white is composed of all the colors, while black is their absence, and grey a blend of the two. White is culturally diverse in meaning: it denotes joy, caregiving, purity, and holiness in Western culture; and death and mourning in African, Asian, and medieval Europe. So too, black is diverse: death, mystery, and the dark side, as well as power, nobility, prestige, and even sensuality. We can try to anticipate the meanings that black and white will assume for the lunar culture, but it might be more interesting to wait and see what lunar inhabitants do with gray.

20.6 Conclusion

It may seem a little odd when summing up projections for how we will be different on the Moon to go back in time. That is exactly what four twenty-first century families were challenged to do, to live on a small Welsh island for a month and be self-sustaining, as in a rural fishing community at the start of the twentieth century. Their fate was documented in a four-episode BBC series, "The 1900 Island." They succeeded, and the secret of their success lay in responding to the necessity of hard work for survival, which built a close-knit community, loath to go back to the twenty-first century at the end of the project.

Lunar pioneers will also find out this secret or they will not survive. In discovering it, they will create a new, environmentally sensitive culture, one that helps them forget about ever living on Earth again.

References

Adee, S. (2019). Kim Stanley Robinson built a moon base in his mind. *IEEE Spectrum.* Retrieved December 30, 2020, from https://spectrum.ieee.org/aerospace/space-flight/kim-sta nley-robinson-built-a-moon-base-in-his-mind.

Allison, P. R. (2015). This is why lunar colonies will need to live underground. *BBC Future.* Retrieved December 30, 2020, from https://www.bbc.com/future/article/20151218-how-to-sur vive-the-freezing-lunar-night.

Benvenuti, S., Ceccanti, F., & De Kestelier, X. (2013). Living on the moon: Topological optimization of a 3D-printed lunar shelter. *Nexus Network Journal, 15*(2), 285–302. https://doi.org/10.1007/ s00004-01.

Bukley, A. P. (2020). *Center for space policy and strategy space agenda 2021 to the moon and beyond: Challenges and opportunities for NASA's artemis program.* El Segundo, CA. https://aer ospace.org/sites/default/files/2020-10/Bukley_TheMoon_20201027.pdf.

Campa, R., Szocik, K., & Braddock, M. (2019). Why space colonization will be fully automated. *Technological Forecasting and Social Change, 143*(October 2018), 162–171. https://doi.org/10. 1016/j.techfore.2019.03.021.

Dieleman, H. (2008). Sustainability, art and reflexivity: Why artists and designers may become key change agents in sustainability. In S. Kagan & V. Kirchberg (Eds.), *Sustainability: A new frontier for the arts and cultures* (pp. 1–26). Verlag für Akademische Schriften.

Dieleman, H. (2017). Arts-based education for an enchanting, embodied and transdisciplinary sustainability. *Artizein: Arts and Teaching Journal, 2*(2), 1–16. https://opensiuc.lib.siu.edu/atj/vol2/iss2/4.

Dolgin, E. (2019). Moondust, radiation, and low gravity: The health risks of living on the moon. *IEEE Spectrum.* Retrieved December 30, 2020, from https://spectrum.ieee.org/aerospace/spaceflight/moondust-radiation-and-low-gravity-the-health-risks-of-living-on-the-moon.

Drake, B. G. (1998). Reference mission version 3.0, addendum to the human exploration of mars: The reference mission of the NASA mars exploration study team. *NASA Special Publication,* (June).

ESA. (2014). *3D-printing a lunar base.* https://www.esa.int/ESA_Multimedia/Videos/2014/11/3D-printing_a_lunar_base. Accessed 11 November 2020

ESA. (2020). *Moon village: Conceptual design of a lunar habitat.* http://esamultimedia.esa.int/docs/cdf/Moon_Village_v1.1_Public.pdf.

Jagoda, J. A. (2020). *An analysis of the viability of 3D-printed construction as an alternative to conventional construction methods in the expeditionary environment.* Air Force Institute of Technology. https://scholar.afit.edu/etd/3240.

Jorgensen, J. (2010). Humans: The strongest and the weakest joint in the chain.*Lunar Settlements,* 247–259. https://doi.org/10.1201/9781420083330.

Kading, B., & Straub, J. (2015). Utilizing in-situ resources and 3D printing structures for a manned mars mission. *Acta Astronautica, 107,* 317–326. https://doi.org/10.1016/j.actaastro.2014.11.036

Kelly, B. E., Bhattacharya, I., Heidari, H., Shusteff, M., Spadaccini, C. M., & Taylor, H. K. (2019). Volumetric additive manufacturing via tomographic reconstruction. *Science, 363*(6431), 1075–1079. https://doi.org/10.1126/science.aau7114

Kohler, F., & Stoukalov, G. (2019). Moon mission I DW documentary. Germany: ZDF. https://www.youtube.com/watch?v=y0mGJNq-bHk.

Kramer, W. R. (2020). A framework for extraterrestrial environmental assessment. *Space Policy, 53,* 101385. https://doi.org/10.1016/j.spacepol.2020.101385.

Labeaga-Martínez, N., Sanjurjo-Rivo, M., Díaz-Álvarez, J., & Martínez-Frías, J. (2017). Additive manufacturing for a moon village. *Procedia Manufacturing, 13,* 794–801. https://doi.org/10.1016/j.promfg.2017.09.186

Leber, J. (2014). How we'll cook breakfast on mars. *Fast Company.* Retrieved December 31, 2020, from https://www.fastcompany.com/3034140/how-well-cook-breakfast-on-mars.

Lynch, S. (2020). Dansaekhwa: Exploring the "Korean Monochrome" art movement. *Singulart.* Retrieved December 31, 2020, from https://blog.singulart.com/en/2020/10/08/dansaekhwa-exploring-the-korean-monochrome-art-movement/.

NASA. (2020a). *Artemis plan: NASA's lunar exploration program overview.* https://www.nasa.gov/sites/default/files/atoms/files/artemis_plan-20200921.pdf.

NASA. (2020b). A next generation spacesuit for the artemis generation of astronauts. Retrieved December 30, 2020, from https://www.nasa.gov/feature/a-next-generation-spacesuit-for-the-artemis-generation-of-astronauts.

Nelson, B. (2020). Scientists create a 'Star Trek'-style replicator. *Treehugger.* Retrieved December 31, 2020, from https://www.treehugger.com/scientists-closer-creating-star-trek-style-replicator-4863606.

Ono, A. (2010). Art: Art as a psychological support for the outer space habitat. In H. Benaroya (Eds.), *Lunar settlements: Advances in engineering* (pp. 197–214). CRC Press. https://doi.org/10.1201/9781420083330.

Parks, J. (2019). Moon village: Humanity's first step toward a lunar colony? *Astronomy,* (May 31), 2019–2020. https://astronomy.com/news/2019/05/moon-village-humanitys-first-step-toward-a-lunar-colony.

Qiu, J., Goode, P. R., Pallé, E., Yurchyshyn, V., Hickey, J., Montanés Rodriguez, P., et al. (2003). Earthshine and the earth's albedo: 1. Earthshine observations and measurements of the lunar phase function for accurate measurements of the earth's bond albedo. *Journal of Geophysical Research: Atmospheres, 108*(22). https://doi.org/10.1029/2003jd003610.

Reuters. (2015). Real-life star trek "replicator" prepares meal in 30 seconds. *Sciencd & Space*. Retrieved December 31, 2020, from https://www.reuters.com/article/us-israel-meals-on-dem and-tracked/real-life-star-trek-replicator-prepares-meal-in-30-seconds-idUSKBN0NQ1PG20 150505.

Robinson, K. S. (1992). *Red Mars* (Mars Trilogy #1). Spectra.

Robinson, K. S. (1993). *Green Mars* (Mars Trilogy #2). Spectra.

Robinson, K. S. (1996). *Blue Mars* (Mars Trilogy #3). Spectra.

Robinson, K. S. (2018). *Red Moon*. Orbit.

Schultz, C. (2014). The first meal eaten on the moon was bacon. *Smithsonian Magazine, SmartNews*, (April 8). https://www.smithsonianmag.com/smart-news/first-meal-eaten-moon-was-bacon-180 950457/.

Szocik, K., Wójtowicz, T., & Braddock, M. (2020). The martian: Possible scenarios for a future human society on mars. *Space Policy, 54*. https://doi.org/10.1016/j.spacepol.2020.101388.

Uri, J. (2020). Space station 20th: Food on ISS. *NASA History*. Retrieved December 31, 2020, from https://www.nasa.gov/feature/space-station-20th-food-on-iss.

Vicens, C., Wang, C., Olabi, A., Jackson, P., & Hunter, J. (2008). Optimized bioregenerative space diet selection with crew choice. *Habitation, 9*(1), 31–39. https://doi.org/10.3727/154296603460 5243.

Index

Lightning Source UK Ltd.
Milton Keynes UK
UKHW020320221221
396048UK00003B/34

9 783030 813871